Monographs in Mathematics
Vol. 81

Birkhäuser
Boston · Basel · Stuttgart

Robert J. Zimmer

Ergodic Theory and Semisimple Groups

1984

Birkhäuser
Boston · Basel · Stuttgart

Library of Congress Cataloging in Publication Data

Zimmer, Robert J., 1947–
Ergodic theory and semisimple groups.

Bibliography: p.
Includes index.
1. Semisimple Lie groups. 2. Ergodic theory.
I. Title.
QA387.Z56 1983 512'.55 84-11127
ISBN 0-8176-3184-4

CIP-Kurztitelaufnahme der Deutschen Bibliothek

Zimmer, Robert J.:
Ergodic theory and semisimple groups / by
Robert J. Zimmer. – Basel ; Boston ; Stuttgart :
Birkhäuser, 1984.
(Monographs in mathematics ; 81)
ISBN 3-7643-3184-4

NE: GT

Printed in Switzerland by Birkhäuser AG, Graphische Unternehmen, Basel
ISBN 0-8176-3184-4
ISBN 3-7643-3184-4

To Terese and David

Table of Contents

Preface

This book is based on a course given at the University of Chicago in 1980–81. As with the course, the main motivation of this work is to present an accessible treatment, assuming minimal background, of the profound work of G. A. Margulis concerning rigidity, arithmeticity, and structure of lattices in semisimple groups, and related work of the author on the actions of semisimple groups and their lattice subgroups. In doing so, we develop the necessary prerequisites from earlier work of Borel, Furstenberg, Kazhdan, Moore, and others.

One of the difficulties involved in an exposition of this material is the continuous interplay between ideas from the theory of algebraic groups on the one hand and ergodic theory on the other. This, of course, is not so much a mathematical difficulty as a cultural one, as the number of persons comfortable in both areas has not traditionally been large. We hope this work will also serve as a contribution towards improving that situation. While there are a number of satisfactory introductory expositions of the ergodic theory of integer or real line actions, there is no such exposition of the type of ergodic theoretic results with which we shall be dealing (concerning actions of more general groups), and hence we have assumed absolutely no knowledge of ergodic theory (not even the definition of "ergodic") on the part of the reader. All results are developed in full detail. On the other hand, there are a number of excellent introductory expositions concerning algebraic groups, semisimple groups, and discrete subgroups. We therefore felt there was little to gain by another reproduction of this material. However, in order to make this work accessible to those without a strong background in algebraic groups, we define all relevant concepts as they arise, and illustrate these by example, usually with $SL(n)$. In the same spirit, we sometimes present a proof in detail for $SL(n)$ where certain structural features are clear, indicating how to extend the argument to the general semisimple case for those with a knowledge of the structure of such groups. In this way, while some knowledge of algebraic groups may technically be called a prerequisite for reading this book, we hope that the presentation is such that without such knowledge one can proceed without much difficulty. A survey of much of the material in this book appears in the author's CIME lectures, "Ergodic theory, group representations, and rigidity" (Bulletin AMS, 1982.)

Our approach to the material has been decidedly non-encyclopedic. We have made no attempt to present all topics of interest which might be included in a work with the present title. For example, we make at most passing reference to the recent work of Dani, Ratner, Sullivan, and our own recent work on actions of semisimple groups and their lattices on compact manifolds. Were the present work to be significantly expanded, it would naturally encompass this material.

Obviously, from our remarks so far, and as will be even more evident to the reader who proceeds beyond this preface, we owe a great intellectual debt to G. A. Margulis which we take pleasure in acknowledging. What is perhaps not so obvious from the approach we have taken is the influence of G. D. Mostow on the ideas developed here. Quoting Margulis (in translation) about his own work: "The present author was under the potent influence of these [Mostow's] ideas." Margulis' remarks apply equally as well to the present work.

We would like to thank those who attended the course in Chicago on which this work is based, and in particular, S. Bloch, D. Ramakrishnan, and V. Srinivas for their perseverence, enthusiasm, and for numerous observations which helped to clarify the exposition. In addition, special thanks are due to G. Prasad. Without his help, section 6.2 could not have been written (lacking a suitable manuscript of Margulis.) We would also like to thank A. Borel for encouraging us to write the notes for the course in a complete, formal form and for numerous suggestions that improved the text; R. Fefferman for some clarifying conversations about section 8.2; D. Witte and S. Adams, who read large portions of the manuscript and noted some errors in its original version; C. C. Moore for his helpful comments on part of the manuscript; R. Fabec for some very helpful conversations about the appendices; A. Connes, J. Feldman, H. Furstenberg, R. Howe, A. Katok, G. W. Mackey, G. A. Margulis, C. C. Moore, G. D. Mostow, G. Prasad, and D. Sullivan for enlightening conversations at various times concerning the material presented here; and the Alfred P. Sloan Foundation and the National Science Foundation for support during the preparation of this work.

Chicago, August 1984

1 Introduction

1.1 Statement of some main results

In this section, we shall briefly describe some of the main results we will be proving in the sequel without striving to state results in their maximal generality. Many of the notions used here will be introduced in more detail in the following chapters.

Let G be a locally compact group. A subgroup $\Gamma \subset G$ is called a lattice subgroup if (i) Γ is discrete, and (ii) G/Γ has a finite G-invariant measure. A subgroup $H \subset G$ is called cocompact if G/H is compact. Any discrete cocompact subgroup of a unimodular group is a lattice.

Example: (a) $Z^n \subseteq \mathbb{R}^n$ is a cocompact lattice.
(b) Let G be the set of $n \times n$ upper triangular real matrices (a_{ij}) with $a_{ii} = 1$ for all i. Let $\Gamma = \{(a_{ij}) \in G \mid a_{ij} \in Z \text{ for all } i, j\}$. Then it is not difficult to verify that Γ is discrete and cocompact, and hence is a lattice. We remark that here G is a nilpotent Lie group.
(c) Let $G = SL(n, \mathbb{R})$ and $\Gamma = SL(n, Z)$. It is then a classical result that Γ is a lattice in G. (See [Borel 2] for example.) In this case Γ is not cocompact. Here, G is a simple Lie group.

Let us make two elementary observations about lattices in $\mathbb{R}^n$.
(1) Suppose $\Gamma_1 \subset \mathbb{R}^n$ and $\Gamma_2 \subset \mathbb{R}^p$ are lattices, and that $\pi: \Gamma_1 \to \Gamma_2$ is an isomorphism of groups. Then π extends to a (continuous) group isomorphism $\pi: \mathbb{R}^n \to \mathbb{R}^p$. (In particular, of course, $n = p$.) We can think of this as a "rigidity" theorem, asserting that the discrete subgroup completely determines the ambient connected group.
(2) If $\Gamma \subset \mathbb{R}^n$ is any lattice, then there is an automorphism $A: \mathbb{R}^n \to \mathbb{R}^n$ such that $A(Z^n) = \Gamma$. We can think of this as an "arithmeticity" theorem, asserting that modulo automorphisms of $\mathbb{R}^n$ any lattice "is" Z^n.

Two of the major results we will prove concern the question as to whether results analogous to these two observations are true for lattices in semisimple Lie groups. Any semisimple Lie group G has an integer attached to it, called the $\mathbb{R}$-rank. For example, $\mathbb{R}$-rank$(SL(n, \mathbb{R})) = n - 1$. More generally, if G is actually a real matrix group, the $\mathbb{R}$-rank of G is the maximal dimension of an abelian subgroup of G which can be diagonalized over $\mathbb{R}$. Some of the results we will prove will require that $\mathbb{R}$-rank$(G) \geqq 2$. This excludes $SL(2, \mathbb{R})$ and the groups of isometries of real or complex hyperbolic space, for example. Our first observation about lattices in $\mathbb{R}^n$ can be generalized to simple Lie groups as follows.

Theorem (*Mostow–Margulis rigidity theorem*): *Let G, G' be connected simple Lie groups with trivial center, and suppose $\Gamma \subset G$ and $\Gamma' \subset G'$ are lattices. Suppose $\mathbb{R}$-rank$(G) \geqq 2$. Then any isomorphism $\pi: \Gamma \to \Gamma'$ extends to an isomorphism $G \to G'$.*

(More generally, this is true for semisimple groups and "irreducible" lattices.) This was first shown by [Mostow 2] for cocompact lattices and then by [Margulis 1] in general. If we exclude $PSL(2, \mathbb{R})$, the result is also true without the $\mathbb{R}$-rank assumption ([Mostow 1, 2] and [Prasad 1]), but the proof if both groups are of $\mathbb{R}$-rank 1 requires different techniques, and we shall not be dealing with that case here. (See also [Munkholm 1] for a nice exposition of a proof due to Gromov in the cocompact real hyperbolic case.)

To describe the analogue of the "arithmeticity" observation above in the context of semisimple groups, we first describe a general arithmetic construction of lattices. Let $H \subset GL(n, \mathbb{R})$ be a semisimple Lie group which is also an algebraic group defined over Q. That is, there exists a polynomial ideal in $n^2 + 1$ variables, $I \subset \mathbb{Q}[a_{ij}, \det(a_{ij})^{-1}]$ such that $H = \{(a_{ij}) \in GL(n, \mathbb{R}) \mid p(a_{ij}, \det(a_{ij})^{-1}) = 0$ for all $p \in I\}$. Clearly $SL(n, \mathbb{R})$ is such an example. Let $H_Z = H \cap GL(n, Z) = \{(a_{ij}) \in H \mid a_{ij}, \det(a_{ij})^{-1} \in Z\}$.

Theorem [*Borel–Harish–Chandra* 1]: *With H as above, H_Z is a lattice in H.*

This is of course a generalization of the fact that $SL(n, Z)$ is a lattice in $SL(n, \mathbb{R})$. If we let H^0 be the connected component of the identity, then $H^0 \subset H$ is a subgroup of finite index, and $H_Z \cap H^0$ will be a lattice in H^0. From the structure theory of semisimple Lie groups, it is known that every connected semisimple Lie group with trivial center is of the form H^0, where H is as above, an algebraic group defined over Q. Thus, a reasonable question to ask is to what extent every lattice in a semisimple Lie group arises via the "arithmetic" construction of taking $H_Z \subset H$. To answer this, we first describe two elementary ways of modifying a given lattice to obtain a new lattice.

Proposition: *Suppose $\varphi: H \to G$ is a surjective homomorphism of locally compact groups so that kernel(φ) is compact. If $\Gamma \subset H$ is a lattice, then $\varphi(\Gamma) \subset G$ is also a lattice.*

To describe the second way of modifying a lattice, we make the following definition.

Definition: Let Γ, Γ' be subgroups of a given group. They are called commensurable if $\Gamma \cap \Gamma'$ is of finite index in both Γ and Γ'.

We then have the following easy proposition.

Proposition: *Suppose $\Gamma, \Gamma' \subset G$ are discrete subgroups of a locally compact group, and that Γ and Γ' are commensurable. If Γ is a lattice in G, so is Γ'.*

Margulis' arithmeticity theorem asserts that modulo these two methods of modifying lattices, every lattice in a suitable semisimple group arises via the above construction.

Theorem [*Margulis* 1]: *Let G be a connected (semi)-simple Lie group with trivial center, and suppose $\Gamma \subset G$ is an (irreducible) lattice. Assume $\mathbb{R}$-rank$(G) \geqq 2$. Then Γ is "arithmetic". That is, there exists a real algebraic semisimple group H defined over $\mathbb{Q}$, and a continuous surjective homomorphism $\varphi: H^0 \to G$ such that*

(*i*) *kernel(φ) is compact,*

and

(*ii*) *$\varphi(H_Z \cap H^0)$ and Γ are commensurable.*

This result was conjectured by Selberg and Piatetski–Shapiro. Earlier progress had been made by [Selberg 1], [Margulis 3], and [Raghunathan 3]. The only groups of $\mathbb{R}$-rank one presently known to have non-arithmetic lattices are (up to local isomorphism) $SO(1, n)$, $n \leqq 5$ [Makarov 1], [Vinberg 1], $SU(1, 2)$ [Mostow 3], and $SU(1, 3)$ [Deligne–Mostow 1]. It is an open problem to determine all such groups.

The Mostow–Margulis rigidity theorem we stated above asserts that the lattice in a sense determines the ambient Lie group. It is then natural to ask as to what extent structural features of the Lie group and those of the lattice reflect each other. The salient feature of simple Lie groups is of course the non-existence of normal subgroups.

Theorem [*Margulis* 6], [*Kazhdan* 1]: *Let G be a connected (semi)simple Lie group* *with finite center, $Z(G)$, and let $\Gamma \subset G$ be an (irreducible) lattice. Suppose $\mathbb{R}$-rank$(G) \geqq 2$. If $N \subset \Gamma$ is a normal subgroup, then either*

(i) $N \subset Z(G)$, and in particular, N is finite;

or

(ii) Γ/N is finite.

In other words, modulo finite groups, Γ is simple.

In the proofs of the rigidity theorem, arithmeticity theorem, and the finiteness theorem for normal subgroups, consistent use is made of various ergodic theoretic properties of group actions. To see why ergodicity is relevant, and in fact to say a word about what it is, let us consider a classical example. Let $G = SL(2, \mathbb{R})$, and let X be the upper half plane, $X = \{z \in \mathbf{C} \mid Im(z) > 0\}$. As is well known, G acts on X via fractional linear transformations, i.e.,

$$g \cdot z = (az + b)/(cz + d) \quad \text{where } g = \begin{pmatrix} a & b \\ c & d \end{pmatrix}.$$

Suppose now that $\Gamma \subset G$ is a lattice, which we assume to be torsion free for simplicity. Since the action of G on X allows an identification of X with G/K, where $K = SO(2)$ (the stabilizer of $i \in X$), and K is compact, it follows that the action of Γ on X is properly discontinuous, and so $\Gamma \backslash X$ will be a manifold, in fact a finite volume Riemann surface. On the other hand, via the same fractional linear formula, G acts on $\bar{\mathbb{R}} = \mathbb{R} \cup \{\infty\}$, and $\bar{\mathbb{R}}$ can be identified with G/P, where P is the group of upper triangular matrices and the stabilizer of $\infty \in \bar{\mathbb{R}}$. Once again, we can consider the action of Γ on $\bar{\mathbb{R}}$, but now the action will be very far from being properly discontinuous. In fact, every Γ-orbit in $\bar{\mathbb{R}}$ will be a (countable) dense set. In particular, if we try taking the quotient $\Gamma \backslash \bar{\mathbb{R}}$, we obtain a space with the trivial topology. On the other hand, $\bar{\mathbb{R}}$ provides a natural compactification of X, and in fact $\bar{\mathbb{R}}$ can be identified with asymptotic equivalence classes of geodesics in X, where X has the essentially unique G-invariant metric. Thus, it is certainly reasonable to expect the action of Γ on $\bar{\mathbb{R}}$ to yield useful information. However, a thorough understanding requires us to come to grips with actions in which the orbits are very complicated (e.g. dense) sets. Ergodic theory is (in large part) the study of complicated orbit structure in the presence of a measure. Not only are there no non-constant Γ-invariant continuous real-valued functions on $\bar{\mathbb{R}}$, but the same is true for measurable functions. This is embodied in the following definition.

Definition: Suppose G acts on a measure space (S, μ) so that the action map $S \times G \to S$ is measurable and μ is quasi-invariant, i.e., $\mu(A) = 0$ if and only if

$\mu(Ag) = 0$. The action is called ergodic if $A \subset S$ is measurable and G-invariant implies $\mu(A) = 0$ or $\mu(S - A) = 0$.

We shall see that the action of Γ on $\bar{\mathbb{R}}$ is ergodic. Analogues, generalizations, and further measure-theoretic features of the action of a lattice in a general semisimple group on an appropriately defined "boundary" will be a basic ingredient in proving the above theorems. In the course of applying ergodic theoretic ideas to problems about discrete groups, Margulis developed powerful techniques and results concerning the ergodic theory (or measure theory) of these boundary actions. Some of these, in turn, played an important role in recent progress on some basic questions in ergodic theory which, at first glance, appear to be unrelated. In particular, Margulis' ideas played an important role in the author's proof of the "rigidity" theorem for ergodic actions, which we briefly describe.

Suppose that for $i = 1, 2$, S_i is an ergodic G_i-space for which the measures are finite and invariant. The actions are called orbit equivalent if there is a measure space isomorphism $\theta: S_1 \to S_2$ so that $\theta(G_1\text{-orbit}) = G_2\text{-orbit}$. We assume the actions are essentially free, i.e., free off some null set.

Theorem [*Zimmer* 8]: *Assume S_i, G_i as above and that G_i are connected simple Lie groups with finite center. (We can take G_i to be semisimple if the actions are "irreducible", which means that the non-trivial normal subgroups still act ergodically.) Assume $\mathbb{R}$-rank $(G_1) \geqq 2$, and suppose the actions are orbit equivalent. Then*

(i) G_1 and G_2 are locally isomorphic.
(ii) In the center-free case, $G_1 \cong G_2$, and identifying G_1 and G_2 via this isomorphism, the actions on S_1 and S_2 are isomorphic.

This result is in sharp contrast to the theory of orbit equivalence for actions of solvable groups. It can also be applied to ergodic actions of lattice subgroups, with (and in some cases without) finite invariant measure.

1.2 Outline of the succeeding chapters

We begin Chapter 2 with a general discussion of the notion of ergodicity. The first major result we prove, to which the bulk of Chapter 2 is devoted, is Moore's ergodicity theorem [Moore 1]. This will imply ergodicity of Γ on the "boundary", but in fact says much more. It also tells us exactly for which homogeneous spaces of G, the restriction of the G-action to Γ will be ergodic.

This is of fundamental importance and these results will be used throughout. The approach we take to proving Moore's theorem is not the same as Moore's original argument, but it is closely related. Namely, Moore's theorem is implied by the vanishing of matrix coefficients of unitary representations of simple Lie groups. This was shown in [Howe and Moore 1] and in [Zimmer 2], the latter proof consisting essentially of some observations on the work of [Sherman 1]. In Chapter 2, we shall follow the argument of [Howe and Moore 1], first developing the prerequisite results of classical representation theory.

The fact that semisimple Lie groups are not only Lie groups but that they are (essentially) algebraic groups will be of basic importance. The main reason for this is that while ergodicity is the study of complicated orbit behavior, many natural actions of algebraic groups exhibit a certain regularity of orbits that is in a sense the opposite of ergodicity. The interplay of ergodicity of some actions and "total non-ergodicity" of others will prove to be a powerful tool. The "total non-ergodicity" condition is that all orbits be locally closed. It is a fundamental fact about algebraic groups that algebraic actions on varieties have locally closed orbits [Borel 1]. We shall review some basic results about algebraic groups at the beginning of Chapter 3. Moreover, other natural actions of algebraic groups, not on varieties but rather on spaces of functions taking values in a variety [Margulis 1], or spaces of measures on a compact homogeneous variety [Zimmer 4], also have the property that all orbits are locally closed. These results, which are used in the proof of the rigidity and arithmeticity theorems, are proved in Chapter 3. The analysis of orbits in spaces of measures mentioned above has as its starting point a lemma of Furstenberg concerning measures on projective space, and this leads to a simple proof [Furstenberg 4] of the Borel density theorem [Borel 3]. This is also basic for further developments and is dealt with in Chapter 3 as well. We also discuss in that chapter one further connection between algebraic structure and measure theory, namely the question of when a measurable map between varieties must be a rational map. This question in fact lies at the heart of the proof of the rigidity and arithmeticity theorems. In Chapter 3, we discuss first results in this direction.

In Chapter 4 we discuss the notion of amenability, which arises in various situations in the following chapters. We begin by developing the basic standard results about amenable groups, and then discuss cocycles of ergodic actions and induced representations. This leads to the notion of an amenable action of a group [Zimmer 3]. A basic observation is that non-amenable groups have amenable actions. In particular, the action of Γ on a suitable boundary, alluded to above (and equal to the action of Γ on $\bar{\mathbb{R}}$ for $\Gamma \subset SL(2, \mathbb{R})$) is not only ergodic but it is also amenable. This is a basic property of the boundary actions that we will use often. For example, in the proof of the rigidity theorems, it will enable us to construct a certain function to which we will apply in various ways the ergodicity results of Chapter 2 and the "non-ergodicity" results of Chapter 3. The notion of amenability, both for groups and actions, also plays an important role in the analysis of normal subgroups of Γ which appears in Chapter 8, and in a variety of other considerations about ergodic actions, some of which are discussed in Chapter 9.

Chapters 2, 3, and 4 form the essential background for the proofs of the rigidity theorems for lattices and ergodic actions which appear in Chapter 5, and the arithmeticity theorems, which appear in Chapter 6. The Mostow–Margulis theorem can also be formulated in more geometric terms, as a result relating the fundamental group and Riemannian structures on certain locally symmetric spaces. The rigidity theorem for actions can also be formulated in geometric terms, as an assertion about foliations, relating the measure theory on a transversal to the Riemannian geometry of the leaves. For these geometric formulations, see [Mostow 2] and [Zimmer 8]. While the arithmeticity theorem we stated above was valid for lattices in groups of $\mathbb{R}$-rank at least 2, Margulis has also given a criterion for a lattice in an arbitrary semisimple Lie group to be arithmetic, in terms of the commensurability group of the lattice. We prove this theorem in Chapter 6 as well.

In Chapter 7 we return to considerations involving unitary representations. Namely, we present background to and a development of Kazhdan's property (T). This is used to deal with the case of amenable quotient groups in considering the normal subgroups of lattices. The case of a non-amenable quotient was dealt with by Margulis by converting it to a (difficult) problem in ergodic theory, and then solving the latter. This work of Margulis is presented in Chapter 8. Kazhdan's property can be defined to be an invariant of an ergodic action as well as a group [Zimmer 7], and this is presented in Chapter 9. Aside from providing an interesting and useful invariant, when combined with the rigidity theorem for actions and certain other considerations, it yields interesting results concerning the entropy of actions of lattices on compact manifolds. This was first observed by [Furstenberg 5] and is also discussed in Chapter 9.

In Chapter 10, we indicate how the main results can be generalized to groups over p-adic fields and to S-arithmetic groups.

2 Moore's Ergodicity Theorem

2.1 Ergodicity and smoothness

Let G be a locally compact second countable group. We shall consider actions of G on a Borel space S so that the action map $S \times G \to S$, $(s, g) \to sg$ is Borel. We shall assume that S is a standard Borel space, i.e., isomorphic as a Borel space to a Borel subset of a complete separable metric space. This includes, of course, many spaces arising naturally in analysis and geometry. If μ is a σ-finite measure on S, μ is called quasi-invariant under the action of G if for all $A \subset S$ and $g \in G$, $\mu(Ag) = 0$ if and only if $\mu(A) = 0$. The measure μ is called invariant if $\mu(Ag) = \mu(A)$ for all A, g. Two measures are said to be in the same measure class if they have the same null sets. Any σ-finite measure is in the same class as a probability measure. An action with quasi-invariant measure can also be thought of as an action with an invariant measure class. If (S, μ), (S', μ') are two such G-spaces, they are called equivalent (or "isomorphic", or "conjugate") if there are conull G-invariant Borel sets $S_0 \subset S$, $S'_0 \subset S'$ and a measure class preserving Borel isomorphism $\varphi: S_0 \to S'_0$ such that $\varphi(sg) = \varphi(s)g$ for all $s \in S_0$, $g \in G$.

2.1.1 Definition: The action of G on (S, μ), with μ quasi-invariant, is called ergodic if every G-invariant measurable set is either null or conull.

2.1.2 Example: If $H \subset G$ is a closed subgroup then it is well known that there is a unique invariant measure class on G/H. The action of G on G/H, being transitive, is clearly ergodic.

An action of G on (S, μ) is called essentially transitive if there is a conull orbit. This again clearly implies ergodicity. An action is called properly ergodic if it is ergodic but not essentially transitive. In this case, every orbit is a null set. [We remark that orbits are always measurable sets. See Corollary 2.1.20.]

2.1.3 Example: Suppose S is a differentiable manifold and that G acts on S by diffeomorphisms. If μ is a measure on S which is of the Lebesgue measure class (i.e., locally in the same class as Lebesgue measure), then μ is clearly quasi-invariant. Of course, it may or may not be ergodic. If G is a Lie group, the

Lebesgue measure class on G/H, where H is closed, is the unique G-invariant measure class of the previous example.

2.1.4 Example: We present a first example of a properly ergodic action. Let $S = \{z \in \mathbb{C} \mid |z| = 1\}$ and let $T: S \to S$ be defined by $Tz = e^{i\alpha}z$ where $\alpha/2\pi$ is irrational. Then T generates an action of Z, the group of integers. This action clearly preserves the arc-length measure on the circle and is not essentially transitive. We claim it is ergodic. If $A \subset S$ is invariant, let $\chi_A(z) = \Sigma a_n z^n$ be the L^2-Fourier expansion of its characteristic function. Then by invariance, $\chi_A(z) = \chi_A(e^{i\alpha}z) = \Sigma a_n e^{in\alpha} z^n$. Thus $a_n e^{in\alpha} = a_n$, and so $a_n = 0$ for $n \neq 0$, since $\alpha/2\pi \notin \mathbb{Q}$. This implies χ_A is constant, verifying ergodicity.

2.1.5 Example: Let $SL(n, Z)$ act on $\mathbb{R}^n$ in the natural way, so in fact $SL(n, Z)$ acts by automorphisms of the group $\mathbb{R}^n$. Then Z^n is left invariant under this action, and so we obtain an induced action of $SL(n, Z)$ on $T^n = \mathbb{R}^n/Z^n$, and this action is by automorphisms of the group T^n. Since Lebesgue measure is preserved by $SL(n, Z)$ acting on $\mathbb{R}^n$ (because the determinants are all 1), the action of $SL(n, Z)$ on T^n will preserve the Haar (= Lebesgue) measure on T^n. [More generally, an automorphism π of a compact group K must be measure preserving. To see this, let μ be Haar probability measure on K. Then $\pi_*(\mu)$ (defined by $\pi_*(\mu)(A) = \mu(\pi^{-1}(A))$ will also be a Haar measure since π is a group automorphism. Thus $\pi_*(\mu)$ and μ must differ by a constant multiple by uniqueness of Haar measure. Since they are both probability measures, they are equal.] We claim that $SL(n, Z)$ is ergodic on T^n. For $z = (z_1, \ldots, z_n) \in T^n$, where z_i is a complex number of modulus 1, and $k = (k_1, \ldots, k_n) \in Z^n$, let us set $z^k = \Pi z_i^{k_i}$. If $A \subset T^n$ is invariant under $SL(n, Z)$, let $\chi_A(z) = \sum_{k \in Z^n} a_k z^k$ be the L^2-Fourier expansion of its characteristic function. It is straightforward to verify that $\chi_A(\gamma^{-1} \cdot z) = \sum_{k \in Z^n} a_{\gamma^*(k)} z^k$ where $\gamma \in SL(n, Z)$ and γ^* is the transposed matrix. Thus, if A is invariant, $a_k = a_{\gamma^*(k)}$ for all $k \in Z^n$ and all $\gamma \in SL(n, Z)$. But it is easy to see that for any $k \in Z^n$, $k \neq 0$, $\{\gamma(k) \mid \gamma \in SL(n, Z)\}$ is infinite. Since $\Sigma |a_k|^2 < \infty$, we have $a_k = 0$ for $k \neq 0$, and this verifies ergodicity.

2.1.6 Example: Let $X = \prod_1^\infty \{\pm 1\}$. Then X is a compact abelian group, being the product of finite abelian groups, and Haar measure is just the product of the Haar measures on each factor. Let $H = \{x \in X \mid x_i = 1$ for all but finitely many $i\}$. Then H is a countable dense subgroup of X. We claim that H acting

$\int_G d\mu = \int_G Jf\, d\mu$

on X by multiplication is ergodic. We proceed as in the previous two examples. Namely, an orthonormal basis for $L^2(X)$ is given by $\{1, p_{i_1} \dots p_{i_n}\}$ where $p_i: X \to \{\pm 1\} \subset \mathbb{C}$ is projection on the i-th factor and $i_1, \dots, i_n$ is an arbitrary sequence of positive integers of finite length without repetitions. It is clear that for any $f = p_{i_1} \dots p_{i_n}$, there is $h \in H$ such that $f(xh) = -f(x)$. Thus, if $A \subset X$ is H-invariant and $\chi_A(x) = c + \Sigma(c_{i_1 \dots i_n} p_{i_1} \dots p_{i_n}(x))$, then $\chi_A(xh) = \chi_A(x)$ a.e., and by the uniqueness of Fourier coefficients, $c_{i_1 \dots i_n} = -c_{i_1 \dots i_n} = 0$. Thus, χ_A is essentially constant, and as in 2.1.4, 2.1.5, this proves ergodicity. (See also Lemma 2.2.13.)

We stated in the introduction that proper ergodicity is a phenomenon of complicated orbits. One elementary reflection of this is the following.

2.1.7 Proposition: *Suppose S is a second countable topological space, that G acts continuously, and that a quasi-invariant μ is positive on open sets. If the action is properly ergodic then for almost every $s \in S$, orbit(s) is a dense null set.*

Proof: If $W \subset S$ is open, then $\bigcup_{g \in G} W \cdot g$ is an open invariant set, and by ergodicity, the complement must have measure 0. Thus, if $\{W_i\}$ is a countable basis for the topology, $\bigcap_i (\bigcup_g W_i \cdot g)$ will be a conull set for which every point has a dense orbit, since any such orbit intersects every W_i.

Our next results will be of constant use and will as well describe another sense in which proper ergodicity is a reflection of complicated orbits. We begin with a definition.

2.1.8 Definition: (i) A Borel space is called countably separated if there is a sequence of Borel sets $\{A_i\}$ which separate points.
(ii) A Borel space is called countably generated if there is a sequence of Borel sets $\{A_i\}$ which separate points and generate the Borel structure.

2.1.9 Definition: Let S be a (Borel) G-space where S is countably separated. The action is called smooth if the quotient Borel structure on S/G is countably separated.

The relevance of smoothness to proper ergodicity is the following.

2.1.10 Proposition: *Suppose G acts smoothly on S. Then every quasi-invariant ergodic measure μ on S is supported on an orbit.*

Proof: We may assume $\mu(S) = 1$. Let $\mathscr{I} = \{A_i\}$ be a sequence of Borel sets separating points in S/G. We can clearly assume $\mathscr{I}$ is closed under taking complements. Let $p: S \to S/G$ be the natural map, and $\nu = p_*(\mu)$. By ergodicity of μ, for every $A \in \mathscr{I}$, $\nu(A) = \mu(p^{-1}(A)) = 0$ or 1. Let $B = \cap\{A \in \mathscr{I} \mid \nu(A) = 1\}$. Then $\nu(B) = 1$ and it suffices to see that B consists of a single point. If B contained two points, these could be separated by an element of $\mathscr{I}$ and either this set or its complement would have measure 1. This would contradict the definition of B.

The proof of Proposition 2.1.10 also shows the following.

2.1.11 Proposition: *Suppose S is an ergodic G-space and that Y is a countably separated space. If $f: S \to Y$ is a G-invariant Borel function (i.e., $f(sg) = f(s)$), then f is essentially constant, i.e., constant on a conull set.*

Proof: If μ is the quasi-invariant ergodic measure on S, it suffices to show that $f_*(\mu)$ is supported on a point. The proof of 2.1.10 shows this.

For continuous actions, smoothness is implied by the condition that all orbits be locally closed. As we shall see below, and in Chapter 3, a number of natural and important actions have this property.

2.1.12 Proposition: *Suppose G acts continuously on a separable metrizable space S. If every G-orbit is locally closed, then the action is smooth.*

Proof: Let $\rho: S \to S/G$ the natural map. Since ρ is open and S has a countable basis for the topology, so does S/G. Thus to see that S/G is a countably separated Borel space, it clearly suffices to show that S/G is T_0; i.e., that any two points are separated by an open set. Suppose $x, y \in S$. If $\rho(x)$ and $\rho(y)$ are not separated by an open set, $yG \subset \overline{xG}$. Similarly, $xG \subset \overline{yG}$. So yG is dense in $\overline{xG}$. But xG is locally closed, so xG is open in $\overline{xG}$. Thus $yG \cap xG \neq \varnothing$, which implies $\rho(x) = \rho(y)$.

This has the following immediate consequence.

2.1.13 Corollary: *Any continuous action of a compact group on a separable metrizable space is smooth.*

This is in fact also true for actions on a countably separated Borel space. See Corollary 2.1.21.

Another fundamental situation to which we shall apply Proposition 2.1.12 is to algebraic groups acting on varieties. A basic result of algebraic geometry asserts that for such actions orbits are locally closed, and hence every ergodic measure is supported in an orbit. We shall discuss this further in Chapter 3.

For continuous actions on complete separable metric spaces, the condition that all orbits be locally closed is equivalent to a large number of other regularity conditions on the orbits, and in particular, is equivalent to smoothness. We now prove this last assertion and present one other equivalent condition we shall need. For $s \in S$, let G_s be the stabilizer of s in G, i.e., $G_s = \{g \in G \mid sg = s\}$.

2.1.14 Theorem [*Glimm* 1] [*Effros* 1]: *Suppose G acts continuously on a complete separable metrizable space S. Then the following are equivalent:*

(i) All orbits are locally closed.
(ii) The action is smooth.
(iii) For every $s \in S$, the natural map $G/G_s \to Orbit(s)$ is a homeomorphism, where Orbit(s) has the relative topology as a subset of S.

Proof: The equivalence of (i) and (iii) for a given orbit follows from the following lemma by passing to the orbit closure.

2.1.15 Lemma: *With S as above, suppose $s \in S$ has a dense orbit. Then Orbit(s) is open if and only if $G/G_s \to Orbit(s)$ is a homeomorphism.*

Proof: Suppose first that the above map is a homeomorphism. Then Orbit(s) satisfies the Baire category theorem with the subspace topology. Since G is σ-compact, some compact set $A \subset \text{Orbit(s)}$ contains an open set, i.e., $A \supset \text{Orbit(s)} \cap U$ for some open set $U \subset S$. Since Orbit(s) is dense, $A \supset \overline{(\text{Orbit(s)} \cap U)} \supset U$. Thus $sG = UG$ which is open. Conversely, suppose sG is open in S. We claim that it suffices to show that for any compact symmetric neighborhood U of $e \in G$, sU contains an open set. Namely, if this holds, let N be any neighborhood of e and choose U with $U^2 \subset N$. If sU is a neighborhood of su, $u \in U$, then sUu^{-1} is a neighborhood of s, and hence so is sN, verifying oppenness of $G/G_s \to sG$. To show that sU contains an open set, choose a countable dense set $g_i \in G$. Then $sG = \bigcup_i sUg_i$, a union of compact sets, so by Baire category, one sUg_i contains an open set, and hence so does sU. This $T_k|S - \cup U(x, k) = \text{identity}$.

The assertion (i)$\Rightarrow$(ii) is Proposition 2.1.12, and hence it remains to show (ii)$\Rightarrow$(i), which is the most difficult part of the theorem.

We suppose $s \in S$ with sG dense in S but that sG contains no open subset of S. We wish to show that S/G is not countably separated. Since a subset of a countably separated space is countably separated, it suffices to show that we can find some other group action, say of H on a space X, which we already know to be non-smooth, and an injective Borel map $\theta: X \to S$ such that $\theta(H\text{-orbit}) \subset G\text{-orbit}$ for every H-orbit, and such that the induced map $X/H \to S/G$ is injective. The reason that this idea is useful is that there is a natural candidate for X, namely the group action described in Example 2.1.6. Here $X = \prod_1^\infty \{\pm 1\}$ is homeomorphic to the Cantor set and it is well known that the classical Cantor set construction on [0, 1] may be generalized in a straightforward manner so that one may construct many injective continuous maps from X into any complete separable metric space. We thus need only verify that we can make this Cantor space construction so as to satisfy the above conditions. Rather than the assertion that the induced map $X/H \to S/G$ is injective, we will establish a somewhat weaker condition which will suffice for our purposes.

2.1.16 Lemma: *Suppose H is a group acting ergodically on (X, μ) where μ has no atoms. If I is a Borel space and $f: X \to I$ is an H-invariant Borel map which is countable-to-one, then I is not countably separated.*

Proof: If I is countably separated, then by Proposition 2.1.11, f would be constant on a conull set. But since μ has no atoms, a conull set cannot be countable, which contradicts our assumption about f.

Thus, it suffices to construct an injective continuous map $\theta: X \to S$ such that

(a) $\theta(xH) \subset \theta(x)G$ for all $x \in X$; and
(b) $\theta(X)$ intersects each G-orbit in at most a countable set.

We now recall the Cantor space construction in S. For $x = (x_i) \in X$, let $p_n(x) = (x_1, \ldots, x_n)$. Suppose that for each $x \in X$ and each $n \geqq 1$, we have a non-empty open set $U(x, n) \subset S$ such that

(i) $\overline{U(x, n+1)} \subset U(x, n)$.
(ii) diameter $U(x, n) \leqq 1/n$.
(iii) If $p_n(x) = p_n(y)$, then $U(x, n) = U(y, n)$.
(iv) If $p_n(x) \neq p_n(y)$, then $U(x, n) \cap U(y, n) = \emptyset$.

Then for each $x \in X$ it follows from (i), (ii) that $\bigcap_n U(x, n)$ contains exactly one point which we denote by $\theta(x)$. From (iv), θ is injective and from (iii) we deduce that θ is continuous. There are of course many such possible choices of $U(x, n)$, and we wish to make such a choice so that (a) and (b) are satisfied. This will hold if we can arrange the following. Let $h_n \in X$ be given by

$$(h_n)_i = \begin{Bmatrix} 1 & \text{if } i \neq n \\ -1 & \text{if } i = n \end{Bmatrix}.$$

(v) For each x, n there is a $g(x, n) \in G$ such that for all $k \leqq n$, $U(xh_k, n) = U(x, n) \cdot g(x, k)$. (Here xh_k is group multiplication.)
(vi) There is a neighborhood N of $e \in G$ such that for all x, y, n with $p_n(x) \neq p_n(y)$, we have $U(x, n)N \cap U(y, n) = \emptyset$.
Then we have:

2.1.17 Lemma: *Condition (v) implies condition (a).*
Condition (vi) implies condition (b).

Proof: It is clear that for $x \in X$, (v) implies that $\theta(xh_n) = \theta(x)g(x, n)$. Since $\{h_n\}$ generates H, we have (v) implies (*a*). If (vi) holds and $x \in X$, it is clear that $\theta(x)N \cap \theta(X) = \theta(x)$. Let M be a symmetric neighborhood of e with $M^2 \subset N$. Let $\{g_i\}$ be a countable dense subset of G. It clearly suffices to see that $\theta(x)g_iM \cap \theta(X)$ has at most one point for each i. But if $\theta(y) = \theta(x)g_im_1$ and $\theta(z) = \theta(x)g_im_2$ for $m_i \in M$, we have $\theta(y) = \theta(z)m_2^{-1}m_1$, and since $m_2^{-1}m_1 \in N$, $\theta(y) = \theta(z)$.

We now show we can choose $U(x, n)$ such that (i)–(vi) hold. The following simple remark is useful.

2.1.18 Lemma: *Let N be a compact symmetric neighborhood of $e \in G$. Then for any $s \in S$ and any decreasing sequence of open sets $W_i \subset S$ with $W_i \downarrow \{s\}$, we have* $\overline{(\cap W_iN)} \subset xG$.

Proof: If $t \in S$ with $t = \lim s_i g_i$, $s_i \in W_i$, $g_i \in N$, then $s_i \to s$, and by passing to a subsequence we can assume $g_i \to g \in N$. Then $t = sg$.

We now construct $U(x, n)$, $g(x, n)$ inductively. Recall we have a dense orbit sG, without interior. Fix a compact symmetric neighborhood N of $e \in G$. We let $I \in X$ denote the identity element of the group X, so $I_i = 1$ for all i. We suppose that for all $x \in X$ and all integers $k \leqq n$, we have constructed $U(x, k)$, $g(x, k)$ such that (i)–(vi) are satisfied, and with the additional assumptions that

(vii) $s \in U(I, k)$, $k \leqq n$.
(viii) as a function of x, $g(x, k)$ depends only on $p_k(x)$.
(ix) $g(xh_k, k) = g(x, k)^{-1}$.
(x) Let $T_k: S \to S$ be defined by $T_k | U(x, k) = g(x, k)$,
$T_k | S - \cup u(x, k) = \text{identity}$.

We assume $g(x, k)$ are chosen so that H_n, the group generated by $\{T_k | k \leqq n\}$ is a finite abelian group of transformations of S, acting simply transitively (via elements of G) on $\{U(x, n) | x \in X\}$.

(For $n = 0$, we take all $g(x, 0) = e$, $U(x, 0) = S$.) Let W_i be a decreasing sequence of open subsets of S such that $W_i \downarrow \{s\}$. Let $G_0 \subset G$ be the set of 2^n-fold products of elements of the form $g(x, k)$, $k \leqq n$.

Then G_0 is a finite set, so $M = \bigcup_{g \in G_0} gNg^{-1}$ is also a compact symmetric neighborhood of $e \in G$. By Lemma 2.1.18, $\bigcap W_i M$ is nowhere dense. Hence, by the Baire category theorem there is some fixed i so that $W_i M$ is not dense in $U(I, n)$. Since sG is dense, we can choose $g(I, n+1) \in G$ such that $sg(I, n+1) \in U(I, n) - \overline{W_i M}$, the latter being an open set in S. Therefore, we can choose an open set $U(I, n+1)$ such that

(a) $s \in U(I, n+1) \subset \overline{U(I, n+1)} \subset U(I, n)$
(b) $U(I, n+1) \subset W_i$
(c) $\overline{U(I, n+1)} g(I, n+1) \subset U(I, n) - W_i M$
(d) diameter $U(I, n+1)g \leqq 1/n+1$ for all $g \in G_1$, where $G_1 = G_0 \cup g(I, n+1)G_0$.

Given x, n, choose $g_0 \in G_0$ such that $U(x, n) = U(I, n)g_0$ and $g_0 = T | U(x, n)$ for some $T \in H_n$. Define

$$U(x, n+1) = \begin{cases} U(I, n+1)g_0 & \text{if } x_{n+1} = 1, \\ U(I, n+1)g(I, n+1)g_0 & \text{if } x_{n+1} = -1, \end{cases}$$

and

$$g(x, n+1) = \begin{cases} g_0^{-1} g(I, n+1) g_0 & \text{if } x_{n+1} = 1, \\ g_0^{-1} g(I, n+1)^{-1} g_0 & \text{if } x_{n+1} = -1. \end{cases}$$

It is then routine to verify that conditions (i)–(x) are satisfied up to $n + 1$. This completes the proof of Theorem 2.1.14.

Remark: The above proof can be extended to show that the condition that every quasi-invariant ergodic measure be supported on an orbit is actually equivalent to the conditions of Theorem 2.1.14. To see this, one need only show that if the orbits are not all locally closed, we can find an ergodic measure not supported on an orbit. However, if we let $\theta: X \to S$ be the map constructed in the above proof, μ the Haar measure on X, and m a probability measure on G in the class of Haar measure, then $\lambda(A) = \int (\theta_* \mu)(Ag) dm(g)$ defines the required measure on S. As we shall not be using this fact, we leave the verification to the reader.

We conclude this section with a result of Varadarajan and some of its consequences.

2.1.19 Theorem [Varadarajan 1]: *Let S be a countably separated Borel G-space. Then there is a compact metric space X on which G acts continuously and an injective Borel G-map $S \to X$.*

2.1.20 Corollary: *Let S be a countably separated Borel G-space. Then orbits are Borel sets and stabilizers of points are closed subgroups.*

Proof: Orbits are Borel sets in a continuous G-space since G is σ-compact.

2.1.21 Corollary: *Any action of a compact group on a countably separated Borel space is smooth.*

Proof: Corollary 2.1.13 and Theorem 2.1.19.

This corollary shows that there is no "proper ergodic theory" for actions of compact groups.

Proof (of Theorem 2.1.19): Let $\{A_i\}$ be a sequence of Borel sets in S separating points, and let χ_i be the characteristic function of A_i. Let B be the unit ball in $L^\infty(G)$, which is a compact metric space with the weak-*-topology. Let $X = \prod_1^\infty B$, which is again a compact metric G-space. Define a map $\varphi: S \to X$ by $\varphi(s) = (\varphi_i(s))$ where $\varphi_i(s) \in B$ is given by $[\varphi_i(s)](g) = \chi_i(sg)$. It is clear that φ is a G-map. To see it is Borel, it suffices to see that each φ_i is Borel, and for this it suffices to see that for all $f \in L^1(G)$,

$$s \to \int f(g)[\varphi_i(s)](g)dg = \int f(g)\chi_i(sg)dg$$

is Borel, which follows from Fubini's theorem. We claim φ is injective. If $s, t \in S$ and $\varphi(s) = \varphi(t)$, then for each i, $\chi_i(sg) = \chi_i(tg)$ for almost all $g \in G$. Thus for some $g_0 \in G$, $\chi_i(sg_0) = \chi_i(tg_0)$ for all i. Since $\{A_i\}$ separates points $sg_0 = tg_0$, so $s = t$.

2.2 Moore's Theorem: Statement and Some Consequences

We have seen some examples of ergodicity above, and we now turn to the problem which is the central focus of this chapter, namely the question as to when certain naturally defined actions are ergodic. For example, we want to establish ergodicity of the "boundary action" of $SL(2, Z)$ on $\bar{\mathbb{R}}$ given by fractional linear transformations, as discussed in 1.1, and to establish various generalizations of this to actions of lattices in general semisimple Lie groups. As we remarked in 1.1., $\bar{\mathbb{R}}$ can be identified with $SL(2, \mathbb{R})/P$ where P is the subgroup of upper triangular matrices. Thus, ergodicity of $SL(2, Z)$ on $\bar{\mathbb{R}}$ is a special case of the following general problem.

2.2.1(a) Problem: *Let G be a semisimple Lie group, $H_1, H_2 \subset G$ closed subgroups. When is H_1 ergodic on G/H_2?*

This is itself a special case of the following.

2.2.1(b) Problem: *Let G be a semisimple Lie group and S an ergodic G-space. If $H \subset G$ is a closed subgroup, when is H ergodic on S?*

If G is transitive on S, this reduces to 2.2.1(a).

The following observation is very useful.

2.2.2 Proposition: *Let G be a locally compact group, S a G-space, and $H \subset G$ a closed subgroup. Then H is ergodic on S if and only if G is ergodic on $S \times G/H$. (Here G acts via the product, $(s, x)g = (sg, xg)$, and the measure class on $S \times G/H$ is the product measure class.)*

Proof: (i) Suppose $A \subset S \times G/H$ is G-invariant and neither null nor conull. For $x \in G/H$, let $A_x = \{s \in S \mid (s, x) \in A\}$. G-invariance of A immediately implies that $A_x \cdot g = A_{xg}$. Since G acts transitively on G/H, it follows that $A_{[e]}$ is neither null nor conull, for then A would be either null or conull by Fubini's theorem. But for $h \in H$, $[e] \cdot h = [e]$, so $A_{[e]} \cdot h = A_{[e]}$, and hence H is not ergodic on S.
(ii) Conversely, suppose $B \subset S$ is H-invariant. We can choose a Borel section $\varphi : G/H \to G$ of the natural projection $G \to G/H$, (see Appendix A), so $[\varphi(x)] = x$ for all $x \in G/H$. Define $A = \{(s, x) \in S \times G/H \mid s \in B \cdot \varphi(x)\}$. Then A is Borel, and since $\varphi(xg) = h\varphi(x)g$ for some $h \in H$, A is clearly G-invariant. If B is neither null nor conull, the same is true for A.

2.2.3 Corollary [*Moore* 1]: *If $H_1, H_2 \subset G$ are two closed subgroups, then H_1 is ergodic on G/H_2 if and only if H_2 is ergodic on G/H_1.*

Proof: By the proposition, both assertions are equivalent to the ergodicity of G on $G/H_1 \times G/H_2$.

Moore's ergodicity theorem is a very general result which completely answers Problem 2.2.1(a) for G a simple Lie group (such as $SL(n, \mathbb{R})$) where H_1 or H_2 is a lattice in G. It also provides an equally complete answer if G is a suitable semisimple Lie group and $\Gamma \subset G$ is an irreducible lattice.

2.2.4 Definition: Suppose G is a connected semisimple Lie group with finite center and $\Gamma \subset G$ is a lattice. Then Γ is called irreducible if for every non-central normal subgroup (equivalently, every normal subgroup of positive dimension) N, Γ is dense when projected onto G/N.

This definition is of course designed to eliminate examples such as $\Gamma_1 \times \Gamma_2 \subset G_1 \times G_2$. A typical example of an irreducible lattice is the following.

2.2.5 Example: Let $G = SL(n, \mathbb{R}) \times SL(n, \mathbb{R})$, and $\mathcal{O} = Z[\sqrt{2}]$. For $a + b\sqrt{2} \in \mathcal{O}$, let $\sigma(a + b\sqrt{2}) = a - b\sqrt{2}$. Let $\Lambda = SL(n, \mathcal{O})$, so that for $\lambda \in \Lambda$, $\sigma(\lambda)$ has an obvious meaning. Then $\Gamma = \{(\lambda, \sigma(\lambda)) \mid \lambda \in \Lambda\}$ is an irreducible lattice in G. (For a discussion, see Corollary 6.1.4.)

Any lattice in a connected semisimple Lie group can essentially be described in terms of irreducible ones. More precisely, if G is any connected semisimple Lie group with finite center and no compact factors and $\Gamma \subset G$ is a lattice then we can write $G = \Pi G_i$ where G_i are normal connected (semisimple) subgroups,

$G_i \cap G_j$ is finite, $\Gamma_i = \Gamma \cap G_i$ is an irreducible lattice in G_i and $\Pi\Gamma_i$ is a normal subgroup of Γ of finite index. [Raghunathan 1].

2.2.6 Theorem: (*Moore's Ergodicity Theorem* [*Moore* 1]): *Let* $G = \Pi G_i$ *be a (finite) product of connected non-compact simple Lie groups with finite center. Let* $\Gamma \subset G$ *be an irreducible lattice. If* $H \subset G$ *is a closed subgroup and* H *is not compact, then* H *is ergodic on* G/Γ.

The proof of this theorem will occupy the final two sections of this chapter. We remark that the converse assertion, that if H is compact then H is not ergodic on G/Γ, is also true, and may be seen in a number of ways. For example, by 2.1.10, 2.1.13 ergodicity of H would mean that some H-orbit is conull, but the H-orbits are closed submanifolds of strictly lower dimension.

2.2.7 Corollary: *With* H, Γ, G *as in* 2.2.6, Γ *is ergodic on* G/H *if and only if* H *is not compact.*

We now consider some examples.

2.2.8 Example: (a) If $\Gamma \subset SL(2, \mathbb{R})$ is a lattice, then Γ acts ergodically on $\bar{\mathbb{R}}$. This follows immediately since $\bar{\mathbb{R}} \cong SL(2, \mathbb{R})/P$ where P is not compact.
(b) More generally, $SL(n, Z)$ (or any lattice in $SL(n, \mathbb{R})$) acts ergodically on $\mathbb{P}^{n-1}$, or any of the Grassmann varieties or flag varieties. This follows immediately by realizing these spaces as homogeneous spaces of $SL(n, \mathbb{R})$.

2.2.9 Example: $SL(n, Z)$ is ergodic on $\mathbb{R}^n$. To see this, we simply note that this is equivalent to ergodicity on $\mathbb{R}^n - \{0\}$, and the latter is a homogeneous space of $SL(n, \mathbb{R})$ with a non-compact stabilizer.

2.2.10 Example: Let H^n be hyperbolic n-space, realized as the unit ball in $\mathbb{R}^n$ with the Poincaré metric. Let G be the group of orientation preserving isometries of H^n. Let $\Gamma \subset G$ be a lattice (so that if Γ is torsion free, H^n/Γ is a finite volume hyperbolic manifold with fundamental group Γ). Every element of G extends to a conformal diffeomorphism of the boundary sphere S^{n-1}, and we can identify $S^{n-1} \cong G/P$ where P is non-compact. Thus Γ acts ergodically on the boundary sphere. When $n = 2$, this is just example 2.2.8(a).

We will actually need a more general version of Theorem 2.2.6. We preface this with a definition.

2.2.11 Definition: Let $G = \Pi G_i$ be a (finite) product, where G_i is a connected simple non-compact Lie group with finite center. Let S be an ergodic G-space with finite invariant measure. Then the action is called irreducible if for every non-central normal subgroup $N \subset G$, N is ergodic on S.

Thus if G is simple, this is no more than ergodicity.

2.2.12 Proposition: *Let G be as in Definition* 2.2.11, *and suppose* $\Gamma \subset G$ *is a lattice. Then* Γ *is an irreducible lattice if and only if the action of* G *on* G/Γ *is irreducible in the sense of* 2.2.11.

Proof: If $N \subset G$ is normal, then N is ergodic on G/Γ if and only if Γ is ergodic on G/N (2.2.3). Thus it clearly suffices to prove the following lemma.

2.2.13 Lemma: *If* $\Gamma \subset H$ *is a subgroup of a locally compact group, then* Γ *is ergodic on* H *(acting by translations) if and only if* Γ is dense.

Proof: If $\bar{\Gamma} \neq H$, $H/\bar{\Gamma}$ will be a non-trivial metrizable space and so there are clearly non-null, non-conull Γ-invariant sets in H.

Conversely, suppose $\bar{\Gamma} = H$, and $A \subset H$ is Γ-invariant. The natural action of H on $L^1(H)$ is well known to be continuous, and so H acts continuously on the unit ball of $L^\infty(H) \cong L^1(H)^*$ with the weak-*-topology. Since χ_A is Γ-invariant, it is also H-invariant, which implies that for all $h \in H$, $ah \in A$ for almost all $a \in A$. By Fubini's theorem, if A is not null there is some $a \in A$ for which $ah \in A$ for almost all $h \in H$, i.e., $a^{-1}A$ is conull. Hence A is either null or conull.

Another example of an irreducible ergodic G-space is the following.

2.2.14 Example: Suppose $G = \Pi G_i$ is as in Definition 2.2.11, and suppose $G \subset H$ is a closed subgroup where H is a simple Lie group (e.g. $SL(n, \mathbb{R})$ for n sufficiently large.) Let $\Gamma \subset H$ be a lattice and $S = H/\Gamma$. Then by Moore's ergodicity theorem (2.2.6), S is an irreducible ergodic G-space.

Here is a more general version of 2.2.6.

2.2.15 Theorem (*Moore's Ergodicity Theorem*): *Let $G = \Pi G_i$ be a finite product where each G_i is a connected non-compact simple Lie group with finite center. Suppose S is an irreducible ergodic G-space with finite invariant measure. If $H \subset G$ is a closed non-compact subgroup, then H is ergodic on S.*

If $S = G/\Gamma$ where Γ is an irreducible lattice, then by 2.2.13 this is equivalent to the first form of the theorem (2.2.6).

Moore's ergodicity theorem follows directly from a general qualitative result about unitary representations of simple Lie groups. To describe this, we begin by discussing the connection between unitary representations and the problem of ergodicity of actions when restricted to subgroups.

Let S be a G-space with finite invariant measure. For each $g \in G$, let $\pi(g): L^2(S) \to L^2(S)$ be given by $(\pi(g)f)(s) = f(sg)$. Thus $\pi: G \to U(L^2(S))$ is a homomorphism where $U(\mathscr{H})$ denotes the unitary group of a Hilbert space $\mathscr{H}$. When endowed with the strong operator topology, $U(\mathscr{H})$ (for $\mathscr{H}$ separable) is a separable metrizable group. (We also recall that on $U(\mathscr{H})$ the weak operator topology and the strong operator topology coincide.) By Fubini's theorem, it follows easily that π is a measurable homomorphism. Since G is locally compact and $U(\mathscr{H})$ is second countable, it follows that π is continuous. (See Appendix B) By a unitary representation of G we shall mean a continuous homomorphism $\pi: G \to U(\mathscr{H})$ where the latter has the strong operator topology.

If $A \subset S$ is G-invariant, then $\chi_A \in L^2(S)$ will be G-invariant, and hence so will $f_A \in L^2(S) \ominus \mathbb{C}$ where f_A is the projection of χ_A onto $L^2(S) \ominus \mathbb{C}$. If A is neither null nor conull, then $f_A \neq 0$. Thus, if G is not ergodic, there exist non-0 invariant vectors in $L^2(S) \ominus \mathbb{C}$. Conversely, suppose G acts ergodically and $f \in L^2(S) \ominus \mathbb{C}$ is G-invariant. This means that for each $g \in G$, $f(sg) = f(s)$ a.e. We call such an f essentially invariant. If G is countable, let $S_0 = \bigcap_{g \in G} \{s \in S \mid f(sg) = f(s)\}$. Then S_0 is a G-invariant Borel set, and defining $\tilde{f}(s) = f(s)$ for $s \in S_0$ and $\tilde{f}(s) = 0$ for $s \notin S_0$, we obtain a G-invariant function $\tilde{f}$ such that $f = \tilde{f}$ a.e. By ergodicity, $\tilde{f}$ is essentially constant (2.1.11), which implies that $f = 0$ in $L^2(S) \ominus \mathbb{C}$. Thus, for G countable, ergodicity is equivalent to there being no non-0 invariant vectors in $L^2(S) \ominus \mathbb{C}$. The following shows that the same is true for general G.

2.2.16 Lemma: *Let S be a G-space with quasi-invariant measure, and Y a countably generated Borel space (Definition 2.1.8). Suppose $f: S \to Y$ is Borel and essentially G-invariant, i.e., for each g, $f(sg) = f(s)$ a.e. Then there is a Borel function $\tilde{f}: S \to Y$ such that $f = \tilde{f}$ a.e. and $\tilde{f}$ is G-invariant.*

Proof: Since Y is countably generated, it is Borel isomorphic to a subset of

[0, 1] (Appendix A) and hence we can assume $Y \subset [0, 1]$. Let $S_0 = \{s \in S \mid g \to f(sg)$ is essentially constant on $G\}$. Then S_0 is conull by Fubini's theorem and S_0 is clearly G-invariant. To see that S_0 is Borel, let λ be a probability measure on G in the class of Haar measure. Define $I(s) = \int f(sg)d\lambda(g)$, so that $I: S \to [0, 1]$ is Borel by Fubini. Let $J(s) = \int |f(sg) - I(s)| d\lambda(g)$. Again by Fubini, J is Borel, and since $S_0 = J^{-1}(0)$, S_0 is Borel. Then define

$$\tilde{f}(s) = \begin{cases} I(s) \text{ for } s \in S_0 \\ y_0 \text{ for } s \in S - S_0 \end{cases}$$

where y_0 is any fixed element of Y. $\tilde{f}$ is readily seen to satisfy the required conditions.

2.2.17 Corollary: *If S is a G-space with finite invariant measure, then G is ergodic on S if and only if there are no non-trivial G-invariant vectors in $L^2(S) \ominus \mathbb{C}$.*

We remark that this result is no longer true if the measure on S is σ-finite and invariant, because for an invariant set A of infinite measure, χ_A will not be in L^2.

We also observe:

2.2.18 Corollary: *If S is an ergodic G-space, Y is countably separated, and $f: S \to Y$ is Borel and essentially G-invariant, then f is essentially constant.*

Proof: This follows from 2.1.11, 2.2.16, and the fact that for any countably separated Y there is an injective Borel map $Y \to [0, 1]$ (Appendix A).

By Corollary 2.2.17, Moore's theorem can be converted to a problem about unitary representations. Namely, if π is a unitary representation of G with no invariant vectors, and $H \subset G$ is a closed subgroup, we want to know when H can have invariant vectors. To prove Theorem 2.2.15 it clearly suffices to prove the following.

2.2.19 Theorem [*Moore* 1]: *Let G_i be a connected non-compact simple Lie group with finite center, $G = \Pi G_i$ (finite product), and suppose π is a unitary representation of G so that for each G_i, $\pi | G_i$ has no invariant vectors. If $H \subset G$ is a closed subgroup and $\pi | H$ has non-trivial invariant vectors, then H is compact.*

This result in turn follows from the following more general theorem. If π is a unitary representation of G on a Hilbert space $\mathscr{H}$, and $v, w \in \mathscr{H}$ are unit vectors let $f(g) = \langle \pi(g)v | w \rangle$. The function f is called a matrix coefficient of π.

2.2.20 Theorem [*Howe-Moore* 1] [*Sherman* 1] [*Zimmer* 2]: *Let G_i, G, π be as in Theorem* 2.2.19. *Then all matrix coefficients vanish at ∞, i.e., $f(g) \to 0$ as g leaves compact subsets of G.*

Theorem 2.2.19 follows from 2.2.20 because if $\pi | H$ has an invariant vector v, the matrix coefficient $\langle \pi(g)v | v \rangle$ is identically 1 along H, and hence H is compact by 2.2.20.

The remaining two sections of this chapter are devoted to the proof of Theorem 2.2.20, thus providing a proof of all the assertions of this section.

2.3 Unitary representations of semi-direct products, I

In this section, we develop some necessary background to the proof of Theorem 2.2.20. We shall first be proving Theorem 2.2.20 in the case $G = SL(2, \mathbb{R})$. This is the simplest example, but moreover, as a general G contains many subgroups (locally) isomorphic to $SL(2, \mathbb{R})$, we will use the result for $SL(2, \mathbb{R})$ in the proof for arbitrary G. To study representations of $SL(2, \mathbb{R})$, we will first study representations of the upper triangular subgroup $P \subset SL(2, \mathbb{R})$, $P = \left\{ \begin{pmatrix} a & b \\ 0 & a^{-1} \end{pmatrix} \right\}$. We will then use the fact that P and $\bar{P}$, the lower triangular group, together generate G. The representation theory of P that we need will follow from its structure as a semi-direct product, and this is what we consider in this section.

We begin by recalling some basic facts about direct integrals. Suppose M is a measure space and that for each $x \in M$ we have a Hilbert space $\mathscr{H}_x$ such that the assignment $x \to \mathscr{H}_x$ is piecewise constant, i.e., there is a disjoint decomposition of M into measurable subsets, $M = \bigcup_{i=1}^{\infty} M_i$, such that for $x, y \in M_i$, $\mathscr{H}_x = \mathscr{H}_y$. By a section we mean an assignment $x \to f(x)$, where $f(x) \in \mathscr{H}_x$. Since $\mathscr{H}_x$ is piecewise constant, the notion of measurability has an obvious meaning, i.e., that it be a measurable function on each M_i into the appropriate Hilbert space. (Here, we give a Hilbert space the Borel structure coming from the weak topology. However, this is actually the same as the Borel structure defined by the norm topology. See Appendix A.) Let $L^2(M, \{\mathscr{H}_x\})$ be the set of sections such that $\int \| f(x) \|^2 < \infty$, identifying two sections if they agree almost everywhere. $L^2(M, \{\mathscr{H}_x\})$ is also denoted by $\int_M^{\oplus} \mathscr{H}_x$. Then $L^2(M, \{\mathscr{H}_x\})$ is a Hilbert space under the inner product $\langle f | g \rangle = \int_M \langle f(x) | g(x) \rangle$.

Suppose now that for each $x \in M$ we have a unitary representation π_x of a group G on $\mathscr{H}_x$. We say that $x \to \pi_x$ is measurable if for $g \in G$, $\pi_x(g)$ is measurable in (x, g) i.e. it is a measurable function on each $M_i \times G$, where unitary groups have the strong operator topology. We can then define a new representation $\pi = \int^{\oplus} \pi_x$ on $L^2(M, \{\mathscr{H}_x\})$ given by $(\pi(g)f)(x) = (\pi_x(g))(f(x))$. (We remark that $g \to \pi(g)$ is clearly measurable, and hence it is also continuous [Appendix B].) If M is an atomic measure space, then π is just the usual direct sum $\sum_{x \in M}^{\oplus} \pi_x$ on $\sum_{x \in M}^{\oplus} \mathscr{H}_x$. A basic result is the following.

2.3.1 Proposition: *Any unitary representation π of a locally compact (second countable) group G is of the form $\int_M^{\oplus} \pi_x$ for some (standard) measure space M, where all π_x are irreducible.*

For a proof, see [Dixmier 1] for example. The point of this proposition, of course, is to reduce a question about an arbitrary unitary representation to one about irreducible representations. Typical examples are the following.

2.3.2 Proposition: *Let $\pi = \int^{\oplus} \pi_x$.*
(i) Suppose all matrix coefficients of all π_x vanish at ∞. Then all matrix coefficients of π vanish at ∞ as well.
(ii) π has a non-trivial invariant vector if and only if for x in a set of positive measure, π_x has a non-trivial invariant vector.

The proof is a straightforward application of the definitions. In (ii), one may wish to use the von Neumann selection theorem (Appendix A).

We now recall the representation theory of $\mathbb{R}^n$. This can be considered as a generalization of the standard spectral theorem for a single unitary operator. Proofs of our assertions about $\mathbb{R}^n$ can be found in any number of standard books, e.g., [Dixmier 1], [Mackey 3], [Loomis 1].

All the irreducible unitary representations of $\mathbb{R}^n$ are one-dimensional and can be described as follows. For $\theta \in \mathbb{R}^n$, $t \in \mathbb{R}^n$, define $\lambda_\theta(t) = e^{i<\theta|t>}$. Then each λ_θ is a one-dimensional representation (or "character") and every character is of this form. In other words, if we let $\hat{\mathbb{R}}^n$ be the characters of $\mathbb{R}^n$, so that $\hat{\mathbb{R}}^n$ forms a group under pointwise multiplication, then $\theta \to \lambda_\theta$ defines a group isomorphism $\mathbb{R}^n \to \hat{\mathbb{R}}^n$. Now let μ be a (σ-finite) measure on $\hat{\mathbb{R}}^n$ and $\{\mathscr{H}_\lambda\}$ a piecewise constant assignment of Hilbert spaces to $\lambda \in \hat{\mathbb{R}}^n$. Form the Hilbert space $L^2(\hat{\mathbb{R}}^n, \mu, \{\mathscr{H}_\lambda\})$ and define a unitary representation $\pi_{(\mu, \mathscr{H}_\lambda)}$ of $\mathbb{R}^n$ on this space by

$$[\pi_{(\mu,\mathcal{H}_\lambda)}(t)f](\lambda) = \lambda(t)f(\lambda).$$

In other words

$$\pi_{\mu,\mathcal{H}_\lambda} = \int^{\oplus}{}_{(\hat{\mathbb{R}}^n,\mu)} \dim(\mathcal{H}_\lambda)\lambda.$$

(If σ is a representation on $\mathcal{H}$ and $n\in\{\infty, 0, 1, \ldots\}$, by $n\cdot\sigma$ we mean the representation $\sum_{i=1}^{n}{}^{\oplus}\sigma_i$ on $\sum^{\oplus}\mathcal{H}_i$ where all $\sigma_i = \sigma$ and $\mathcal{H}_i = \mathcal{H}$.) The representation theory of $\mathbb{R}^n$ is summarized by:

2.3.3 Proposition: (*i*) *For any unitary representation π of $\mathbb{R}^n$, there exist μ, $\mathcal{H}_\lambda$ on $\hat{\mathbb{R}}^n$ such that $\pi \cong \pi_{(\mu,\mathcal{H}_\lambda)}$.*
(*ii*) *$\pi_{(\mu,\mathcal{H}_\lambda)}$ and $\pi_{(\mu',\mathcal{H}'_\lambda)}$ are unitarily equivalent if and only if*
a) $\mu \sim \mu'$, i.e., they are in the same measure class; and
b) $\dim \mathcal{H}_\lambda = \dim \mathcal{H}'_\lambda$ a.e.

We wish to consider groups containing $\mathbb{R}^n$ as a normal subgroup. Therefore, we first consider the effect of an automorphism of $\mathbb{R}^n$ on its representation theory. The knowledgeable reader will recognize the development as the beginnings of the "Mackey machine" [Mackey 3]. Let $A:\mathbb{R}^n \to \mathbb{R}^n$ be a continuous group automorphism. Let α be the adjoint automorphism $\alpha:\hat{\mathbb{R}}^n \to \hat{\mathbb{R}}^n$, given by $[\alpha(\lambda)](t) = \lambda(A^{-1}(t))$. If π is any unitary representation of $\mathbb{R}^n$, we let $\alpha(\pi)$ be the unitary representation $[\alpha(\pi)](t) = \pi(A^{-1}(t))$. If π is given in the form $\pi = \pi_{(\mu,\mathcal{H}_\lambda)}$, we wish to express $\alpha(\pi)$ in a similar form.

2.3.4 Proposition: *Let $\pi = \pi_{(\mu,\mathcal{H}_\lambda)}$, $A\in Aut(\mathbb{R}^n)$, α the adjoint automorphism of $\hat{\mathbb{R}}^n$. Then*
(*i*) *$\alpha(\pi)$ is unitarily equivalent to $\pi_{(\alpha_*\mu,\mathcal{H}_{\alpha^{-1}(\lambda)})}$. (By $\mathcal{H}_{\alpha^{-1}(\lambda)}$ we mean the assignment $\lambda \to \mathcal{H}_{\alpha^{-1}(\lambda)}$.)*
(*ii*) *If $V: L^2(\hat{\mathbb{R}}^n,\mu,\{\mathcal{H}_\lambda\}) \to L^2(\hat{\mathbb{R}}^n,\alpha_*\mu,\{\mathcal{H}_{\alpha^{-1}(\lambda)}\})$ is any unitary equivalence between $\alpha(\pi)$ and $\pi_{(\alpha_*\mu,\mathcal{H}_{\alpha^{-1}(\lambda)})}$, then for any Borel set $E \subset \hat{\mathbb{R}}^n$, we have*

$$V(L^2(E, \mu, \{\mathcal{H}_\lambda\})) = L^2(\alpha(E), \alpha_*\mu, \{\mathcal{H}_{\alpha^{-1}(\lambda)}\}).$$

Proof: Define $U: L^2(\hat{\mathbb{R}}^n, \mu, \{\mathcal{H}_\lambda\}) \to L^2(\hat{\mathbb{R}}^n, \alpha_*\mu, \{\mathcal{H}_{\alpha^{-1}(\lambda)}\})$ by $(Uf)(\lambda) = f(\alpha^{-1}(\lambda))$. Then one immediately verifies that U is unitary and in fact provides the unitary equivalence required in (i). It is clear from the definition that U satisfies the conclusion of (ii). If V is any unitary equivalence as in (ii), then $U^{-1}V$ commutes with all operators $\alpha(\pi)(t)$, $t\in\mathbb{R}^n$, i.e. with all operators $\pi(t)$. We recall that this is given by $(\pi(t)f)(\lambda) = \lambda(t)f(\lambda)$. But then a standard argument shows that $U^{-1}V$ commutes with all operators of the form $(\pi_\varphi f)(\lambda) = \varphi(\lambda)f(\lambda)$ where $\varphi \in L^\infty(\hat{\mathbb{R}}^n)$. (Namely, one sees easily that if $\|\varphi_N\|_\infty, \|\varphi\|_\infty \leqq M$, and $\varphi_N \to \varphi$ pointwise a.e., then $\pi_{\varphi_N} \to \pi_\varphi$ in the weak operator topology. Thus, if π_{φ_N} commutes with an operator A for all N, then π_φ commutes with A as well. Thus, it suffices to prove our assertion if φ is smooth with compact support. Identify $\hat{\mathbb{R}}^n$ with $\mathbb{R}^n$. For N sufficiently large, $[-N, N]^n \supset \operatorname{supp}(\varphi)$ and we let φ_N be the $(N \cdot Z)^n$-invariant function that is equal to φ on $[-N, N]^n$. Since $\varphi_N \to \varphi$ pointwise it suffices to prove our assertion for each φ_N. However φ_N is the uniform limit of its Fourier series, and hence the assertion for φ_N follows from the validity of the assertion for $\lambda \to \lambda(t)$ for all t.) Letting φ be the characteristic function of the set E, we obtain that $U^{-1}V$ leaves $L^2(E, \mu, \{\mathcal{H}_\lambda\})$ invariant. Thus

$$V(L^2(E, \mu, \{\mathcal{H}_\lambda\})) \subset U(L^2(E, \mu, \{\mathcal{H}_\lambda\})) = L^2(\alpha(E), \alpha_*\mu, \{\mathcal{H}_{\alpha^{-1}(\lambda)}\}).$$

Applying the same argument to $V^{-1}U$ completes the proof.

We now apply this proposition to representations of groups containing $\mathbb{R}^n$ as a normal subgroup. Suppose G is such a group. Then for each $g \in G$, conjugation by g is an automorphism of $\mathbb{R}^n$. Thus, we have an action of G on $\hat{\mathbb{R}}^n$ given by $(g \cdot \lambda)(t) = \lambda(g^{-1}tg)$, and similarly if π is any unitary representation of $\mathbb{R}^n$, we define $(g \cdot \pi)(t) = \pi(g^{-1}tg)$. Of course, the conclusions of 2.3.4 hold for each g.

2.3.5 Proposition [*Mackey* 3]: *Suppose* $\mathbb{R}^n \subset G$ *is a normal subgroup and* π *is a unitary representation of* G. *Write* $\pi|\mathbb{R}^n \cong \pi_{(\mu, \mathcal{H}_\lambda)}$ *for some* $(\mu, \mathcal{H}_\lambda)$ *by* 2.3.3. *Then*
(i) μ *is quasi-invariant under the action of* G *on* $\hat{\mathbb{R}}^n$.
(ii) If $E \subset \hat{\mathbb{R}}^n$ *is measurable, let* $\mathcal{H}_E = L^2(E, \mu, \{\mathcal{H}_\lambda\})$. *Then* $\pi(g)\mathcal{H}_E = \mathcal{H}_{g \cdot E}$.
(iii) If π *is irreducible, then* μ *is ergodic and* $\dim \mathcal{H}_\lambda$ *is constant on a* μ*-conull set.*

Proof: For $g \in G$ and $t \in \mathbb{R}^n$, we have $\pi(g^{-1}tg) = \pi(g)^{-1}\pi(t)\pi(g)$. In other words $\pi(g)$ provides a unitary equivalence of $\pi|\mathbb{R}^n$ and $g \cdot (\pi|\mathbb{R}^n)$, i.e., of $\pi_{(\mu, \mathcal{H}_\lambda)}$ and $g \cdot \pi_{(\mu, \mathcal{H}_\lambda)}$. From 2.3.4 (i) and 2.3.3 (ii), we deduce that μ is quasi-invariant and $\dim \mathcal{H}_\lambda$ is an essentially invariant function under G. (We may therefore assume $\lambda \to \mathcal{H}_\lambda$ is essentially invariant under G.) Thus, if we fix $g \in G$, we can define an operator $T: L^2(\hat{\mathbb{R}}^n, \mu, \{\mathcal{H}_\lambda\}) \to L^2(\hat{\mathbb{R}}^n, g*\mu, \{\mathcal{H}_{g^{-1}\lambda}\})$ by $Tf = (d\mu/dg_*\mu)^{1/2}f$. This is unitary and intertwines the representations $\pi_{(\mu, \mathcal{H}_\lambda)}$ and $\pi_{(g_*\mu, \mathcal{H}_{g^{-1}\lambda})}$. It follows that $T \circ \pi(g): L^2(\hat{\mathbb{R}}^n, \mu, \{\mathcal{H}_\lambda\}) \to L^2(\hat{\mathbb{R}}^n, g_*\mu, \{\mathcal{H}_{g^{-1}\lambda}\})$ is a unitary equivalence of $g \cdot \pi_{(\mu, \mathcal{H}_\lambda)}$ and $\pi_{(g_*\mu, \mathcal{H}_{g^{-1}\lambda})}$). By 2.3.4 (ii), if $E \subset \hat{\mathbb{R}}^n$ is measurable, $(T\pi(g))(\mathcal{H}_E) = L^2(gE, g_*\mu, \{\mathcal{H}_{g^{-1}\lambda}\})$. Thus (ii) follows from the observation that $T^{-1}(L^2(gE, g_*\mu, \{\mathcal{H}_{g^{-1}\lambda}\})) = \mathcal{H}_{gE}$.

If μ is not ergodic, let $E \subset \hat{\mathbb{R}}^n$ be a G-invariant set which is neither null nor conull. Then by (ii), $\mathscr{H}_E$ is a proper G-invariant subspace, showing that π is reducible. Finally, if μ is ergodic, the fact that $\dim \mathscr{H}_\lambda$ is essentially constant follows from the fact that it is essentially invariant.

We can now easily deduce the information concerning representations of

$$P = \left\{ \begin{pmatrix} a & b \\ 0 & a^{-1} \end{pmatrix} \middle| a, b \in \mathbb{R}, a \neq 0 \right\}$$

that we shall need. Let N be the set of matrices in P with $a = 1$, and A the set of matrices with $b = 0$. Then $N \subset P$ is normal, $N \cong \mathbb{R}$, and P is a semi-direct product $P = AN$. By pursuing the development of the above construction, one can specifically list all the unitary representations of P. (In fact, we will discuss this in Chapter 7.) However, here we will simply observe the following qualitative result which will be sufficient for our present purposes.

2.3.6 Theorem: *Let π be a unitary representation of $P = AN$. Then either*
(i) $\pi|N$ has non-trivial invariant vectors or
(ii) for $g \in A$, and any vectors v, w, $\langle \pi(g)v|w\rangle \to 0$ as $g \to \infty$ (in A).

2.3.7 Corollary: *Let π be a unitary representation of P. Then any A-invariant vector is also P-invariant.*

Proof (of corollary): If π is a representation on $\mathscr{H}$, let $W = \{v \in \mathscr{H} | v$ is $\pi(N)$-invariant$\}$. Since N is normal, W is a P-invariant subspace, and hence so is $W^\perp$. Since $W^\perp$ has no N-invariant vectors, we are in case (ii) in the above theorem for the representation on $W^\perp$, and in particular, there are no A-invariant vectors in $W^\perp$. Since the projection onto $W^\perp$ of an A-invariant vector in $\mathscr{H}$ is also A-invariant, this shows that all A-invariant vectors are in W, and hence that all A-invariant vectors are P-invariant.

Proof of Theorem 2.3.6: We apply 2.3.5, identifying $N \cong \mathbb{R}$. Let $\pi|N = \pi_{(\mu, \mathscr{H}_\lambda)}$. If $\mu(\{0\}) > 0$, then $\pi|N$ has invariant vectors (namely $\mathscr{H}_0$). We now show that if $\mu(\{0\}) = 0$, then assertion (ii) in the theorem is satisfied. To see this, consider the action of P on $\hat{N}$. An elementary calculation shows that

$$\begin{pmatrix} a & 0 \\ 0 & a^{-1} \end{pmatrix} \in A$$

acts on $\hat{N} \cong \hat{\mathbb{R}}$ via multiplication by a^2. Hence, given any compact subsets $E, F \subset \hat{\mathbb{R}} - \{0\}$, for $g \in A$ outside a sufficiently large compact set we have $\mu(gE \cap F) = 0$. Given any two unit vectors $f, h \in L^2(\hat{\mathbb{R}}, \mu, \{\mathscr{H}_\lambda\})$, and $\varepsilon > 0$ we can choose compact subsets $E, F \subset \hat{\mathbb{R}} - \{0\}$ such that

$$\| \chi_E \cdot f - f \| \leqq \varepsilon \text{ and } \| \chi_F \cdot h - h \| \leqq \varepsilon.$$

Then

$$|\langle \pi(g) f | h \rangle| \leqq 2\varepsilon + |\langle \pi(g)(\chi_E \cdot f) | (\chi_F \cdot h) \rangle|.$$

But $\pi(g)(\chi_E \cdot f) \in \mathscr{H}_{gE}$ by 2.3.5 (ii) and by our above remark, choosing $g \in A$ outside a sufficiently large compact subset of A we can ensure $\mathscr{H}_{gE} \perp \mathscr{H}_F$, and hence that $|\langle \pi(g) f | h \rangle| \leqq 2\varepsilon$. This completes the proof of the theorem.

Theorem 2.3.6 gives a vanishing theorem for the matrix coefficients of representations of P. In the next section we will see how to use this to prove Moore's theorem.

2.4 Vanishing of matrix coefficients for semisimple groups

We begin with the proof of Theorem 2.2.20 for $G = SL(2, \mathbb{R})$. As we indicated in the previous section, this result will be used in the proof for general G.

To see the idea of the proof, let us recall the polar decomposition of a matrix. If $T \in SL(n, \mathbb{R})$, then we can write $T = U \cdot S$, where U is orthogonal and S is a positive definite symmetric matrix. However S has an orthonormal basis of eigenvectors, i.e., $S = U_0 D U_0^{-1}$ where D is a positive diagonal matrix and U_0 is orthogonal. Thus, any $T \in SL(n, \mathbb{R})$ has an expression $T = U_1 D U_2$ where $U_i \in SO(n, \mathbb{R})$ and D is a positive diagonal matrix. Thus, we can write $SL(n, \mathbb{R}) = KAK$, where $K = SO(n, \mathbb{R})$ is compact and A is the group of positive diagonals. This is called the Cartan decomposition of $SL(n, \mathbb{R})$.

2.4.1 Lemma: *If π is a unitary representation of a group G, and we can write $G = KAK$ where K is a compact group, to see that all matrix coefficients of π vanish at ∞ in G, it suffices to see $f(a) \to 0$ for $a \in A$, $a \to \infty$ in A, for all matrix coefficients f of π.*

Proof: We observe that if v, w are vectors and $g \in G$ is expressed as $g = k_1 a k_2$, $k_i \in K$, $a \in A$, then $\langle \pi(g) v | w \rangle = \langle \pi(a)(\pi(k_2) v) | (\pi(k_1)^{-1} w \rangle$. If this fails to vanish at

∞, we can find $g_n \to \infty$ such that $|\langle \pi(g_n)v|w\rangle| \geqq \varepsilon$ for some $\varepsilon > 0$. Let $g_n = k_{1n}a_nk_{2n}$. By passing to a subsequence we can suppose $k_{2n} \to k$ and $k_{1n}^{-1} \to k'$. Then an elementary argument shows that for n sufficiently large $|\langle \pi(a_n)(\pi(k)v)|\pi(k')w\rangle| \geqq \varepsilon/2$. But since K is compact and $g_n \to \infty$, we must have $a_n \to \infty$. This shows there is a matrix coefficient of $\pi|A$ that fails to vanish at ∞, completing the proof.

We now prove 2.2.20 for $SL(2, \mathbb{R})$.

2.4.2 Theorem: *If π is a unitary representation of $G = SL(2, \mathbb{R})$ with no invariant vectors, then all matrix coefficients of π vanish at ∞.*

Proof: By the preceding lemma, it suffices to see that the matrix coefficients vanish at infinity along A and by Theorem 2.3.6, it suffices to see that there are no N invariant vectors. Suppose to the contrary that $v \neq 0$ is N-invariant. Let $f(g) = \langle \pi(g)v|v\rangle$. Then f is continuous and bi-invariant under N, i.e., f lifts from a continuous N-invariant function on G/N. Now N is exactly the stabilizer of a vector (namely (1, 0)) in $\mathbb{R}^2$ under the natural $SL(2, \mathbb{R})$ action. Thus, we can identify G/N with $\mathbb{R}^2 - \{0\}$. The action of N on G/N is therefore identified with the action on $\mathbb{R}^2 - \{0\}$ given by ordinary matrix multiplication. Thus there are two types of orbits, namely all horizontal lines except the x-axis, and each point on the x-axis (except the origin, of course). Clearly any continuous function on $\mathbb{R}^2 - \{0\} \cong G/N$ which is constant along these orbits must actually be constant on the x-axis. But the x-axis is identified with $P/N \subset G/N$ under the identification of G/N with $\mathbb{R}^2 - \{0\}$. Hence $f(g)$ is constant on P. However, since π is unitary, if $f(g) = \langle \pi(g)v|v\rangle$ is constant on P, it follows that v must be P-invariant. Therefore f is actually bi-invariant under P. But P has a dense orbit in G/P. (For example, identify G/P with projective space of $\mathbb{R}^2$ under ordinary matrix multiplication.) Thus f is actually a constant function, and as above, this implies that v is G-invariant.

We are now ready to prove 2.2.20.

Proof of Theorem 2.2.20: (Vanishing of matrix coefficients). Following our remark in the preface, we shall prove this in detail for $G = SL(n, \mathbb{R})$, and then indicate how the proof carries over to general G.

Let $A \subset SL(n, \mathbb{R})$ be the group of diagonal matrices. We denote an element $a \in A$ by $(a_1, \ldots, a_n)$, where these are to be interpreted as the diagonal elements of a matrix. We note $\Pi a_i = 1$. Let B be the set of matrices (c_{ij}) with $c_{ii} = 1$, and $c_{ij} = 0$ for $i \neq j$ and $i \geqq 2$. We denote an element $b \in B$ by $b = (1, b_2, \ldots, b_n)$ where this is to be interpreted as the first row of the corresponding matrix. Then

$aBa^{-1} = B$ for $a \in A$, and hence $H = AB$ is a subgroup of G, and $B \subset H$ is normal. We observe $B \cong \mathbb{R}^{n-1}$. As with $SL(2, \mathbb{R})$, by Lemma 2.4.1, it suffices to show that the matrix coefficients of $\pi|A$ vanish at ∞. For $SL(2, \mathbb{R})$ we obtained this using knowledge of the representation of P. In our more general situation, we will examine the representation of H. (Note that $H = P$ for $n = 2$.)

Express $\pi|B \cong \pi_{(\mu, \mathscr{H}_\lambda)}$ (by 2.3.3) via the above identification of B with $\mathbb{R}^{n-1}$. Matrix multiplication shows that for $a \in A$, $b \in B$, $aba^{-1} = (1, a_1 a_2^{-1} b_2, \ldots, a_1 a_n^{-1} b_n) \in B$. The adjoint action on $\hat{\mathbb{R}}^{n-1}$ will be given by the same expression, replacing b_i by the dual variables λ_i, $i = 2, \ldots, n$. Therefore, if E, $F \subset \hat{\mathbb{R}}^{n-1}$ are compact subsets which are disjoint from the union of the hyperplanes $\lambda_i = 0$, $i = 2, \ldots, n$ then for $a \in A$ outside a sufficiently large compact set, we have $a \cdot E \cap F = \varnothing$. Therefore, arguing exactly as in the proof of Theorem 2.3.6, we deduce that if μ assigns measure 0 to the union of the hyperplanes $\lambda_i = 0$, then all matrix coefficients vanish along A, and by our comments above, this suffices to prove the theorem. Therefore, it remains to show that $\mu(\{\lambda_i = 0\}) > 0$ is impossible.

If $\mu(\{\lambda_i = 0\}) > 0$, then by definition of $\mu_{(\mu, \mathscr{H}_\lambda)}$, the subgroup $B_i \subset B$, $B_i = \{b \in B | b_j = 0 \text{ for } i \neq j\}$ leaves non-trivial vectors invariant (namely, the subspace $\mathscr{H}_{\{\lambda_i = 0\}}$.) However $B_i \subset H_i \subset G$ where $H_i \cong SL(2, \mathbb{R})$ and is defined as follows

$$H_i = \{(c_{jk}) \in SL(n, \mathbb{R}) | c_{jj} = 1 \text{ for } j \neq 1, i, \text{ and for } j \neq k \text{ and } \{1, i\} \neq \{j, k\}, c_{jk} = 0\}.$$

From the vanishing of matrix coefficients for $SL(2, \mathbb{R})$, (2.4.2), the existence of a B_i-invariant vector implies the existence of a H_i-invariant vector (since B_i is clearly non-compact). In particular, $A_i = H_i \cap A$ has non-trivial invariant vectors. Let $W = \{v \in \mathscr{H} | \pi(a)v = v \text{ for all } a \in A_i\}$. It suffices to show that W is G-invariant. For then the representation π_W of G on W has kernel $(\pi_W) \supset A_i$ which by simplicity of G implies that kernel$(\pi_W) = G$, so that G itself leaves all vectors in W fixed, contradicting our assumptions. (For the analogous argument in the semisimple case the fact that dim(kernel $\pi_W) > 0$ contradicts the assumption that no simple factor of G leaves vectors invariant.)

We now turn to G-invariance of W. For $k \neq j$, let $B_{kj} \subset G$ be the one-dimensional subgroup defined by $B_{kj} = \{(c_{rs}) | c_{rr} = 1, \text{ and for } r \neq s \text{ and } (r, s) \neq (k, j), c_{rs} = 0\}$. We consider two possibilities.

(i) $k \neq i$ or 1 and $j \neq i$ or 1. Then B_{kj} commutes with A_i, and hence B_{kj} leaves W invariant.

(ii) If $\{k, j\} \cap \{i, 1\} \neq \varnothing$ then A_i normalizes B_{kj}. Hence $A_i B_{kj}$ is a 2-dimensional subgroup and is isomorphic to P in such a way that $A_i \leftrightarrow$(diagonal matrices

in P), $B_{kj} \leftrightarrow N$. By Corollary 2.3.7, all A_i-invariant vectors are also B_{kj} invariant. Hence in this case, too, B_{kj} leaves W invariant.
Finally, we remark that since $A_i \subset A$, A abelian, A also leaves W invariant. However, A and all B_{kj} together generate G. Therefore G leaves W invariant, completing the proof.

In concluding this section, we indicate the modifications necessary in the above argument for a general semisimple G. Let $A \subset G$ be a maximal $\mathbb{R}$-split torus. Then $A \subset G' \subset G$ where G' is semisimple and split over $\mathbb{R}$, and A is the maximal $\mathbb{R}$-split torus of G'. Choose a maximal linearly independent set S of positive roots of G' relative to A such that for $\alpha, \beta \in S$, $\alpha + \beta$ is not a root. Then the direct sum of the root spaces is the Lie algebra of an abelian subgroup $B \subset G'$, with $\dim B = \dim A$, and B is normalized by A. The representations of AB can be analyzed exactly as in the case of $SL(n, \mathbb{R})$, and since the relevant copies of $sl(2, \mathbb{R})$ are present, we deduce that either we are done, or some one-dimensional subgroup $A_0 \subset A$ leaves a non-trivial vector fixed. (Actually to obtain this we may need to use the universal covering $\tilde{G}$ of $SL(2, \mathbb{R})$ rather than $SL(2, \mathbb{R})$ itself. Namely, we need that for $N \subset SL(2, \mathbb{R})$ as in the proof of 2.4.2, $\tilde{N} \subset \tilde{G}$ the connected component of the lift of N to $\tilde{G}$ (so that $\tilde{N} \cong N$), that $\tilde{N}$ invariant vectors are $\tilde{G}$-invariant. However, this follows by elementary covering space arguments applied to the picture in the proof of 2.4.2. If G is algebraic, which will be our main concern, consideration of $SL(2, \mathbb{R})$ suffices.) The proof then proceeds as in the case of $SL(n, \mathbb{R})$; G is generated by elements that either commute with A_0 or lie in a suitable copy of the group P.

3 Algebraic Groups and Measure Theory

3.1 Review of algebraic groups

In this section, we shall review without proof some basic notions concerning algebraic groups. For proofs and further details, we refer to the reader to the standard works on the subject, e.g. [Borel 1], [Humphreys 1], [Chevalley 1], [Borel-Tits 1], [Springer 1], [Steinberg 1].

Let $k \subset K$ be fields, with K algebraically closed. For convenience, and as it is all we shall require in the sequel, we assume the fields are of characteristic 0. By an affine variety we mean a subset $V \subset K^n$ which is the set of zeroes of some polynomial ideal in $K[X_1, \ldots, X_n]$. V is said to be defined over k, or to be an affine k-variety if it is actually the set of zeroes of a polynomial ideal in $k[X_1, \ldots, X_n]$. The affine (k-) varieties contained in V are the closed subsets of a topology on V, called the Zariski ($k-$) topology. Let $I(V)$ be the ideal of all polynomials over K vanishing on V, and $I_k(V)$ the corresponding polynomial ideal over k. Then $K[V] = K[X_1, \ldots, X_n]/I(V)$ can be identified with a set of K-valued functions on V, called the ring of regular functions on V. Similarly $k[V] = k\ [X_1, \ldots, X_n]/I_k\ (V)$ is the ring of regular functions defined over k (and we can clearly identify $k[V] \subset K[V]$). Any affine variety V can be written as a finite union $V = \cup V_i$, where V_i are irreducible, i.e., not the union of two proper closed subsets. If V is irreducible, we set $K(V) =$ field of fractions of $K[V]$, called the field of rational functions on V. Similarly, if V is an irreducible k-variety, we have the field $k(V)$ of rational functions defined over k. A rational function is defined at all points except on a closed subvariety. One can show that a rational function defined at all points of V is actually a regular function. If V is an affine (k-) variety and $p \in K[V]$ ($p \in k[V]$), then $\{a \in V \mid p(a) \neq 0\}$ is called a principal (k-) open set in V. Then $(x_1, \ldots, x_n) \to (x_1, \ldots, x_n, 1/p(x_1, \ldots, x_n))$ defines a bijection of this open set onto an affine (k-) variety U in K^{n+1}, and $K[U] \cong (K[V])[1/p]$. We can clearly speak of regular and rational mappings (over k) defined between affine k-varieties by examining the coordinate functions. If V is an affine variety and $A \subset K$ is a subring, then we set $V_A = V \cap A^n$. For example, if $A = k$, then V_k can alternatively be described as exactly those elements of V left fixed by the Galois group $\mathrm{Gal}(K/k)$. If $f\colon V \to W$ is a rational mapping of affine k-varieties and is defined over k, then $f(V_k) \subset W_k$. By a (k-) isomorphism of affine (k-) varieties, we mean a bijective bi-regular map (defined over k).

We enlarge the spaces under consideration by defining a variety in general to be a space which locally looks like an affine variety. More precisely, by a pre-variety V we mean a topological space V which is a finite union $V = \cup U_i$ such that

(i) each U_i is an open set which is itself an affine variety;
(ii) $U_i \cap U_j$ is a principal open set in both U_i and U_j, and the two affine structures they induce on $U_i \cap U_j$ are equal.

One can form products of affine varieties and pre-varieties, and finally one calls a pre-variety a variety if it satisfies the technical condition that the diagonal is closed in $V \times V$. By insisting that everything be defined over k, we can clearly speak of a k-variety, and of $V_k \subset V$.

Examples of varieties:

(i) affine varieties.
(ii) open sets in affine varieties (being a finite union of principal open subsets).
(iii) projective space.
(iv) open and closed sets in a variety.

A closed subvariety of projective space is called a projective variety and an open subvariety of a projective variety is called quasi-projective. Thus the class of quasi-projective varieties contains both the affine and projective varieties. Regular functions into K and regular maps between varieties are defined to be continuous functions that are regular when restricted to mappings between open affine subvarieties. A rational map is one that is regular on an open set. Clearly, all of these notions can be defined over k if the varieties are, and if f is defined over k, we still have $f(V_k) \subset W_k$.

The group $GL(n, K)$ is a principal open set in K^{n^2}, and hence is a variety. (In fact, since it is defined by the non-vanishing of det, it is defined over $\mathbb{Q}$. Recall $\mathrm{char}(K) = 0$, so $\mathbb{Q} \subset K$.) By an algebraic (k-) group we mean a Zariski (k-) closed subgroup of $GL(n, K)$. For example, $SL(n, K)$ is an algebraic $\mathbb{Q}$-group. For any ring $A \subset K$, let $GL(n, A) = \{(a_{ij}) \in GL(n, K) | a_{ij} \in A \text{ and } \det (a_{ij})^{-1} \in A\}$. For any algebraic group G, let $G_A = G \cap GL(n, A)$. By an algebraic action of an algebraic group G we mean an action of G on a variety V such that $V \times G \to V$, $(x, g) \to xg$ is regular. In particular, for each $x \in V$, $g \to xg$ is regular. It is a fundamental fact about regular maps that if $f: V \to W$ is regular, then $f(V)$ contains an open set in its closure. (See the above references, for example.) Applying this to the case of actions, we deduce that we have $U \subset xG \subset \overline{xG}$, and so $UG = xG$, and hence xG is open in $\overline{xG}$. Therefore, we deduce:

3.1.1 Theorem: *If an algebraic group acts algebraically on a variety, then every orbit is locally closed.*

The following is a straightforward consequence.

3.1.2 Corollary: *If $\varphi: G \to H$ is a regular homomorphism of algebraic groups, then $\varphi(G)$ is an algebraic group.*

If $K = \mathbb{C}$, then any variety also has a locally compact Hausdorff topology coming from the usual topology on $\mathbb{C}^n$. Clearly, the map (V, Hausdorff topology) $\to$ (V, Zariski topology) is continuous. Hence, for an algebraic action the orbits are locally closed in the Hausdorff topology as well. The following fundamental fact generalizes this to local fields. We remark that if k is a local field (i.e., locally compact, non-discrete, and with our standing assumption about all our fields, of characteristic 0), and V is a variety defined over k, then V_k also has a locally compact topology, which we call the Hausdorff topology.

3.1.3 Theorem: [*Borel-Serre* 1] *If k is a local field of characteristic* 0, *and a k-group G acts k-algebraically on a k-variety V, then every G_k-orbit in V_k is locally closed in the Hausdorff topology.*

The proof of Borel and Serre depends upon showing that a certain Galois cohomology group is finite. A good introduction to Galois cohomology for the purposes of this theorem is [Springer 2].

If G and H are k-groups and $\varphi: G \to H$ is a regular homomorphism defined over k, then $\varphi(G)$ is a k-group, but it is not necessarily true that $\varphi(G_k) = \varphi(G)_k$. (For example, consider the map $\varphi(z) = z^2$ on $\mathbb{C}^*$. Then $\varphi(\mathbb{R}^*) \neq \mathbb{R}^*$.) However, with k as in 3.1.3 it is always true that $\varphi(G_k) \subset \varphi(G)_k$ is a subgroup of finite index.

If $H \subset G$ is an inclusion of k-groups, then G/H has a natural quotient structure as a k-variety (quasi-projective) on which G acts k-algebraically. We have an injection $G_k/H_k \to (G/H)_k$, although this is not always a surjection. With k as in 3.1.3 the image is locally closed, by the above theorem. Furthermore, there are only finitely many G_k-orbits in $(G/H)_k$ [Borel–Serre 1]. The existence of such a structure on G/H (for arbitrary k, not just for local fields) is usually proven using the following theorem of Chevalley.

3.1.4 Proposition: *If $H \subset G$ is a k-subgroup of G, then there is a rational representation (i.e., a regular homomorphism $G \to GL(n, K)$) defined over k and a point $x \in \mathbb{P}^{n-1}(k)$ such that H is the stabilizer of x in G.*

As an example of why the condition that groups be algebraic rather than just Lie is very useful in questions related to ergodic behavior, let us observe the following consequence of Theorem 3.1.3 and Proposition 2.1.12.

3.1.5 Corollary: *Let G be as in Theorem 3.1.3 and H, $J \subset G$ k-subgroups. Then every G_k-orbit on $G_k/H_k \times G_k/J_k$ is locally closed. In particular, every ergodic measure under the action of G_k is supported on an orbit.*

The analogue of this statement for Lie groups is of course very far from true. In fact, we saw many examples in the preceding chapter. (Although the examples in Chapter 2 considered the case in which one of the subgroups is discrete, it is easy to give examples of connected subgroups H, J of a connected Lie group G so that the orbit space of G acting on $G/H \times G/J$ is not countably separated. For example taking the product of 2 copies of the usual action of the circle on $\mathbb{R}^2$ gives us a representation of the torus T^2 on $\mathbb{R}^4$. There is a homomorphism $\mathbb{R} \to T^2$ with dense image (namely, a line with irrational slope), and this gives a representation of $\mathbb{R}$ on $\mathbb{R}^4$. Let G be the associated semi-direct product $\mathbb{R} \ltimes \mathbb{R}^4$. Then the orbit space of the action of G on $G/\mathbb{R} \times G/\mathbb{R}$, being the same as the orbit space of $\mathbb{R}$ acting on $G/\mathbb{R} \cong \mathbb{R}^4$, will not be countably separated. This is the well-known Mautner group.)

An algebraic group is called connected if there are no proper open subgroups. If we let G^0 be the largest connected subgroup of G, then G/G^0 is finite. If G is defined over k, so is G^0. *If* $K = \mathbb{C}$, then G^0 is just the connected component of the identity in the Hausdorff topology. If G is an $\mathbb{R}$-group and is connected, it is not necessarily true that $G_{\mathbb{R}}$ is connected in the Hausdorff topology (for example $GL(n, \mathbb{R})$ is not connected), but the connected component $G^0_{\mathbb{R}}$ (Hausdorff topology) will be of finite index in $G_{\mathbb{R}}$. Groups of the form $G^0_{\mathbb{R}}$ in fact include all semi-simple Lie groups with trivial center.

3.1.6 Proposition: *Let H be a connected semisimple Lie group with trivial center. Then there is a connected (semi-simple) algebraic group $G \subset GL(n, \mathbb{C})$ defined over $\mathbb{Q}$ such that as Lie groups, H and $G^0_{\mathbb{R}}$ are isomorphic.*

Proof: Let $L(H)_{\mathbb{C}}$ be the complexified Lie algebra of H. Via the complexified adjoint representation, H acts as automorphisms of the Lie algebra $L(H)_{\mathbb{C}}$. If we let $G = \mathrm{Aut}(L(H)_{\mathbb{C}})^0$, then G is easily seen to be an algebraic group over $\mathbb{R}$ and it will be a k-group ($k \subset \mathbb{R}$) if there is a basis with respect to which the structural constants of the Lie algebra lie in k. It is well known from Lie theory that $H = G^0_{\mathbb{R}}$, and that the structural constants can be chosen to lie in $\mathbb{Q}$.

The following basic result of Borel and Harish-Chandra then shows that lattices exist in any semi-simple Lie group.

3.1.7 Theorem: [*Borel–Harish–Chandra* 1]: *Let* $G \subset GL(n, \mathbb{C})$ *be a semisimple algebraic group defined over* $\mathbb{Q}$. *Then* $G_{\mathbb{Z}}$ *is a lattice in* $G_{\mathbb{R}}$.

In fact, in [Borel 4] it is shown that any connected semisimple non-compact Lie group admits a cocompact lattice. Small modifications in Borel's proof show that any such group also has a non-cocompact lattice [Raghunathan 1]. For a complete proof of Theorem 3.1.7 see [Borel 2] or [Borel–Harish–Chandra 1].

It will be useful to have the following criterion for determining when an algebraic group is defined over a fixed $k \subset K$.

3.1.8 Proposition: *Suppose* $G \subset GL(n, K)$ *is an algebraic group and that* $G \cap GL(n, k)$ *is Zariski dense in* G *for some subfield* $k \subset K$. *Then* G *is defined over* k.

Proof: [Margulis 3] Let I_d be the polynomials of degree $\leqq d$ annihilating G. Since $G \cap GL(n, k)$ is Zariski dense, a polynomial p of degree $\leqq d$ is in I_d if and only if $p(g) = 0$ for all $g \in G \cap GL(n, k)$. We can consider this as a system of homogeneous linear equations, one equation for each $g \in G \cap GL(n, k)$, where the unknowns are the r coefficients of p, and the coefficients of the linear equations are in k. Since there are r unknowns, we can find r of the equations so that the solutions of the original system are the same as the simultaneous solutions to the r equations. But the kernel of an $r \times r$ matrix with coefficients in k considered as a linear transformation on K^r has a basis of elements in k^r. Thus, I_d has a basis of elements with coefficients in k, and since this is true for all d, it is clear that G is defined over k.

In the converse direction we have the following theorem of Rosenlicht and Chevalley.

3.1.9 Theorem: [*Rosenlicht* 1], [*Chevalley* 1]. *Suppose* G *is a connected algebraic* k*-group (where* $char(k) = 0$*). Then* G_k *is Zariski dense in* G. *If* $k = \mathbb{R}$, *and* $G^0_{\mathbb{R}}$ *is the connected component in the Hausdorff topology, then* $G^0_{\mathbb{R}}$ *is Zariski dense in* G.

If G is an $\mathbb{R}$-group, then G is a complex Lie group. $G_{\mathbb{R}}$ is a real Lie group, and we have $\dim_{\mathbb{R}}(G_{\mathbb{R}}) = \dim_{\mathbb{C}}(G)$. (See [Chevalley 1].) From this, 3.1.9 follows for $k = \mathbb{R}$. For the proof in general, see [Rosenlicht 1] or [Borel–Springer 1].

The next proposition gives a useful criterion for determining when a regular map is defined over $k \subset K$.

3.1.10 Proposition: *Suppose* V, W *are* k*-varieties and that* $f: V \to W$ *is a regular* *map. Suppose there is a set $A \subset V_k$ which is Zariski dense in V such that $f(A) \subset W_k$. Then f is defined over k.*

Proof: It suffices to consider the case in which both V, W are affine, and by considering coordinate functions, the case in which $W = K$. Then we can write $f = f_0 + \sum_{i=1}^{r} \alpha_i f_i$ where f_0, f_i are polynomials with coefficients in k and $\alpha_i \in K$ with $\{1, \alpha_1, \ldots, \alpha_r\}$ linearly independent over k. Since $f(A) \subset k$, it follows from the linear independence of $\{1, \alpha_1, \ldots, \alpha_r\}$ that $f(a) = f_0(a)$ for all $a \in A$. Since A is Zariski dense in V, $f = f_0$ on V, verifying the proposition.

If $f: G \to G'$ is a regular map of algebraic groups which is an isomorphism of abstract groups, then f^{-1} is also regular, i.e., f is an isomorphism of algebraic groups. From 3.1.10 we obtain:

3.1.11 Corollary: *Suppose G, G' are k-groups and $f: G \to G'$ is an isomorphism. If f is defined over k, so is f^{-1}.*

We recall that a connected algebraic group G is called semisimple if the radical of G (i.e. the maximal normal connected solvable subgroup) is trivial. If G is a k-group, G is called k-simple if every proper normal k-subgroup is trivial and almost k-simple if every proper normal k-subgroup is finite. (If G is a k-group, so is its radical, and hence semisimplicity does not depend upon a field of definition $k \subset K$.) A regular (k-) homomorphism $f: G \to H$ is called a (k-) isogeny if it is surjective with finite kernel. For every connected semisimple G there exist algebraic groups $\tilde{G}$, G' and isogenies $\tilde{G} \xrightarrow{\pi_1} G \xrightarrow{\pi_2} G'$ with the following properties. For every isogeny $H \to G$, there is an isogeny $\tilde{G} \to H$ such that the composition $\tilde{G} \to H \to G$ is π_1. For every isogeny $G \to H$, there is an isogeny $H \to G'$ such that the composition $G \to H \to G'$ is π_2. $\tilde{G}$ is called the (algebraic) universal covering of G and G' the adjoint of G. If G is a k-group, so are $\tilde{G}$ and G' and π_1 and π_2 are defined over k. G is called (algebraically) simply connected if $G = \tilde{G}$ and is called an adjoint group if $G = G'$. For example, $SL(n, \mathbb{C})$ is (algebraically) simply connected and $PSL(n, \mathbb{C})$ is an adjoint group. If G is a connected semisimple k-group which is either simply connected or adjoint, then $G \cong \Pi G_i$ as k-groups where G_i is an almost k-simple k-group. If G is adjoint, then G_i are k-simple. (See [Tits 2] for further details.)

Finally, suppose $k \subset K$ are fields of characteristic 0, K algebraically closed. Let $G \subset GL(n, K)$ be a k-group. Suppose now that we also have $k \subset \Omega$ where Ω is algebraically closed. Then taking the Zariski closure of G_k in $GL(n, \Omega)$ we obtain a k-subgroup G_Ω of $GL(n, \Omega)$. Since the connected component of the identity and the radical of G are both defined over k, G_Ω will be connected or semisimple if and only if G has the same property. For example, suppose $G \subset GL(n, \mathbb{C})$ is a connected semisimple $\mathbb{Q}$-group. If $\mathbb{Q}_p$ is the field of p-adic numbers with algebraic closure $\bar{\mathbb{Q}}_p$, then $G_{\bar{\mathbb{Q}}_p}$ will be a connected semisimple $\mathbb{Q}$-group, and in particular, a connected semisimple $\mathbb{Q}_p$-group. This construction will arise in Chapter 6.

In this context, the following result is useful.

3.1.12 Theorem: *Let G be a semisimple $\mathbb{Q}$-group. Then for all but finitely many primes p, $G_{\mathbb{Q}_p}$ is not compact.*

For a proof, see [Springer 3, lemma 4.9].

3.2 Orbits of measures on projective varieties and the Borel density theorem

Suppose X is a compact metric space. Let $M(X)$ be the space of probability measures on X, and $C(X)$ the continuous functions on X. Then we can identify $M(X)$ as a closed subset of the unit ball in the dual space $C(X)^*$, where the latter is given the weak-*-topology. Thus, $M(X)$ is a compact metrizable space with this topology. If $\mu_n, \mu \in M(X)$, we have $\mu_n \to \mu$ if and only if $\int f d\mu_n \to \int f d\mu$ for all $f \in C(X)$. If G acts continuously on X (i.e., the action map $X \times G \to X$ is continuous), then G will act continuously on $M(X)$, via the action $(\mu \cdot g)(A) = \mu(Ag^{-1})$ for $A \subset X$ Borel, or alternatively, via the equation $\int f d(\mu \cdot g) = \int (g \cdot f) d\mu$, for $f \in C(X)$, where $(g \cdot f)(x) = f(xg)$.

We shall be particularly interested in this action when G is a semisimple Lie group and $X = G/H$ for some algebraic subgroup H, or more generally for groups of the form G_k (G algebraic over k) where $X = G_k/H_k$ for some k-subgroup $H \subset G$. The space $M(X)$ will of course in general be infinite dimensional. Nevertheless, the main results of this section will show that in certain respects, the action of G_k on $M(G_k/H_k)$ behaves much like the action of G_k on a variety. Namely, the stabilizers are close to being algebraic and in fact are algebraic for $k = \mathbb{R}$, (first shown when H is minimal parabolic in [Moore 2]), and the orbits are locally closed [Zimmer 4]. For $G = PGL(n, k)$ acting on $M(\mathbb{P}^{n-1})$, an examination of the stabilizers will lead to an easy proof, due to [Furstenberg 4], of the Borel density theorem [Borel 3].

Let k be a local (i.e., locally compact, non-discrete) field of characteristic 0. If $V \subset k^n$ is a non-zero linear subspace, we denote by $[V] \subset \mathbb{P}^{n-1}(k)$ its image in $\mathbb{P}^{n-1}$. We let $PGL(n, k)$, as usual, be the projective general linear group, i.e., $GL(n, k)/k^* \cdot I$, where k^* is the set of non-zero elements of k. If $g \in GL(n, k)$, we denote by $[g]$ its image in $PGL(n, k)$. There is a natural action of $PGL(n, k)$ on $\mathbb{P}^{n-1}(k)$, and hence on $M(\mathbb{P}^{n-1}(k))$. The following will be very useful.

3.2.1 Lemma [*Furstenberg* 1]: *Suppose* $[g_m] \in PGL(n, k)$, μ, $\nu \in M(\mathbb{P}^{n-1}(k))$ *and that* $\mu \cdot [g_m] \to \nu$. *Then either (i)* $\{[g_m]\}$ *is bounded, i.e., has compact closure in* $PGL(n, k)$*; or (ii) there exist linear subspaces* $V, W \subset k^n$, $1 \leqq \dim V, \dim W \leqq n-1$, *such that* ν *is supported on* $[V] \cup [W]$.

Proof: We can choose an absolute value on k, and then a norm on the k-linear space $M(n \times n, k)$ of $n \times n$ matrices over k. Since we are free to multiply g_m by a non-0 scalar, we can assume that $\| g_m \|$ is bounded and bounded away from 0. Let us assume that $\{[g_m]\}$ is not bounded. Then by passing to a subsequence, we can assume $[g_m] \to \infty$ in $PGL(n, k)$, i.e., eventually leaves every compact set in $PGL(n, k)$. Since $\| g_m \|$ is bounded, again by passing to a subsequence we can assume $g_m \to g \in M(n \times n, k)$. Since $[g_m] \to \infty$, $g \notin GL(n, k)$. On the other hand, since $\| g_m \|$ is bounded away from 0, $g \neq 0$. Thus, if we let $N = \text{kernel}(g)$ and $R = \text{image}(g)$, then $1 \leqq \dim N, \dim R \leqq n-1$. Since the Grassmann varieties are compact, again by passing to a subsequence, we can assume $[N] \cdot g_m \to [V]$ for some V, $\dim V = \dim N$. We also observe that if $x \in \mathbb{P}^{n-1}$, $x \notin [N]$ then $\lim x \cdot g_m = x \cdot g \in [R]$. We can write $\mu = \mu_1 + \mu_2$ where $\mu_1(\mathbb{P}^{n-1} - [N]) = 0$ and $\mu_2([N]) = 0$. We can assume by compactness of $M(\mathbb{P}^{n-1})$ that $\mu_i \cdot g_m \to \nu_i$, $i = 1, 2$, and we have $\nu = \nu_1 + \nu_2$. Since μ_1 is supported on $[N]$ and $[N] \cdot g_m \to [V]$, it follows that ν_1 is supported on $[V]$. Thus to prove the lemma, it suffices to show that ν_2 is supported on $[R]$. Suppose $f \in C(\mathbb{P}^{n-1})$ and that $f = 0$ on $[R]$. Then

$$\int \mathrm{fd}\nu_2 = \lim \int f d(\mu_2 \cdot g_m) = \lim \int_{\mathbb{P}^{n-1}} f(x g_m) d\mu_2(x).$$

However $\mu_2[N] = 0$, so this is $\int_{\mathbb{P}^{n-1}-[N]} f(x g_m) d\mu_2(x)$. But since $f = 0$ on $[R]$, $f(x g_m) \to 0$ pointwise on $\mathbb{P}^{n-1} - [N]$, and hence the integral converges to 0 by the dominated convergence theorem. This completes the proof of the lemma.

3.2.2 Corollary: *For* $\mu \in M(\mathbb{P}^{n-1})$, *let* $PGL(n, k)_\mu$ *be the stabilizer of* μ *in* $PGL(n, k)$. *Then either*

(i) μ *is not supported on a finite union of proper projective subspaces, in which case* $PGL(n, k)_\mu$ *is compact; or*

(ii) There is a proper subspace $V_0 \subset k^n$ *such that*

(a) $\mu[V_0] > 0$, *and*

(*b*) *$PGL(n, k)_\mu([V_0])$ is a finite union of projective subspaces $[V_0], \ldots, [V_r]$.*
In particular, $PGL(n, k)_\mu$ has a subgroup of finite index leaving V_0 invariant.

Proof: (i) is an immediate consequence of Lemma 3.2.1. To see (ii), let V_0 be a subspace with minimal dimension so that $\mu([V_0]) > 0$. For $g \in PGL(n, k)_\mu$, $\mu(g \cdot [V_0]) > 0$. Furthermore, if $\mu(g[V_0] \cap [V_0]) > 0$, it follows by the minimality of dimension property of V_0 that $g[V_0] = [V_0]$. Since μ is a probability measure, it follows immediately that $\{g \cdot [V_0] | g \in PGL(n, k)_\mu\}$ must be a finite collection of projective subspaces. Thus, (ii) follows as well.

From this corollary, we can see that the stabilizers of measures are almost algebraic. More precisely, let us make the following definition.

3.2.3 Definition: Suppose k is as above, a local field of characteristic 0. Suppose $G \subset GL(n, k)$. We call G k-almost algebraic if there is an algebraic k-group H such that $H_k \subset G$ is a cocompact normal subgroup.

We recall that compact real matrix groups are $\mathbb{R}$-points of algebraic $\mathbb{R}$-groups. (See, e.g., [Baily 1].) Thus, for $k = \mathbb{R}$, any k-almost algebraic group is actually (the $\mathbb{R}$-points of) an algebraic group (over $\mathbb{R}$). For $k = \mathbb{R}$, a result of the following type was first observed in [Moore 2].

3.2.4 Theorem: *For any measure $\mu \in M(\mathbb{P}^{n-1}(k))$, the stabilizer $PGL(n, k)_\mu$ has a normal subgroup of finite index which is k-almost algebraic. In particular, for $k = \mathbb{R}$, $PGL(n, \mathbb{R})_\mu$ is itself the real points of an $\mathbb{R}$-group.*

Proof: As it should cause no confusion, for this proof we shall refer to the k-points of a k-group simply as an algebraic group.

For any $\mu \in M(\mathbb{P}^{n-1}(k))$ we can find a countable family of (positive) measures $\{u_i\}$ on $\mathbb{P}^{n-1}(k)$ of total measure at most 1 such that:

(a) for $i \neq j$, $\mu_i \perp \mu_j$;

(b) $\sum_i \mu_i = \mu$;

(c) if V_i is of minimal dimension among all linear subspaces with $\mu_i([V]) > 0$, then $\text{supp}(\mu_i) \subset [V_i]$;

(d) with $[V_i]$ as in (c), if $i \neq j$, then $[V_i] \neq [V_j]$.

As in 3.2.2, each μ_i must have a finite orbit under $PGL(n, k)_\mu$. Replacing each μ_i by $\Sigma \mu_j$ where $\mu_j \in PGL(n, k)_\mu \mu_i$, and eliminating repetitions, we obtain a countable family of measures $\{v_i\}$ such that:

(1) $v_i \perp v_j$ for $i \neq j$;
(2) $\Sigma v_i = \mu$;
(3) $PGL(n, k)_\mu = \bigcap_i PGL(n, k)_{v_i}$;
(4) if V_i is of minimal dimension among all linear subspaces for which $v_i([V_i]) > 0$, then $PGL(n, k)_\mu([V_i])$ is a finite union $\{[W_{ij}], j = 1, \ldots, n_i\}$ of projective subspaces of $\mathbb{P}^{n-1}(k)$, and v_i is supported on $PGL(n, k)_\mu([V_i])$.

For each i, let $H_i = \{g \in PGL(n, k) | g[W_{ij}] \subset PGL(n, k)_\mu([V_i])$ for all $j\}$. Let $N_i = \{g \in PGL(n, k) | g | W_{ij}$ is scalar for all $j\}$. Then H_i, N_i are algebraic, $N_i \subset H_i$ is a normal subgroup, $N_i \subset PGL(n, k)_{v_i}$, and $PGL(n, k)_\mu \subset H_i$. Thus $\bigcap_i N_i \subset PGL(n, k)_\mu \subset \bigcap H_i$. Since N_i, H_i are algebraic, by the descending chain condition on algebraic groups we can find a finite set of i, say F, such that $\bigcap_i N_i = \bigcap_{i \in F} N_i$ and $\bigcap_i H_i = \bigcap_{i \in F} H_i$. For $i \in F$, let $H'_i = \{g \in PGL(n, k) | g[W_{ij}] = [W_{ij}]$ for all $j\}$. Then $N_i \subset H'_i \subset H_i$, and H'_i is a normal algebraic subgroup of finite index in H_i. Letting $G = PGL(n, k)_\mu \cap \bigcap_{i \in F} H'_i$, we have that G is a normal subgroup of finite index in $PGL(n, k)_\mu$. Therefore, it suffices to show that $G/\bigcap_{i \in F} N_i$ is compact.

We have a natural homomorphism $\pi: \bigcap_{i \in F} H'_i / \bigcap_{i \in F} N_i \rightarrow \prod_{i \in F, j} PGL(W_{ij})$ which is regular and hence has a closed image. Since it is also injective, it is a homeomorphism onto its image. Thus, to see that $G/\bigcap_{i \in F} N_i$ is compact, it suffices to see that the projection of $\pi(G/\bigcap_{i \in F} N_i)$ into each factor $PGL(W_{ij})$ is precompact. But this image leaves the measure $v_i|[W_{ij}]$ on $\mathbb{P}(W_{ij})$ invariant, and by conditions (3), (4), and Furstenberg's lemma (3.2.1), this image is precompact.

We now turn to Furstenberg's proof of the Borel density theorem, first proved in [Borel 3].

3.2.5 Theorem (*Borel density theorem*, [*Borel* 3]): *Let* **G** *be a connected semi-*

simple algebraic $\mathbb{R}$*-group*, $G = \mathbf{G}^0_{\mathbb{R}}$, *and assume* G *has no compact factors. Let* $\Gamma \subset G$ *be a (topologically) closed subgroup such that* G/Γ *has a finite* G*-invariant measure. Then*

(i) Γ *is Zariski dense in* $\mathbf{G}$.
(ii) Γ^0 *(the connected component in the Hausdorff topology) is normal in* G. *In particular, if* G *is simple and* Γ *is a proper subgroup, then* Γ *is discrete.*

Proof: Let $\mathbf{H}$ be the Zariski closure of Γ in $\mathbf{G}$ and $H = \mathbf{H} \cap G$. By 3.1.9, it suffices to see that $H = G$. Since $\Gamma \subset H$, we have a G-map $G/\Gamma \to G/H$, and so G/H also has a finite G-invariant measure. By Chevalley's theorem (3.1.4), there is a representation $G \to SL(n, \mathbb{R})$ for some n, and a point $x \in \mathbb{P}^{n-1}$ such that $H = G_x$, i.e. the stabilizer of x in G. By passing to a subspace, we can clearly suppose that the linear span of $x \cdot G$ is $\mathbb{R}^n$. If $n = 1$, then the representation of G is trivial, $H = G$, and we are done. If $n \geqq 2$, the image of G in $SL(n, \mathbb{R})$ does not have compact closure. (This follows from the well-known fact that G does not have non-trivial finite-dimensional unitary representations. This assertion follows from 2.2.20 for example, although simpler proofs are available.) However, the map $G \to \mathbb{P}^{n-1}$, $g \to x \cdot g$, induces a map $G/H \to \mathbb{P}^{n-1}$, and since G/H has a finite invariant measure, there is a measure $\mu \in M(\mathbb{P}^{n-1})$ left invariant by G. Choose V_0 as in Corollary 3.2.2. By connectedness of G, G leaves $[V_0]$ invariant. Since μ is supported on $x \cdot G$ and $\mu([V_0]) > 0$, it follows that $x \cdot G \cap [V_0] \neq \emptyset$, and by invariance, that $x \cdot G \subset [V_0]$. Since V_0 is a proper subspace, this contradicts the fact that $x \cdot G$ spans $\mathbb{R}^n$. Thus, (i) is established.
(ii) let $L(G)$ be the Lie algebra of G, and $L(\Gamma^0) \subset L(G)$ the Lie algebra of Γ^0. Since $\Gamma^0 \subset \Gamma$ is normal, $L(\Gamma^0)$ is $Ad(\Gamma)$-invariant. But Ad is a rational representation of G, so by (i), $L(\Gamma^0)$ must also be $Ad(G)$-invariant. Thus, Γ^0 is normal.

Remark: The same proof shows the following result due to [Wang 2]. Let k be a local field of characteristic 0, G a connected semisimple k-group such that for every simple k-group H is which a factor of G, H_k is non-compact. Let $\Gamma \subset G_k$ be a closed subgroup (in the Hausdorff topology) such that G_k/Γ has a finite G-invariant measure. Then Γ is Zariski dense in G.

Thus far, our results in this section have concerned the stabilizers of measures in $M(\mathbb{P}^{n-1}(k))$. We now turn to consideration of the orbits as subsets of $M(\mathbb{P}^{n-1}(k))$. Our next goal is the following.

3.2.6 Theorem [*Zimmer* 4]: *Let* k *be a local field of characteristic* 0. *Then every*

$PGL(n, k)$-orbit in $M(\mathbb{P}^{n-1}(k))$ is locally closed, and hence the action is smooth.

We begin the proof by establishing some notation. Let d be a metric on $\mathbb{P}^{n-1}(k)$. Let $\mathscr{C}$ be the space of closed non-empty subsets of $\mathbb{P}^{n-1}(k)$. Since $\mathbb{P}^{n-1}(k)$ is compact, $\mathscr{C}$ will itself be a compact metric space with the Hausdorff metric $d(A, B) = \sup_{x\in A,\, y\in B} \{d(x, B), d(y, A)\}$. If $A\in\mathscr{C}$ we have a natural identification of $M(A)$ with the measures on $\mathbb{P}^{n-1}(k)$ which are supported on A. With this identification, we shall write $M(A) \subset M(\mathbb{P}^{n-1}(k))$.

3.2.7 Lemma: *If $A, A_i\in\mathscr{C}$ with $A_i \to A$, and $\mu_i\in M(A_i)$ with $\mu_i \to \mu\in M(\mathbb{P}^{n-1}(k))$, then $\mu\in M(A)$.*

The proof of the lemma is a straightforward exercise.

If $\mathscr{A} \subset \mathscr{C}$, let $M_{\mathscr{A}} = \bigcup_{A\in\mathscr{A}} M(A)$. From Lemma 3.2.7, we have:

3.2.8 Lemma: *If $\mathscr{A} \subset \mathscr{C}$ is closed, then $M_{\mathscr{A}} \subset M(\mathbb{P}^{n-1}(k))$ is closed.*

We shall now fix a choice of $\mathscr{A} \subset \mathscr{C}$. Namely, let $\mathscr{A} = \{A\in\mathscr{C} \mid A = \bigcup_{i=1}^{\ell} [V_i]$ for some linear subspaces $V_i \subset k^n$, such that for all $i \neq j$, $V_i \not\subset V_j$, and $\Sigma \dim V_i \leqq n\}$. Then it is straightforward that $\mathscr{A} \subset \mathscr{C}$ is a closed subset. For $A\in\mathscr{A}$, define: $n(A) = \ell$, $d(A) = \Sigma \dim V_i$, and $D(A) = \dim(\Sigma V_i)$. Then we have $1 \leqq n(A), d(A), D(A) \leqq n$. The following two lemmas are again straightforward to verify.

3.2.9 Lemma: *Let*

$$\mathscr{A}_{\ell} = \{A\in\mathscr{A} \mid n(A) \leqq \ell\},$$
$$\mathscr{A}^{d} = \{A\in\mathscr{A} \mid d(A) \leqq d\},$$
$$\mathscr{A}(D) = \{A\in\mathscr{A} \mid D(A) \leqq D\}.$$

Then these are all closed subsets of $\mathscr{A}$.

3.2.10 Lemma: *Let $\mathscr{B}^{d}_{\ell} = \{A\in\mathscr{A} \mid n(A) = \ell, d(A) = d\}$. If $A_j\in\mathscr{B}^{d}_{\ell}$, and $A_j \to A\in\mathscr{C}$, then $A\in\mathscr{B}^{d}_{\ell} \cup \mathscr{A}^{d-1}$. Hence $\mathscr{B}^{d}_{\ell} \cup \mathscr{A}^{d-1}$ is closed.*

We are now ready to prove the theorem.

Proof (of Theorem 3.2.6): Let $\mu \in M(\mathbb{P}^{n-1}(k))$. Define

$$d(\mu) = \min\{d(A) \mid A \in \mathscr{A} \text{ and } \mu \in M(A)\}.$$

Let

$$n(\mu) = \max\{n(A) \mid A \in \mathscr{A},\ \mu \in M(A), \text{ and } d(A) = d(\mu)\}.$$

Fix $A \in \mathscr{A}$ with $\mu \in M(A)$, $d(A) = d(\mu)$ and $n(\mu) = n(A)$. Then define $D(\mu) = D(A)$. We observe that if $A = \bigcup_{j=1}^{\ell} [V_j]$, then μ is also supported on $[\Sigma V_j]$. Thus, by the definition of $d(\mu)$ and the choice of A, we have $d(A) = D(A)$, i.e., the subspaces $V_1, \ldots, V_\ell$ are independent.

Now let $K(\mu) \subset \mathscr{C}$ be defined by

$$K(\mu) = \mathscr{B}^{d(\mu)}_{n(\mu)+1} \cup \mathscr{A}^{d(\mu)-1} \cup \mathscr{A}(D(\mu) - 1).$$

By Lemmas 3.2.9, 3.2.10, $K(\mu)$ is closed. Finally, we define $\mathscr{U}(\mu) = M(\mathbb{P}^{n-1}(k)) - M_{K(\mu)}$. Then $\mu \in \mathscr{U}(\mu)$, $\mathscr{U}(\mu) \subset M(\mathbb{P}^{n-1}(k))$ is open by Lemma 3.2.8, and $\mathscr{U}(\mu)$ is readily seen to be $PGL(n, k)$-invariant. To show that Orbit(μ) is locally closed, it suffices to see that $\overline{\text{Orbit}(\mu)} \cap \mathscr{U}(\mu) = \text{Orbit}(\mu)$. By our observations of the preceding sentence, $\text{Orbit}(\mu) \subset \overline{\text{Orbit}(\mu)} \cap \mathscr{U}(\mu)$. To establish the reverse inclusion, let $v \in \mathscr{U}(\mu)$ and assume there is a sequence $[g_i] \in PGL(n, k)$ such that $[g_i] \cdot \mu \to v$. We want to show $v \in \text{Orbit}(\mu)$.

Recall the set $A = \cup [V_j]$ defined above. By passing to a subsequence, we can assume that for each j, $[g_i] \cdot [V_j] \to [W_j]$ for some W_j with $\dim V_j = \dim W_j$. Thus by Lemma 3.2.7, $v \in M(\cup [W_j])$. Since $v \in \mathscr{U}(\mu)$,

$$\begin{aligned}\dim(\Sigma W_j) &= D(\cup [W_j]) \\ &\geqq D(\mu) \text{ (by definition of } K(\mu)) \\ &= d(\mu) \\ &= \Sigma \dim V_j = \Sigma \dim W_j.\end{aligned}$$

In other words, the subspaces W_j are independent as well. Let $\mu_j = \mu \mid [V_j]$. Again by passing to a subsequence, we can assume that for each j, $[g_i] \cdot \mu_j \to v_j$, where $v_j \in M([W_j])$. For each i, j, let $g_{ij}: V_j \to k^n$ be $g_{ij} = g_i \mid V_j$. We will apply the following variant of Furstenberg's lemma (3.2.1) to the sequence $\{g_{ij} \mid i \in Z\}$.

3.2.11 Lemma: *Let $M(n \times r, k)$ be the set of $n \times r$ matrices over k, and let $I(n \times r, k)$ be the matrices corresponding to injective linear maps $k^r \to k^n$. Then $I(n \times r, k)$ is an open subset of $M(n \times r, k)$, and the non-0 scalars k^* acts by* *multiplication on* $I(n \times r, k)$. *For* $f \in I(n \times r, k)$, *let* $[f]: \mathbb{P}^{r-1}(k) \to \mathbb{P}^{n-1}(k)$ *be the induced map. If* $\mu \in M(\mathbb{P}^{r-1}(k))$ *and* $f_i \in I(n \times r, k)$ *with* $[f_i]_* \mu \to \nu \in M(\mathbb{P}^{n-1}(k))$, *then either (i)* $\{[f_i]\}$ *is bounded in* $I(n \times r, k)/k^*$ *or (ii)* ν *is supported on* $[V] \cup [W]$ *where* $1 \leq \dim V$, $\dim W \leq n-1$ *and* $\dim V + \dim W = r$.

(We recall that if $f: X \to Y$ and $\mu \in M(X)$, then $(f_*\mu)(B) = \mu(f^{-1}(B))$.)

The proof of Lemma 3.2.11 is virtually identical to that of Lemma 3.2.1.

We now complete the proof of the theorem. We claim that for each j, $\{[g_{ij}] \mid i \in Z\}$ is bounded in $I(n \times \dim V_j, k)/k^*$. Suppose this fails for some j. Then by Lemma 3.2.11, ν_j is supported on $[Y_j] \cup [Z_j]$ where $\dim Y_j + \dim Z_j = \dim W_j$. If $Y_j \cap Z_j \neq 0$, then replacing $[W_j]$ by $[Y_j + Z_j]$ in the support of ν, we see that $\nu \in M_{\mathscr{A}(D(\mu)-1)}$ which is impossible by the definition of $K(\mu)$ (recalling that $\nu \in \mathscr{U}(\mu)$). On the other hand, if $Y_j \cap Z_j = 0$, a similar replacement implies that ν is supported on an element of $\mathscr{B}^{d(\mu)}_{n(\mu)+1}$, which is likewise impossible. Thus, the boundedness of the above sequences is verified.

Therefore, by passing to a subsequence we can assume there are $c_{ij} \in k^*$ such that for each j, $\lim_i c_{ij} g_{ij} = h_j$ where $h_j: V_j \to k^n$ is an injection. In this case, we clearly have $[h_j]\mu_j = \nu_j$ and $h_j(V_j) = W_j$. Since $\{V_j\}$ and $\{W_j\}$ are both independent collections of subspaces, we can clearly find $h \in GL(n, k)$ such that $h \mid V_j = h_j$. Then $[h] \cdot \mu = \nu$, and $\nu \in \text{Orbit}(\mu)$ as required. This completes the proof of Theorem 3.2.6.

More generally, we can consider the G_k-orbits in $M(\mathbb{P}^{n-1}(k))$ where G is an algebraic k-group. We then have the following consequence of Theorem 3.2.6.

3.2.12 Corollary [*Zimmer* 4]: *Suppose* k *is a local field of characteristic* 0 *and* $G \subset GL(n, K)$ *is a* k*-group. Then the action of* G_k *on* $M(\mathbb{P}^{n-1}(k))$ *is smooth.*

We preface the proof with the following lemma.

3.2.13 Lemma: *Let* S *be a countably separated Borel* G*-space and* $H \subset G$ *a closed subgroup. Then* H *acts smoothly on* S *if and only if* G *acts smoothly on* $S \times G/H$. *In particular, if* $H_1, H_2 \subset G$ *are two closed subgroups, then* H_1 *acts smoothly on* G/H_2 *if and only if* H_2 *acts smoothly on* G/H_1.

Proof: The second statement follows from the first, because the first implies that both are equivalent to G acting smoothly on $G/H_1 \times G/H_2$. (Cf. 2.2.2, 2.2.3.) It is straightforward to see that the inclusion map $S \to S \times G/H$, $s \to (s, [e])$ defines a Borel bijection $i: S/H \to (S \times G/H)/G$. It suffices to show that the inverse map is also Borel. Choose a Borel section $\varphi: G/H \to G$ of the natural projection $p: G \to G/H$ with $\varphi([e]) = e$. Note that $p(\varphi(xg)) = p(\varphi(x)g)$. Define $\lambda: S \times G/H \to S$ by $\lambda(s, x) = s\varphi(x)^{-1}$. Then one easily checks that $\lambda(sg, xg) = \lambda(s, x)h$ for some $h \in H$, and hence λ induces a Borel map $(S \times G/H)/G \to S/H$ which is clearly the inverse of i.

Proof (of Corollary 3.2.12): Let $\mu \in M(\mathbb{P}^{n-1}(k))$. Then $\mu \cdot G_k \subset \mu \cdot PGL(n, k)$ and by Theorem 3.2.6 and 2.1.14, we can identify $\mu \cdot PGL(n, k)$ with $PGL(n, k)/PGL(n, k)_\mu$. Thus, it suffices to see that the G_k orbits on $PGL(n, k)/PGL(n, k)_\mu$ are locally closed. If $PGL(n, k)_\mu$ were algebraic (e.g., $k = \mathbb{R}$ by Proposition 3.2.4), this would follow immediately from Theorem 3.1.3. In general, by Proposition 3.2.4, we can find subgroups $H_k \subset Q \subset PGL(n, k)_\mu$ where H is algebraic over k, both inclusions are as normal subgroups, and both quotient groups are compact. By Lemma 3.2.13, G_k is smooth on $PGL(n, k)/PGL(n, k)_\mu$ if and only if $PGL(n, k)_\mu$ is smooth on $PGL(n, k)/G_k$. Since H_k is smooth on $PGL(n, k)/G_k$, it clearly suffices to verify the following.

3.2.14 Lemma: *Suppose S is a countably separated G-space and $H \subset G$ is a cocompact normal subgroup. If H acts smoothly on S, so does G.*

Proof: G/H acts on the countably separated space S/H, and hence the result follows from Proposition 2.1.21.

Just as Corollary 3.2.12 gives us information on the orbit space of $M(\mathbb{P}^{n-1}(k))$ under a general G_k, the following result gives us information on the stabilizers.

3.2.15 Proposition: *Suppose $G \subset SL(n, K)$ is a connected k-group and that the action of G_k on k^n is irreducible. Then for any measure $\mu \in M(\mathbb{P}^{n-1}(k))$, we have either (i) $(G_k)_\mu$ is compact; or*
(ii) there is an algebraic k-group $L \subset G$ with $\dim L < \dim G$ and $L_k \supset (G_k)_\mu$.

Proof: If $(G_k)_\mu$ is not compact, we can choose V_i as in Corollary 3.2.2. Let M be the algebraic k-group with $M_k = \{g \in SL(n,k) \mid g(\cup V_i) = \cup V_i\}$, and $L = G \cap M$. By construction, $(G_k)_\mu \subset L_k$.

If $\dim G = \dim L$, then $L = G$ by connectedness, and so $G \subset M$ and hence

$G \subset M^0$. Therefore, G_k leaves each V_i invariant, contradicting irreducibility.

We can also generalize the results we have obtained so far concerning the action of G_k on $M(\mathbb{P}^{n-1}(k))$ to actions on the space of measures on more general varieties.

3.2.16 Definition: Suppose G is an algebraic k-group, and $H \subset G$ is a k-subgroup. We call H k-cocompact in G if G_k/H_k is compact.

For example, all the Grassmann spaces and the various flag spaces over k arise as G_k/H_k where $G_k = SL(n, k)$. We also recall that if G is an algebraic group, an algebraic subgroup $H \subset G$ is called a parabolic subgroup if G/H is a projective variety. If G is a k-group and H is a parabolic k-subgroup, then H is k-cocompact.

By Chevalley's theorem (3.1.4), there is a rational representation of G defined over k such that H_k is the stabilizer in G_k of a point $x \in \mathbb{P}^{n-1}(k)$. Thus, we can identify G_k/H_k as a closed subset of $\mathbb{P}^{n-1}(k)$, and hence we also have an identification of $M(G_k/H_k)$ as a closed G_k-invariant subspace of $M(\mathbb{P}^{n-1}(k))$.

3.2.17 Corollary [*Zimmer* 4]: *If H is k-cocompact in G, then G_k acts smoothly on $M(G_k/H_k)$.*

3.2.18 Corollary: *Suppose $k = \mathbb{R}$ and $H \subset G$ is $\mathbb{R}$-cocompact. Let $\mu \in M(G_{\mathbb{R}}/H_{\mathbb{R}})$. Then $(G_{\mathbb{R}})_\mu$ is the set of real points of an algebraic $\mathbb{R}$-group.*

When H is a minimal $\mathbb{R}$-parabolic subgroup, 3.2.18 is due to [Moore 2].

3.2.19 Corollary: *Suppose G is a connected k-group, almost simple over k, and that $H \subset G$ is a proper k-cocompact subgroup. Then for any measure $\mu \in M(G_k/H_k)$, either*

(i) $(G_k)_\mu$ is compact; or
(ii) there is an algebraic k-group $L \subset G$ with dim $L <$ dim G and $L_k \supset (G_k)_\mu$.

Proof: Identify G_k/H_k as an orbit in $\mathbb{P}^{n-1}(k)$ for some rational representation $\pi: G_k \to GL(n, k)$. We may clearly assume that the orbit G_k/H_k linearly spans k^n. Since π is a rational representation, $\pi: G_k/\ker(\pi) \to \pi(G_k)$ is an isomorphism (of topological groups) onto a topologically closed subgroup of $GL(n, k)$ and since G is almost k-simple, $\ker(\pi)$ is finite. Thus, $(G_k)_\mu$ is compact if and only if its image in $PGL(n, k)$ is compact. If $[\pi(G_k)]_\mu$ is not compact, we argue as in 3.2.15. Namely, choose V_i as in Corollary 3.2.2, let M be the k-group with $M_k = \{g \in GL(n, k) | g(\cup V_i) = \cup V_i\}$, and $L = \pi^{-1}(M)$. By construction $(G_k)_\mu \subset L_k$, and hence it suffices to see that $\dim L < \dim G$. If they were equal then by connectedness of G, $L = G$ and again by connectedness, $\pi(G_k)$ leaves each V_i invariant. Since $\mu([V_i]) > 0$, $(G_k/H_k) \cap [V_i] \neq \emptyset$, which by invariance of V_i implies $G_k/H_k \subset [V_i]$. This contradicts the assumption that G_k/H_k linearly spans k^n.

It is natural to enquire as to the relation of the algebraic groups $(G_{\mathbb{R}})_\mu$ and $H_{\mathbb{R}}$ in Corollary 3.2.18, and although we shall not actually require this information in the sequel, we include some observations for completeness. The following result shows that the non-compact semisimple part of $(G_{\mathbb{R}})_\mu$ can only be as large as the non-compact semisimple part of $H_{\mathbb{R}}$.

3.2.20 Theorem: *With notation as in Corollary* 3.2.18, *let us further assume that* $L \subset G$ *is an algebraic* $\mathbb{R}$*-group with* $L_{\mathbb{R}} \subset (G_{\mathbb{R}})_\mu$ *and* $L_{\mathbb{R}}$ *a semisimple Lie group with no compact factors. Then* $L^0_{\mathbb{R}}$ *is contained in a conjugate of* $H_{\mathbb{R}}$.

We begin the proof with the following useful general observation.

3.2.21 Lemma: *Let G be a locally compact group acting continuously on a compact metrizable space X. Let $M_0 = \{\mu \in M(X) | \mu$ is G-invariant$\}$, so that M_0 is a compact convex subset of $M(X)$. Assume $M_0 \neq \emptyset$. Then each extreme point of M_0 is ergodic under G.*

Proof: If μ is not ergodic let $A \subset X$ be G-invariant with A neither μ-null nor μ-conull. Let $\mu_1(B) = \mu(B \cap A)/\mu(A)$, $\mu_2(B) = \mu(B \cap (X - A))/\mu(X - A)$. Then $\mu_i \in M_0$, $\mu_i \neq \mu$, $\mu = \mu(A) \cdot \mu_1 + \mu(X - A) \cdot \mu_2$, showing that μ is not extreme.

Proof (of Theorem 3.2.20): Since $L^0_{\mathbb{R}}$ leaves a measure invariant, by the lemma, there is an ergodic invariant measure ν. However, $L_{\mathbb{R}}$ (and hence $L^0_{\mathbb{R}}$) acts smoothly on $G_{\mathbb{R}}/H_{\mathbb{R}}$, and hence ν is supported on an $L^0_{\mathbb{R}}$-orbit. Thus, replacing $H_{\mathbb{R}}$ by some conjugate subgroup (in $G_{\mathbb{R}}$) ν is an $L^0_{\mathbb{R}}$-invariant measure on $L^0_{\mathbb{R}}/L^0_{\mathbb{R}} \cap H_{\mathbb{R}}$. By the Borel density theorem (3.2.5), $L^0_{\mathbb{R}} \cap H_{\mathbb{R}} = L^0_{\mathbb{R}}$, completing the proof.

In particular, if $H_{\mathbb{R}}$ is solvable, or more generally, has a cocompact solvable normal subgroup, then $H_{\mathbb{R}}$ has no non-compact connected simple subgroups, and hence neither does $(G_{\mathbb{R}})_\mu$ for any measure μ. This implies from the structure theory of Lie groups that $(G_{\mathbb{R}})_\mu$ has a cocompact solvable normal subgroup. For example, this will be the case for $SL(n, \mathbb{R})$ acting on the space of full flags in $\mathbb{R}^n$. More generally, if G is semisimple and $P \subset G$ is a minimal parabolic $\mathbb{R}$-subgroup, then $P_{\mathbb{R}}$ has a cocompact solvable normal subgroup. Hence, we have the following, first obtained by [Moore 2].

3.2.22 Corollary [*Moore* 2]: *Let G be a semisimple $\mathbb{R}$-group and $P \subset G$ a minimal $\mathbb{R}$-parabolic subgroup. If $\mu \in M(G_{\mathbb{R}}/P_{\mathbb{R}})$, then $(G_{\mathbb{R}})_\mu$ is (the real points of) an algebraic group and has a cocompact solvable normal subgroup.*

Finally, we observe that since a connected semisimple Lie group of trivial center is of finite index in the $\mathbb{R}$-points of an algebraic $\mathbb{R}$-group (3.1.6) from Corollaries 3.2.17, 3.2.18, 3.2.22 we easily deduce:

3.2.23 Corollary: *Let G be a connected semisimple Lie group with trivial center, and $P \subset G$ a parabolic subgroup. (Recall for example, that if $G = SL(n, \mathbb{R})$ the parabolic subgroups are the stabilizers of flags.) Then every G-orbit in $M(G/P)$ is locally closed and the stabilizers are algebraic. (Recall that we call $M \subset G$ algebraic if, writing $G = H^0_{\mathbb{R}}$ as in 3.1.6, we have $M = G \cap L_{\mathbb{R}}$ for some $\mathbb{R}$-group $L \subset H$.) If P is a minimal parabolic, the stabilizers are compact extensions of solvable groups.*

3.3 Orbits in function spaces

In the previous section we saw that the action of an algebraic group on certain (infinite dimensional) spaces of measures behaved in some respects like the action on a variety. In this section we wish to observe the same phenomenon for actions on certain spaces of functions. We will actually be considering two different types of function spaces, the first being measurable functions into a variety.

Let (X, μ) be a finite measure space and suppose V is a complete separable metrizable space. Let $F(X, V)$ be the space of measurable functions $X \to V$, two functions being identified if they agree almost everywhere. We recall that if f_n, $f \in F(X, V)$, we say that $f_n \to f$ in measure if for every $\varepsilon > 0$,

$$\mu\{x \mid d(f_n(x), f(x)) \geqq \varepsilon\} \to 0$$

as $n \to \infty$, where d is a metric on V. It is easy to see that this does not depend on the metric d, and depends only on the measure class of μ, not on μ itself. If μ is σ-finite, we say that $f_n \to f$ in measure if it does so with respect to one (and hence any) finite measure in the same measure class as μ. This is easily seen to be equivalent to convergence in measure on all subsets of X of finite μ-measure. If (X, μ) is σ-finite, convergence in measure in $F(X, V)$ defines a topology on $F(X, V)$ which is metrizable by a complete separable metric. Furthermore, if $f_n \to f$ in measure, then by passing to a subsequence we have pointwise convergence almost everywhere. (The reader is referred to any standard book on measure theory for these assertions, e.g. [Berberian 1].)

Suppose now that G is a locally compact group acting continuously on a locally compact V. Then it is easy to check that G also acts continuously on $F(X, V)$ via the action $(f \cdot g)(x) = f(x) \cdot g$, for $f \in F(X, V)$, $g \in G$. If X is finite, this is of course just the product action of G on $V \times \ldots \times V$, and as such may or may not be smooth. In particular, if G is the set of k-points of an algebraic k-group and V is the set of k-points of a k-variety on which G acts k-regularly, then for X finite, the action will be smooth by Theorem 3.1.3 (for k a local field of characteristic 0). Margulis has observed that the action is still smooth for an arbitrary (σ-finite) measure space.

3.3.1 Proposition: [*Margulis* 1]: *Suppose* $\mathbf{G}$ *is a* k*-group,* $\mathbf{V}$ *a* k*-variety, and that* $\mathbf{G}$ *acts* k*-regularly on* $\mathbf{V}$. *Let* $G = \mathbf{G}_k$, $V = \mathbf{V}_k$, *and assume* k *is a local field of characteristic* 0. *Let* X *be a (standard* σ*-finite) measure space. Then the action of* G *on* $F(X, V)$ *is smooth and the stabilizers are* k*-points of algebraic* k*-groups.*

Proof [Zimmer 8]: By Theorem 2.1.14 (or more easily by Lemma 2.1.15), it suffices to show that for each $f \in F(X, V)$, $G/G_f \to f \cdot G$ is a homeomorphism. That is, it suffices to show that if $f \cdot g_n \to f$, then $[g_n] \to [e]$ in G/G_f. For this it suffices to show that each subsequence of $[g_n]$ has a sub-subsequence converging to $[e]$, and hence it suffices to consider the case in which $(f \cdot g_n)(x) \to f(x)$ on a conull set $X_1 \subset X$. It is clear that $g \in G_f$ if and only if $g \in G_{f(x)}$ for almost all $x \in X$. Let H_f be a countable dense subgroup of G_f. Then for some conull set $X_0 \subset X_1$, we have $H_f \subset \bigcap_{x \in X_0} G_{f(x)}$. Since the latter group is closed, we have $G_f \subset \bigcap_{x \in X_0} G_{f(x)}$ and since the reverse inclusion is clear, we have $G_f = \bigcap_{x \in X_0} G_{f(x)}$. The groups $\mathbf{G}_{f(x)}$ are all algebraic over k since $\mathbf{G}$ acts k-regularly on $\mathbf{V}$, and by the descending chain condition on algebraic subgroups, there exists $x_1, \ldots, x_n \in X_0$ such that $G_f = \bigcap_{i=1}^{n} G_{f(x_i)}$. By Theorem 3.1.3, the action of G on $\prod_{i=1}^{n} G/G_{f(x_i)}$ is smooth, and hence by 2.1.14, the map $G/G_{([e],\ldots,[e])} \to ([e],\ldots,[e])\cdot G$ is a homeomorphism. But clearly $G_{([e],\ldots,[e])} = \bigcap_i G_{f(x_i)}$, and hence we have a homeomorphism

$$G/G_f \to ([e],\ldots,[e])\cdot G \subset \Pi G/G_{f(x_i)}.$$

Therefore, to see that $[g_n] \to [e]$ in G/G_f, it suffices to see $[g_n] \to [e]$ in each $G/G_{f(x_i)}$. This follows directly from Theorem 2.1.14 and the fact that G acts smoothly on V.

The second type of action on a function space we will consider is an action on the space of rational functions between varieties. Namely, assume now that V is a (complex) variety defined over $\mathbb{R}$, such that $V_{\mathbb{R}}$ is Zariski dense in V, and let $F(V_{\mathbb{R}}, \mathbb{P}^m(\mathbb{R}))$ be as above. Define $\mathrm{Rat}(V_{\mathbb{R}}, \mathbb{P}^m(\mathbb{R})) = \{f \in F(V_{\mathbb{R}}, \mathbb{P}^m(\mathbb{R})) \mid f$ is the restriction to $V_{\mathbb{R}}$ of a rational function $f: V \to P^m(\mathbb{C})$ defined over $\mathbb{R}\}$.

The result we will need is the following.

3.3.2 Proposition: *Let G be an algebraic $\mathbb{R}$-group acting on $\mathbb{P}^n(\mathbb{C})$ via a rational representation defined over $\mathbb{R}$, and suppose H and an H-action on $\mathbb{P}^m(\mathbb{C})$ are similarly defined. Suppose $V \subset \mathbb{P}^n(\mathbb{C})$ is a closed G-invariant $\mathbb{R}$-subvariety, with $V_{\mathbb{R}}$ Zariski dense in V. (In particular, $G_{\mathbb{R}}$ leaves the Lebesgue measure class on $V_{\mathbb{R}}$ invariant.) Define a $G_{\mathbb{R}} \times H_{\mathbb{R}}$-action on $F(V_{\mathbb{R}}, \mathbb{P}^m(\mathbb{R}))$ by $(f\cdot(g,h))(x) = f(xg^{-1})\cdot h$, so that $Rat(V_{\mathbb{R}}, \mathbb{P}^m(\mathbb{R}))$ is an invariant set. Then*
(i) The $G_{\mathbb{R}}$, $H_{\mathbb{R}}$, and $G_{\mathbb{R}} \times H_{\mathbb{R}}$ actions on $Rat(V_{\mathbb{R}}, \mathbb{P}^m(\mathbb{R}))$ are smooth.
(ii) The stabilizers of points in $Rat(V_{\mathbb{R}}, \mathbb{P}^m(\mathbb{R}))$ are real points of algebraic $\mathbb{R}$-groups.

Proof: The idea of the proof is that rational functions can be identified with a class of closed subvarieties of a certain projective space, and that the space of subvarieties of projective space has the structure of a countable union of varieties. Thus, the action in question can be reduced to a countable disjoint union of algebraic actions, for which the desired conclusions are known. We now sketch some further details.

If X is a compact metric space, let $\mathscr{C}(X)$ be the space of closed subsets, which is compact metric with the Hausdorff metric. (See e.g., the paragraph following the statement of Theorem 3.2.6.) If X is a complex projective variety, we let $\mathrm{Var}(X) \subset \mathscr{C}(X)$ be the set of closed subvarieties, and if X is defined over $\mathbb{R}$, let $\mathrm{Var}_{\mathbb{R}}(X)$ be the closed $\mathbb{R}$-subvarieties. Let $R(V, \mathbb{P}^m(\mathbb{C}))$ be the rational maps $V \to P^m(\mathbb{C})$ defined over $\mathbb{R}$. For $f \in R(V, \mathbb{P}^m(\mathbb{C}))$, the set of points at which f is not defined is a closed $\mathbb{R}$-subvariety of $\mathbb{P}^n(\mathbb{C})$, which we denote by W_f. Hence f is regular on the quasi-projective variety $V - W_f$. We recall that this implies that the graph of f, $\mathrm{gr}(f) \subset (V - W_f) \times \mathbb{P}^m(\mathbb{C})$ is closed. (See, e.g. [Shafarevich 1, p. 45].) Hence, if we let $Z_f \subset P^n(\mathbb{C}) \times \mathbb{P}^m(\mathbb{C})$ be the Zariski closure of $gr(f)$, the map $R(V, \mathbb{P}^m(\mathbb{C})) \to \mathrm{Var}_{\mathbb{R}}(\mathbb{P}^n(\mathbb{C})) \times \mathrm{Var}_{\mathbb{R}}(\mathbb{P}^n(\mathbb{C}) \times \mathbb{P}^m(\mathbb{C}))$ given by $f \to (W_f, Z_f)$ is injective. Give $R(V, \mathbb{P}^m(\mathbb{C}))$ the subspace topology for the topology defined by the Hausdorff metric on $\mathscr{C}(\mathbb{P}^m(\mathbb{C})) \times \mathscr{C}(\mathbb{P}^n(\mathbb{C}) \times \mathbb{P}^m(\mathbb{C}))$. We also have the natural restriction mapping $R(V, \mathbb{P}^m(\mathbb{C})) \to \mathrm{Rat}(V_{\mathbb{R}}, \mathbb{P}^m(\mathbb{R}))$, which is injective since $V_{\mathbb{R}}$ is Zariski dense in V (and surjective by definition). It is easy to check that this map is continuous, and hence a Borel isomorphism (Appendix A). Thus, it suffices to see that our results are true for the actions on $\mathrm{Var}_{\mathbb{R}}(\mathbb{P}^n(\mathbb{C})) \times \mathrm{Var}_{\mathbb{R}}(\mathbb{P}^n(\mathbb{C}) \times \mathbb{P}^m(\mathbb{C}))$. We can choose an isomorphism of $\mathbb{R}$-varieties of $\mathbb{P}^n(\mathbb{C}) \times \mathbb{P}^n(\mathbb{C}) \times \mathbb{P}^m(\mathbb{C})$ with a closed $\mathbb{R}$-subvariety of a higher dimensional projective space $\mathbb{P}^r(\mathbb{C})$ in such a way that $G \times H$ has a compatible rational representation into $GL(r+1, \mathbb{C})$ defined over $\mathbb{R}$. Therefore, it suffices to verify our results for the $G_{\mathbb{R}} \times H_{\mathbb{R}}$ action on $\mathrm{Var}_{\mathbb{R}}(\mathbb{P}^r(\mathbb{C}))$. The stabilizers are clearly $\mathbb{R}$-algebraic, and hence only smoothness is at issue. However, it is well known, using the "Chow coordinates" of a projective variety (see e.g. [Shafarevich 1], or [Samuel 1]) that there is a disjoint decomposition $\mathrm{Var}(\mathbb{P}^r(\mathbb{C})) = \cup Y_i$ into a countable union of $GL(r+1, \mathbb{C})$ invariant Borel sets such that each Y_i is a subspace of a projective $\mathbb{R}$-variety W_i on which $GL(r+1, \mathbb{C})$ acts $\mathbb{R}$-regularly. Furthermore $Y_i \cap \mathrm{Var}_{\mathbb{R}}(\mathbb{P}^r(\mathbb{C}))$ is contained in $(W_i)_{\mathbb{R}}$. To see smoothness of $G_{\mathbb{R}}$, $H_{\mathbb{R}}$, $G_{\mathbb{R}} \times H_{\mathbb{R}}$ on $\mathrm{Var}_{\mathbb{R}}(\mathbb{P}^r(\mathbb{C}))$, it suffices to see smoothness on $(W_i)_{\mathbb{R}}$, which in turn follows from Theorem 3.1.3.

3.4 Rationality of measurable mappings – first results

In this section we consider a different question relating measure theory and algebraic geometry, namely the question as to when one can deduce that a measurable map is rational.

We begin by recalling one situation in which measurability implies a much stronger regularity property, namely in the case of homomorphisms of groups. More precisely, if $\pi: G \to H$ is a measurable homomorphism of second countable groups, and G is locally compact, then π is continuous. (For a proof, see Appendix B). Furthermore, if G, H are both Lie groups, it is well known that any continuous homomorphism is C^∞. It is not true, however, that if G, H are (real points of) algebraic groups then the homomorphism must be rational. A simple example is the map $\mathbb{R} \to \mathbb{R}^*$, $t \to \exp(t)$, where $\mathbb{R}^*$ is the multiplicative group of non-zero real numbers. The first result in this section is to describe a simple situation in which one can deduce that a homomorphism of algebraic groups is rational.

Let k be a field of charactistic 0. A subgroup $U \subset GL(n, k)$ is called unipotent if it is conjugate to a subgroup of $N(n, k)$, where $N(n, k) = \{(a_{ij}) | a_{ii} = 1$ and

$a_{ij} = 0$ if $i > j\}$. A matrix $A \in GL(n, k)$ is called unipotent if all eigenvalues of A are equal to 1. If $U \subset GL(n, k)$ is the k-points of an algebraic k-group, then U is unipotent if and only if every element of U is a unipotent matrix. (For a proof, see any of the standard references on algebraic groups.) For $k = \mathbb{R}$, a measurable representation of a unipotent group need not be rational. For example,

$$\begin{pmatrix} 1 & a \\ 0 & 1 \end{pmatrix} \to (e^a)$$

is a non-rational one-dimensional representation. However, we do have the following.

3.4.1 Proposition: *Suppose $k = \mathbb{R}$ or $\mathbb{C}$, and U is a k-group. Suppose U is unipotent. If $\pi : U_k \to GL(m, k)$ is a measurable unipotent representation, i.e., $\pi(U_k)$ is a unipotent group, then π is k-regular.*

Proof: Let $N = N(n, \mathbb{C})$ be defined as above, and let $L(N)$ be the Lie algebra of N. That is, $L(N)$ is the linear subspace of all $n \times n$ matrices given by $L(N) = \{(a_{ij}) \mid a_{ij} = 0 \text{ if } i \geqq j\}$. We remark that for any $A \in L(N)$, $A^n = 0$. The map $\exp : L(N) \to N$ given by $\exp(A) = \sum_{j=0}^{n-1} A^j/j!$ is a regular mapping, and has an inverse $\log : N \to L(N)$ given by $\log(B) = \sum_{j=0}^{n-1} (-1)^j (B - I)^j/j!$. (Note that for $B \in N$, $(B - I)^n = 0$.) Thus, exp is a biregular $\mathbb{R}$-isomorphism of varieties. If $U \subset N$ is an ($\mathbb{R}$-) algebraic subgroup, and $L(U) \subset L(N)$ is the Lie algebra of U, then $\exp : L(U) \to U$ will also be a biregular ($\mathbb{R}$-) isomorphism.

Since π is measurable, by our above remarks it is C^∞. Let $d\pi : L(U_k) \to L(N(m, \mathbb{C}))$ be the derivative and, $d\pi_{\mathbb{C}} : L(U) \to L(N(m, \mathbb{C}))$ its extension to $L(U)$ (which only differs from $L(U_k)$ if $k = \mathbb{R}$). Then by well-known properties of the exponential map, the following diagram commutes:

$$\begin{array}{ccc} U_k & \xrightarrow{\quad \pi \quad} & N(m, \mathbb{C}) \\ \log \downarrow & & \uparrow \exp \\ L(U) & \xrightarrow{\quad d\pi_{\mathbb{C}} \quad} & L(N(m, \mathbb{C})). \end{array}$$

Since log, exp are regular, and $d\pi_{\mathbb{C}}$ is linear, it follows that π is regular.

In light of Proposition 3.4.1, it is useful to have some criteria for determining when a representation of a unipotent group is unipotent. In this direction, we have the following.

3.4.2 Proposition: *Suppose $G \subset GL(n, \mathbb{R})$ is a semisimple Lie group and $\pi: G \to GL(m, \mathbb{R})$ is a measurable (equivalently, C^∞) representation. If $U \subset G$ is a connected unipotent subgroup, then $\pi|U$ is a unipotent representation.*

Proof: By letting $\lambda = d\pi$, it suffices to prove the following fact concerning Lie algebras: if $\mathscr{G} \subset \mathscr{G}\ell(n, \mathbb{R})$ is a semisimple Lie algebra, $\lambda: \mathscr{G} \to \mathscr{G}\ell(m, \mathbb{R})$ is a representation, and $A \in \mathscr{G}$ is a nilpotent matrix, then $\lambda(A) \in \mathscr{G}\ell(m, \mathbb{R})$ is also a nilpotent matrix.

To see this, first recall that for any linear transformation $A \in \text{End}(V)$, V a finite-dimensional vector space, we can write $A = A_s + A_n$ in a unique way so that A_n is nilpotent, A_s is semisimple (i.e., is diagonalizable over an algebraically closed extension field), and A_s and A_n commute. A basic fact about semisimple Lie algebras $\mathscr{G}$ is that if $A \in \mathscr{G}$, then $A_s, A_n \in \mathscr{G}$. (For a proof, see [Jacobson 1] or [Bourbaki 1].) It is straightforward to check that if $\mathscr{G} \subset \mathscr{G}\ell(n, \mathbb{R})$ is a general Lie algebra and $A \in \mathscr{G}$, then A semisimple (resp. nilpotent) implies $ad(A): \mathscr{G} \to \mathscr{G}$ is also semisimple (resp. nilpotent). If $\mathscr{G}$ is semisimple, the converse is also true. For if $A = A_s + A_n$, then $ad(A) = ad(A_s) + ad(A_n)$ and by the preceding sentence, this is just the decomposition $ad(A)_s + ad(A)_n$. Thus, for example, if $ad(A)$ is nilpotent, $ad(A)_s = ad(A_s) = 0$, and since $\mathscr{G}$ is semisimple, ad is faithful and hence $A_s = 0$.

With these observations, we now prove our required assertion. Since $\mathscr{G}$ is semisimple, so is the subalgebra $\lambda(\mathscr{G}) \subset \mathscr{G}\ell(m, \mathbb{R})$. Thus, if $A \in \mathscr{G}$ is a nilpotent matrix, to see $\lambda(A)$ is as well, it suffices by the previous paragraph to see that $ad_{\lambda(\mathscr{G})}(\lambda(A))$ is nilpotent. However, the following diagram commutes:

$$\begin{array}{ccc}
\mathscr{G} & \xrightarrow{\quad ad_{\mathscr{G}}(A) \quad} & \mathscr{G} \\
\lambda \downarrow & & \downarrow \lambda \\
\lambda(\mathscr{G}) & \xrightarrow{\quad ad_{\lambda(\mathscr{G})}(\lambda(A)) \quad} & \lambda(\mathscr{G})
\end{array}$$

Since $ad_{\mathscr{G}}(A)$ is nilpotent, $ad_{\lambda(\mathscr{G})}(\lambda(A))$ clearly is as well.

We now turn to a different result, due to [Margulis 2], asserting rationality of certain measurable maps. This result will in fact be used in conjunction with

3.4.1 and 3.4.2 in Chapters 5 and 6 to establish rationality of a specific measurable map.

3.4.3 Definition: Let V be a complex ($\mathbb{R}$-) variety, W an $\mathbb{R}$-variety, and $A \subset W_{\mathbb{R}}$ of positive (Lebesgue) measure. A measurable function $f: A \to V$ is called essentially ($\mathbb{R}$-) rational if there is an ($\mathbb{R}$-) rational function $R: W \to V$ such that $f = R$ a.e. on A.

3.4.4 Theorem [*Margulis* 2]: *Suppose $f: \mathbb{R}^k \times \mathbb{R}^n \to V$ is a measurable function such that for almost all $x \in \mathbb{R}^k$, $f_x: \mathbb{R}^n \to V$ given by $f_x(y) = f(x, y)$ is essentially* ($\mathbb{R}$-) *rational, and similarly, for almost all $y \in \mathbb{R}^n$, $f^y: \mathbb{R}^k \to V$ given by $f^y(x) = f(x, y)$ is essentially* ($\mathbb{R}$-) *rational. Then f is essentially* ($\mathbb{R}$-) *rational.*

Proof: By induction, we can assume $n = 1$. For simplicity, let us first assume that V is affine. Then by taking coordinate functions, it suffices to assume $V = \mathbb{C}$. We first claim it suffices to prove the following: there is a set $A \subset \mathbb{R}^k$ of positive measure and a rational function R such that $R = f$ a.e. on $A \times \mathbb{R}$. For if we have this condition, then by Fubini's theorem, for almost all $y \in \mathbb{R}$, $R^y = f^y$ on a set of positive measure in $\mathbb{R}^k$. Since both R^y and f^y are essentially rational for almost all y, this implies that for almost all y, $R^y = f^y$ a.e. on $\mathbb{R}^k$. Then Fubini's theorem implies $R = f$ a.e. on $\mathbb{R}^k \times \mathbb{R}$.

We now show that such a set A exists. For each pair of non-negative integers r, s, let $A_{r,s} \subset \mathbb{R}^k$ be defined by $A_{r,s} = \{x \in \mathbb{R}^k \mid f_x: \mathbb{R} \to \mathbb{R}$ is equal a.e. to a rational function p_x/q_x where degree$(p_x) = r$ and degree$(q_x) = s\}$. Then $A_{r,s}$ are measurable sets and $\bigcup_{r,s} A_{r,s}$ is conull in $\mathbb{R}^k$. Therefore, some $A_{r,s}$ has positive measure, and we let $A = A_{r,s}$ for such a pair r, s. Then $B = \{y \in \mathbb{R} \mid f_x(y) = p_x(y)/q_x(y)$ for almost all $x \in A_{r,s}\}$ is conull in $\mathbb{R}$ by Fubini's theorem. Therefore, we may choose $r + s + 1$ distinct points $\lambda_0, \ldots, \lambda_{r+s} \in \mathrm{B}$ with the further property that for each j, $f_{\lambda_j}: \mathbb{R}^k \to \mathbb{C}$ is essentially rational. Set $c_j(x) = f_{\lambda_j}(x)$. For $x \in A = A_{r,s}$, we can write

$$p_x(y)/q_x(y) = (y^r + \sum_{i=0}^{r-1} a_i(x)y^i)/\sum_{i=0}^{s} b_i(x)y^i.$$

Since $f(x, y) = p_x(y)/q_x(y)$ a.e. on $A \times \mathbb{R}$, it clearly suffices to see that $a_i(x)$, $b_i(x)$ are essentially rational. However, a rational function p/q from $\mathbb{R}$ into $\mathbb{C}$ with degree$(p) = r$, degree$(q) = s$, is determined by its values at any $r + s + 1$ distinct points. Therefore, for almost all x, $a_i(x)$, $b_i(x)$ are the unique solutions to the system of equations

$$c_j(x) = (\lambda_j^r + \sum_{i=0}^{r-1} a_i(x)\lambda_j^i)/\sum_{i=0}^{s} b_i(x)\lambda_j^i, \; 0 \leqq j \leqq r+s.$$

For each x, this is a system of $r+s+1$ linear equations in the $r+s+1$ unknowns $a_i(x)$, $b_i(x)$. Because $c_j(x)$ is essentially rational in x for each j, the coefficients of this linear system will be rational in x and hence so will its solutions (by Cramer's rule for inverting a matrix). It follows that $a_i(x)$, $b_i(x)$ are essentially rational, and by our observations above, so is f. It is clear that f will be essentially $\mathbb{R}$-rational if f_x and f^y were assumed to be so.

Finally, we reduce the case of a general V to an affine V. Write $V = \cup V_i$ (finite union) where V_i are affine and Zariski open. It clearly suffices to see that some $f^{-1}(V_i)$ is conull. Clearly, some $f^{-1}(V_i)$ is of positive measure. However, if $y \in \mathbb{R}$ such that f^y is essentially rational and $(f^y)^{-1}(V_i)$ is of positive measure, $(f^y)^{-1}(V_i)$ must be conull in $\mathbb{R}^k$. A similar statement holds for f_x, $x \in \mathbb{R}^k$. A straightforward application of Fubini's theorem then shows that $f^{-1}(V_i)$ must be conull.

3.5 A homomorphism theorem

To describe the main result of this section, we begin with some general observations about abstract groups. Suppose A, B, C are groups with $A \subset B$. If $h: C \to B$ is a homomorphism, then of course there is a map $f: C \to B/A$ such that $f(xy) = f(x)h(y)$. There is also a type of converse assertion. Namely, suppose there is a map $f: C \to B/A$ and a map (not necessarily a homomorphism) $h: C \to B$ such that $f(xy) = f(x)h(y)$. Let $B_0 \subset B$ be the subgroup that pointwise fixes $f(C)$, and $N(B_0)$ the normalizer of B_0 in B. Then for all $y \in C$, $h(y) \in N(B_0)$, and the induced map $h: C \to N(B_0)/B_0$ is a homomorphism. The group $N(B_0)/B_0$ acts on the fixed point set of B_0 which contains $f(C)$. If we are given a map $f: C \to B/A$, the existence of such a map $h: C \to B$ can be described as follows. For each $y \in C$, let $f_y: C \to B/A$ be $f_y(x) = f(xy)$. Then the existence of such an h is equivalent to the statement that all f_y, $y \in C$, are in the same B-orbit in $F(C, B/A)$ (the set of functions $C \to B/A$). We summarize these remarks in the following:

3.5.1 Proposition: *Let A, B, C be abstract groups, $A \subset B$, and let $f \in F(C, B/A)$. Suppose all f_y, $y \in C$ are in the same B-orbit in $F(C, B/A)$. Then there is*
(i) a subset $W \subset B/A$ with $f(C) \subset W$;
(ii) a group Q which acts on W; and

(iii) a homomorphism $h: C \to Q$;
such that $f(xy) = f(x) \cdot h(y)$. In particular, there is a point $a \in W$ such that $f(y) = a \cdot h(y)$.

The point of this section is to prove an analogous result where A, B, C are not abstract groups but are locally compact and all our hypotheses are assumed to hold only almost everywhere. In a sense, this is basically a technical result combining the observations of Proposition 3.5.1. with a repeated use of Fubini's theorem. However, although it is technical, this result plays a basic role in the proof of the theorems of Chapters 5 and 6. $F(C, B/A)$ will now be taken in the sense of sections 3.3 and 3.4, namely functions are identified if they agree almost everywhere. The generalization of 3.5.1 we will require is the following.

3.5.2 Proposition [*Margulis* 1]: *Let C be a locally compact group, and k a local field of characteristic 0. Suppose H is an algebraic k-group, and $L \subset H$ a k-subgroup. Let $\varphi \in F(C, H_k/L_k)$ and for $g \in C$, let $\varphi_g \in F(C, H_k/L_k)$ be $\varphi_g(c) = \varphi(cg)$. Suppose that almost all φ_g lie in a single H_k-orbit in $F(C, H_k/L_k)$. (This is equivalent to the existence of a map $h: C \to H_k$ such that $\varphi(cg) = \varphi(c)h(g)$ for almost all (c, g).) Then there exist:*
(i) a k-subvariety $W \subset H/L$ such that $\varphi(c) \in W$ for almost all c;
(ii) an algebraic k-group Q which acts k-regularly on W;
(iii) a measurable homomorphism $h: C \to Q_k$; and
(iv) a point $x \in W \cap H_k/L_k$;
such that $\varphi(c) = x \cdot h(c)$ for almost all $c \in C$.

Proof: The essential range of $\varphi: C \to H_k/L_k$ will be a subset of H/L and we let $B \subset H$ be the subgroup pointwise fixing the essential range. Then B is a k-group. Let $W \subset H/L$ be the k-variety of all fixed points of B. Let $N(B)$ be the normalizer of B in H, so that $N(B)$ leaves W invariant, and hence the algebraic k-group $Q = N(B)/B$ acts k-regularly on the k-variety W.

Since almost all φ_g lie in a single H_k-orbit, it follows that for each fixed $a \in C$, we have φ_{ag} and φ_g are in this H_k-orbit for almost all $g \in C$. Thus, for any $a \in C$, we can write $\varphi_{ag} = \varphi_g \cdot \mathrm{h}_g(\mathrm{a})$ for almost all $g \in C$, where $h_g(a) \in H_k$ and $h_g(a)$ is measurable in (g, a). (To see this last measurability assertion, we use the existence of measurable sections (Appendix A).) The essential range of φ_g for any $g \in C$ is identical to the essential range of φ, and it follows that for each $a \in C$, $h_g(a) \in N(B)_k$ for almost all g. Furthermore, we claim that for each $a, b \in C$ and almost all $c, g \in C$,

(*) $\varphi_g(c)\cdot h_g(ab) = \varphi_g(c)\cdot h_g(a)h_g(b)$.

To see this, we simply observe that for each a, b and almost all c, g,

$$\begin{aligned}\varphi_g(c)\cdot h_g(ab) &= \varphi_{abg}(c)\\ &= \varphi(cabg)\\ &= \varphi_{bg}(ca)\\ &= \varphi_g(ca)\cdot h_g(b)\\ &= \varphi(cag)h_g(b)\\ &= \varphi_{ag}(c)h_g(b)\\ &= \varphi_g(c)h_g(a)h_g(b).\end{aligned}$$

In other words, using Fubini's theorem, we have for almost all $g \in C$, $h_g(ab)h_g(b)^{-1}h_g(a)^{-1} \in B_k$ for almost all a, $b \in C$. In conclusion, for almost all $g \in C$, we have an almost everywhere defined measurable map

$$h_g: C \to N(B)_k \to N(B)_k/B_k \to Q_k,$$

and as a map into Q_k, h_g satisfies $h_g(ab) = h_g(a)h_g(b)$ a.e.

It follows (Appendix B) that for almost all g, there is a measurable homomorphism $\bar{h}_g: C \to Q_k$ such that $h_g = \bar{h}_g$ a.e. The definition of $h_g(a)$ and Fubini's theorem imply that for almost all c, a, $g \in C$, $\varphi(cag) = \varphi(cg)h_g(a)$, and hence $\varphi(cag) = \varphi(cg)\bar{h}_g(a)$ a.e. By Fubini, we can fix $c, g \in C$ such that this last equation holds for almost all $a \in C$, and such that $\varphi(cg)$ is in the essential range of φ, and in particular, $\varphi(cg) \in W \cap H_k/L_k$. Thus, for almost all $a \in C$,

$$\varphi(a) = \varphi(cg)\bar{h}_g(c^{-1}ag^{-1}) = \varphi(cg)\bar{h}_g(c^{-1}g^{-1})\bar{h}_g(gag^{-1}).$$

Then $x = \varphi(cg)\bar{h}_g(c^{-1}g^{-1})$, and $a \to \bar{h}_g(gag^{-1})$ are the required point in $W \cap H_k/L_k$ and measurable homomorphism $C \to Q_k$. This completes the proof.

4 Amenability

4.1 Amenable groups

In this section we develop the notion of amenability for a locally compact group. Here, we shall be concentrating on a fixed point property of these groups, but we shall give other equivalent characterizations of amenability in Chapter 7.

We recall that if X is compact and metrizable, then $M(X)$, the space of probability measures on X, is a compact convex subset of the unit ball in $C(X)^*$, where the latter has the weak-*-topology. If G is locally compact (and, as usual, second countable) and acts continuously on X, then it acts continuously on $M(X)$.

4.1.1 Definition: G is called amenable if for every continuous G-action on a compact metrizable space X, there is a G-invariant probability measure on X, i.e., a G-fixed point in $M(X)$.

The following is a classical result of functional analysis.

4.1.2 Theorem (*Kakutani-Markov fixed point theorem*): *If G is abelian, then G is amenable.*

Proof: For each $g \in G$, let $A_n(g): M(X) \to M(X)$ be given by

$$A_n(g)(\mu) = \Big(\sum_{j=0}^{n} \mu \cdot g^j\Big)/(n+1).$$

We remark that $A_n(g)(\mu) \in M(X)$ because $M(X)$ is convex. Let $\tilde{G}$ be the semigroup of transformations of $M(X)$ generated by $\{A_n(g) \mid {\scriptstyle \xi} \in G,\ n \geqq 0\}$. For each $\gamma \in \tilde{G}$, $\gamma(M(X))$ is compact. We claim $\bigcap_{\gamma \in \tilde{G}} \gamma(M(X))$ is non-empty. By compactness, we need only check $\bigcap_{i=1}^{n} \gamma_i(M(X)) \neq \varnothing$ for $\gamma_1, \ldots, \gamma_n \in \tilde{G}$. Let $\gamma = \gamma_1 \circ \gamma_2 \circ \cdots \circ \gamma_n$. Then, $\gamma(M(X)) \subset \gamma_1(M(X))$. But G is abelian, and hence all γ_i commute. Therefore, $\gamma(M(X)) \subset \gamma_i(M(X))$ for all i, verifying our assertion. Let $\mu_0 \in \bigcap_{\gamma \in \tilde{G}} \gamma(M(X))$. We claim μ_0 is G-invariant. If $g \in G$, and $n \geqq 0$, then

$\mu_0 = (A_n(g))(\mu)$ for some μ. Therefore,

$$\mu_0 - \mu_0 \cdot g = \frac{\mu - \mu \cdot g^{n+1}}{n+1},$$

and hence $\| \mu_0 - \mu_0 \cdot g \| \leqq 2/n + 1$, for all n. This proves the theorem.

We caution the reader that the condition of amenability does not assert that one can find an invariant measure in any predetermined measure class. For example, putting a probability measure in the class of Haar measure along a given orbit will in general give a counterexample. One can in fact have homeomorphisms of compact metric spaces with a quasi-invariant measure which will not even have a σ-finite invariant measure in the same measure class. Such an example was first constructed by [Ornstein 3].

In the above proof, the only properties of $M(X)$ that we used were compactness and convexity. It will be useful to consider actions on more general compact convex sets.

4.1.3 Definition: By an affine action of a group G on a space A we mean a continuous action where
(i) A is a compact convex set in the unit ball of E^*, where E is some separable Banach space, and E^* has the weak-*-topology.
(ii) The action of G comes from A being a G-invariant set in E^* under the adjoint representation of some continuous (strong operator topology) isometric representation of G on E.

Remark: One can of course weaken the definition, e.g. by eliminating separability, allowing A to be a compact convex set in an arbitrary locally convex linear space, or only assuming the action of G on A to be affine (i.e., $(tx + (1-t)y)g = t(xg) + (1-t)yg$, $0 \leqq t \leqq 1$) rather than linear on the ambient linear space. However, the above situation will suffice for our purposes.

The proof of Theorem 4.1.2 shows that if G is abelian, then G has a fixed point in every affine action. In fact, we have:

4.1.4 Proposition: *G is amenable if and only if there is a fixed point in every compact affine G-space.*

The proof of Proposition 4.1.4 follows easily from properties of the barycenter construction on compact convex sets, which we briefly recall. See [Phelps 1] or [Choquet 1] for a full discussion. If μ is a probability measure on a compact convex set A in a locally convex linear space, there is a unique point $b(\mu) \in A$, called the barycenter of μ, such that for each continuous linear functional λ on the linear space, $\lambda(b(\mu)) = \int_A \lambda(a) d\mu(a)$. For example, if μ is an atomic measure, say $\mu = \sum_{i=1}^{n} c_i \delta_{a_i}$, where $a_i \in A$, δ_a is the point measure at a, and $0 \leqq c_i \leqq 1$, $\Sigma c_i = 1$, then $b(\mu) = \sum_{i=1}^{n} c_i a_i$. Since the set of atomic measures is dense, this defines the barycenter map $b: M(A) \to A$ on a dense subset, and one shows (in [Phelps 1] for example) that this extends continuously to all of $M(A)$. If $T: A \to A$ is any affine transformation, then T commutes with b; more precisely, $b(T_* \mu) = T(b(\mu))$.

Proof: (of Proposition 4.1.4): If X is a compact metrizable G-space, then $M(X)$ is an affine G-space, so one direction is clear. On the other hand, suppose G is amenable and A is an affine G-space. Since A is compact and metrizable, there is a G-invariant measure $\mu \in M(A)$. But then by the above remarks, $b(\mu)$ is a G-fixed point.

4.1.5 Proposition: *Compact groups are amenable.*

Proof: If X is a compact G-space, choose $x \in X$, and let $f: G \to X$ be given by $f(g) = xg$. If μ is Haar measure on G, then $f_*(\mu)$ is a G-invariant measure on X.

We now wish to describe the connected Lie groups which are amenable. We begin with some general properties.

4.1.6 Proposition: (*a*) *If G is amenable and $H \subset G$ is closed, then H is amenable.* (*b*) *If $\varphi: G \to H$ is a continuous surjective homomorphism with kernel K, then G is amenable if and only if both K and H are amenable.*

Proof: (a) It will be convenient to postpone the proof of (a) until the following section (see Proposition 4.2.20) where it will be more transparent.
(b) Suppose H, K are amenable. Let A be an affine G-space. Let A_0 be the set

of K-fixed points in A, which is non-empty since K is amenable. Because K is normal, A_0 is G-invariant, and hence is an affine G/K-space. Hence, there is an H-fixed point in A_0 which must be a G-fixed point. Conversely, suppose G is amenable. Then K is amenable by (a). If A is an affine H-space, it is also an affine G-space, and a G-fixed point will be H-fixed.

4.1.7 Corollary: *Solvable groups are amenable. Compact extensions of solvable groups are amenable, i.e., groups with a cocompact solvable normal subgroup are amenable.*

In particular, the subgroup of $SL(n, \mathbb{R})$ consisting of upper triangular matrices, being solvable, is amenable. More generally, if G is a connected semisimple Lie group and $P \subset G$ is a minimal parabolic subgroup, then P is a compact extension of a solvable group and hence amenable.

We thus have a significant collection of amenable groups. On the other hand, Furstenberg's Lemma (3.2.1) shows that there are no invariant measures on $\mathbb{P}^{n-1}$ under the action of $SL(n, \mathbb{R})$. Thus, $SL(n, \mathbb{R})$ is not amenable ($n \geqq 2$). More generally, we have:

4.1.8 Proposition [*Furstenberg* 1]: *Suppose G is a connected semisimple Lie group with finite center. Then G is amenable if and only if G is compact.*

Proof: If G is not compact then G admits an irreducible representation with a non-relatively compact projective image. Then by 3.2.2, G will not leave a measure on projective space invariant, showing non-amenability.

For connected Lie groups, we then have the converse to 4.1.7.

4.1.9 Corollary: *If G is a connected Lie group, then G is amenable if and only if G is a compact extension of a solvable group.*

Proof: Let R be the radical of G, i.e., the maximal normal connected solvable subgroup. Then G/R is a semisimple Lie group. Let $p: G \to G/R$ the natural map and $R' = p^{-1}(Z(G/R))$. Then R' is also solvable normal, and G/R' is a connected, semisimple Lie group with trivial center. If G is amenable, so is G/R', and the result follows from 4.1.8.

Remark: Using the structure theory for connected groups (for example, that they are almost Lie groups), one can see that 4.1.9 is true if we assume G is connected but not necessarily Lie. See [Greenleaf 1] for details.

We now consider amenability for some discrete groups.

4.1.10 Example: Let F_n be the free group on n generators. Then if $n \geqq 2$, F_n is not amenable. For example, let $\varphi:[0, 1] \to [0,1]$ be $\varphi(x) = x^2$. Then any invariant measure is easily seen to be supported on $\{0, 1\}$. (This follows from the fact that for any $a < 1$, $\varphi([0, a]) = [0, a^2]$, and $a^n \to 0$.) We can consider φ as a map on the circle. If we let ψ be a rotation of the circle by any angle which is not a multiple of 2π, then there are no probability measures on the circle invariant under both φ and ψ. But φ, ψ generate an F_2 action, showing non-amenability of F_2, from which the non-amenability of a general F_n follows.

The following observation tells us when lattices in Lie groups are amenable.

4.1.11 Proposition: *Suppose $\Gamma \subset G$ is a closed subgroup such that G/Γ has a finite G-invariant measure. Then G is amenable if and only if Γ is amenable.*

Proof: In light of 4.1.6 it remains to show that Γ amenable implies G amenable. Let A be an affine G-space. Then there is a Γ-fixed point in A, say a_0. Then $g \to a_0 g$ induces a G-map $f: G/\Gamma \to A$, and if μ is the G-invariant measure on G/Γ, then $f_*(\mu)$ will be a G-invariant measure on A. Thus the barycenter $b(f_*(\mu))$ will be a G-fixed point.

In particular, lattices in non-compact semisimple Lie groups are not amenable. The following result is useful.

4.1.12 Proposition: *Every locally compact G has a unique maximal normal amenable subgroup.*

We begin with two easy lemmas.

4.1.13 Lemma: *If H is amenable and $\varphi: H \to G$ a continuous homomorphism with $\varphi(H)$ dense in G, then G is amenable.*

Proof: If A is an affine G-space, it is also an affine H-space. Since $\varphi(H)$ is dense, an H-fixed point will be G-fixed.

4.1.14 Lemma: *Suppose $H_\alpha \subset G$ where $\{H_\alpha\}$ is a directed set of amenable closed subgroups. Then $\overline{\cup H_\alpha}$ is amenable.*

Proof: This is a straightforward finite intersection argument.

Proof (of 4.1.12): By Lemma 4.1.14, there exists a maximal normal amenable subgroup. Suppose $H_1, H_2 \subset G$ are two such groups. Then H_1 acts by conjugation on H_2, so we can form the (external) semi-direct product $H_1 \ltimes H_2$. This group is amenable by 4.1.6. We have a continuous homomorphism $H_1 \ltimes H_2 \to H_1 H_2 \subset G$, and hence $\overline{H_1 H_2}$ is amenable by Lemma 4.1.13 and is clearly normal. By maximality of H_1, H_2, we have $H_1 = H_2$.

Perhaps surprisingly, the conclusion of Lemma 4.1.13 is also true if G is the real points of an $\mathbb{R}$-algebraic group and we only assume $\varphi(H)$ is Zariski dense.

4.1.15 Theorem [*Moore* 2]: *Suppose G is the set of real points of an $\mathbb{R}$-algebraic group and $\varphi: H \to G$ is a (topologically) continuous homomorphism with $\varphi(H)$ Zariski dense in G. If H is amenable, so is G.*

Proof: It clearly suffices to consider $\varphi: H \to G = GL(n, \mathbb{R})$, and to show that $\overline{\varphi(H)}$ (Zariski closure) is amenable. Let P be the subgroup of G consisting of upper triangular matrices so that G/P is the space of full flags. H acts on G/P and since H is amenable, it leaves a measure $\mu \in M(G/P)$ invariant. However, from Corollary 3.2.22, we see that G_μ is both amenable and algebraic. Therefore, $\overline{\varphi(H)} \subset G_\mu$ is amenable by 4.1.6.

We conclude this section by mentioning without proof (we will not be using this result) a striking theorem of J. Tits. To put this in some perspective, we recall that any discrete group containing the free group F_2 cannot be amenable. A conjecture of von Neumann, which had remained open for many years, was that the converse is also true, namely that any non-amenable discrete group contains a non-abelian free group. A counterexample to this conjecture has recently been constructed by [Olshanski 1]. However, Tits' result proves the von Neumann conjecture for linear groups.

4.1.16 Theorem [*Tits 3*]: *Suppose k is a field of characteristic* 0, *and* $\Gamma \subset GL(n, k)$ *is a subgroup. Then either* Γ *contains a non-abelian free group or* Γ *has a solvable subgroup of finite index* (*and hence is amenable*).

4.2 Cocycles

Cocycles of a group action are a fundamental tool for understanding the action, and in this section we present an introduction to this important concept. In the next section we will use cocycles to extend the notion of an amenable group to that of an amenable action. We shall be seeing many other uses as well.

To motivate the discussion, suppose that (S, μ) is a G-space and that T is some countably separated Borel space. As usual, we let $F(S, T)$ be the space of measurable functions $S \to T$, two functions being identified if they agree almost everywhere. Then G acts on $F(S, T)$ by translation, i.e. $(g \cdot f) = f(sg)$. Suppose now that T is also a (Borel) H-space for some other group H. (For notational convenience, assume H acts on the left.) Then we can define new actions of G on $F(S, T)$ which consist of translation together with some "twisting" by H. More precisely, suppose that for each $s \in S$, $g \in G$, we have an element $\alpha(s, g) \in H$. Then we define $(g \cdot f)(s) = \alpha(s, g) f(sg)$ – a translation and a twist by $\alpha(s, g)$. For this to define an action, α must satisfy some compatibility conditions, namely:

4.2.1 Definition: Suppose (S, μ) is a G-space and H is a second countable group. A Borel function $\alpha: S \times G \to H$ is called a cocycle if for all $g, h \in G$, $\alpha(s, gh) = \alpha(s, g)\alpha(sg, h)$ for almost all $s \in S$. The cocycle is called strict if this equation holds for all s, g, h.

If α is a cocycle, then the above definition of $g \cdot f$ yields an action of G on $F(S, T)$ which we call the α-twisted action. Suppose now that $\alpha, \beta: S \times G \to H$ are both cocycles. What relation between α and β will ensure that the twisted actions on $F(S, T)$ will be equivalent?

4.2.2 Definition: Two cocycles $\alpha, \beta: S \times G \to H$ are called equivalent, or cohomologous, if there is a Borel function $\varphi: S \to H$ such that for each g, $\beta(s, g) = \varphi(s)\alpha(s, g)\varphi(sg)^{-1}$ for almost all s. We then write $\alpha \sim \beta$. If α, β are strict, and the above equation holds for all s, g, we say that α and β are strictly equivalent.

If $\alpha \sim \beta$, then the map $\Phi: F(S, T) \to F(S, T)$, given by $\Phi(f)(s) = \varphi(s) f(s)$ is a Borel isomorphism which is a G-map, where the domain has the α-twisted action and the range has the β-twisted action.

Pictorially, we think of cocycles and cohomology as follows. Consider elements of $F(S, T)$ as sections of the bundle $S \times T \to S$, and consider $\alpha(s, g) \in H$ as giving a map from the fiber over sg to the fiber over s. In the twisted action, $\alpha(s, g)$ then tells you how to bring $f(sg)$ back to an element of the fiber over s. Then the cocycle identity is simply commutativity of the following picture:

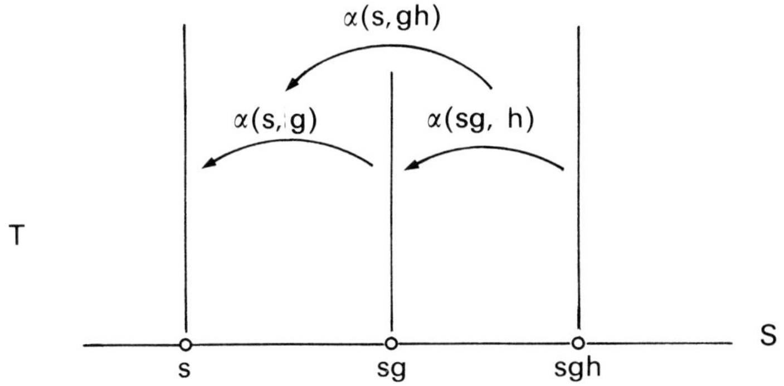

The relation of cohomology is simply the existence of a bundle self-equivalence, i.e., for each s, a self-map $\varphi(s)$ of the fiber, so that the following picture commutes:

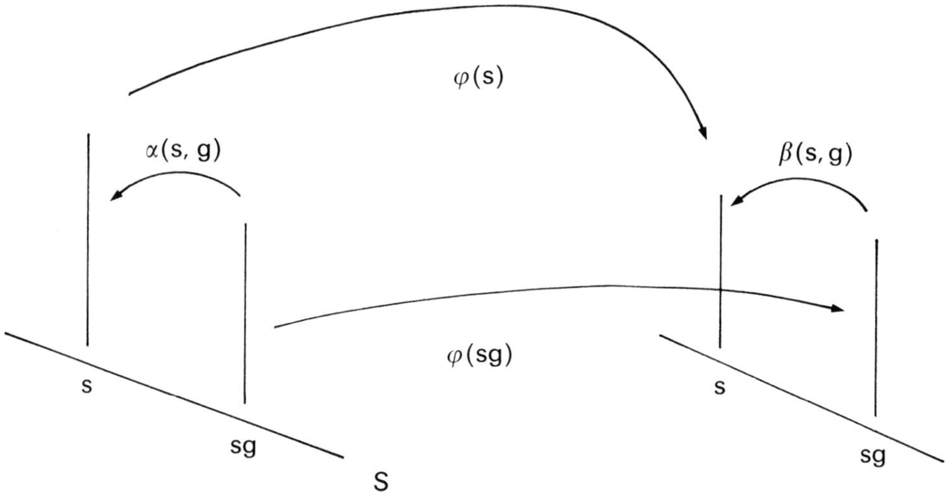

Remarks: (1) By considering $\alpha: S \times G \to H$ as an element of $F(G, F(S, H))$, we see that the cocycle identity is exactly the condition that α be a Borel 1-cocycle on G (in the sense of Eilenberg-MacLane) with values in the G-module $F(S, H)$,

where the latter action is ordinary translation. Similarly, cohomology in the sense of Definition 4.2.2 is cohomology in the sense of Eilenberg-MacLane (using only Borel cochains). Although this point of view is quite useful for certain considerations, and raises many interesting questions (e.g. concerning higher cohomology), we shall not pursue this connection here. See [Feldman-Moore 1], [Zimmer 14].
(2) The question as to how close an arbitrary cocycle is to a strict cocycle is a technical point which is of considerable importance for certain considerations. Although we will be using cocycles defined on an arbitrary G-space, the only actions for which we will need to deal with the above question are transitive actions. Thus, we will deal with this point for transitive actions in this section, and discuss the more technical general situation in Appendix B. We remark, however, that if G is countable, it is easy to see that for any α, there is a strict cocycle α' such that for all g, $\alpha(s, g) = \alpha'(s, g)$ a.e. Namely, for $g, h \in G$, let $A_{g,h} = \{s | \alpha(s, gh) = \alpha(s, \vec{g})\alpha(sg, h)\}$, $A = \bigcap_{g,h} A_{g,h}$ and $S_0 = \bigcap_{g \in G} A \cdot g$. Then S_0 is an invariant conull Borel set, and hence we can take $\alpha'(s, g) = \alpha(s, g)$ for all $s \in S_0$, $\alpha'(s, g) = e$ for $s \notin S_0$.

We now consider some cocycles that arise naturally.

4.2.3 Example: Suppose M is a manifold of dimension n and G acts on M by diffeomorphisms. For each $s \in M$, $g \in G$, let $\alpha(s, g) = (dg^{-1})_{sg}: TM_{sg} \to TM_s$, where TM is the tangent bundle. Then the chain rule for differentiation is exactly the cocycle identity of Definition 4.2.1. We can always choose enough Borel sections of the tangent bundle, and hence measurably TM is trivial. Under any trivialization, α will be identified with a strict cocycle $M \times G \to GL(n, \mathbb{R})$. Different trivializations of TM will yield strictly equivalent cocycles. The α-twisted action is just the natural action of G on measurable vector fields on M.

4.2.4 Example: Suppose (S, μ) is a G-space with quasi-invariant measure μ. Let $r_\mu: S \times G \to \mathbb{R}^+$ be defined by $r_\mu(s, g) = d\mu(sg)/d\mu(s)$, the Radon-Nikodym derivative. Once again, the chain rule is exactly the cocycle identity and r_μ is called the Radon-Nikodym cocycle. If μ and ν are equivalent measures, so that $d\mu = \varphi d\nu$, $\varphi > 0$, then $r_\mu(s, g) = \varphi(sg)d\nu(sg)/\varphi(s)d\nu(s)$, i.e., $\varphi(s)r_\mu(s, g)\varphi(sg)^{-1} = r_\nu(s, g)$. Thus, $r_\mu \sim r_\nu$. Conversely, any cocycle $r: S \times G \to \mathbb{R}^+$ with $r_\mu \sim r$ is of the form r_ν, where ν is a measure equivalent to μ. Since $r_\nu \equiv 1$ (a.e.) if and only if ν is G-invariant, we have:

4.2.5 Proposition: *If (S, μ) is a G-space, then there is a (σ-finite) G-invariant measure equivalent to μ if and only if the Radon-Nikodym cocycle r_μ is equivalent to the identity cocycle, i.e., $i(s, g) = 1$ for all s, g.*

4.2.6 Example: Suppose $\pi: G \to H$ is a (continuous) homomorphism. Then $\alpha_\pi: S \times G \to H$ given by $\alpha_\pi(s, g) = \pi(g)$ is a cocycle. The cocycles of the form α_π are exactly those that are independent of s.

An important example of a cocycle is one coming from an orbit equivalence of actions.

4.2.7 Definition: Suppose (S, μ) is a G-space, (S', μ') a G'-space. The actions are called orbit equivalent if there are conull Borel sets $S_0 \subset S$, $S'_0 \subset S'$ and a measure class preserving Borel isomorphism $\varphi: S_0 \to S'_0$ such that if $s, t \in S_0$, then s and t are in the same G-orbit if and only if $\theta(s)$ and $\theta(t)$ are in the same G'-orbit.

4.2.8 Example: If θ is an orbit equivalence, and the action of G' on S' is essentially free (i.e., almost all stabilizers are trivial), then for each g and almost all s, there is a unique $\alpha(s, g) \in G'$ such that $\theta(s)\alpha(s, g) = \theta(sg)$. It is straightforward to check that α is a cocycle. Some of the significance of its cohomology class can be seen from the following.

4.2.9 Proposition: *Suppose $G = G'$ in Example 4.2.8. As above, let $\alpha: S \times G \to G$ be the cocycle corresponding to an orbit equivalence of essentially free G-spaces, $\theta: S \to S'$. If there is an inner automorphism $\pi: G \to G$ such that $\alpha \sim \alpha_\pi$, (the latter as in Example 4.2.6), then the actions of G on S and S' are isomorphic.*

We shall indicate the idea of the proof, which in fact is essentially a complete proof if G is discrete. A full treatment in the general case becomes somewhat technical and we shall refer the reader to [Zimmer 9] for details.

Proof: Suppose that for each g, $\alpha(s, g) = \varphi(s)\alpha_\pi(s, g)\varphi(sg)^{-1}$ where $\varphi: S \to G$ is Borel. We have $\alpha_\pi(s, g) = hgh^{-1}$ for some $h \in G$. Let $\psi(s) = \varphi(s)h$, so $\alpha(s, g) = \psi(s)g\psi(sg)^{-1}$ (for each g, and almost all s). Define $\lambda: S \to S'$ by $\lambda(s) = \theta(s) \cdot \psi(s)$. Then for each g, and almost all s, we have

$$\lambda(sg) = \theta(sg)\psi(sg) = \theta(sg)\alpha(s, g)^{-1}\psi(s)g = \theta(s)\psi(s)g = \lambda(s) \cdot g.$$

Because θ is a measure space isomorphism, one can deduce by a technical argument that (modulo null sets in S, S') λ is as well. (See [Zimmer 9] for details.) Thus, the proposition will follow from the following generally useful fact.

4.2.10 Lemma: *If S, S' are G-spaces and $\lambda: S \to S'$ is a measure space isomorphism (modulo null sets), and for each g, $\lambda(sg) = \lambda(s)g$ for almost all s, then S and S' are isomorphic G-spaces (in the sense of the beginning of section 2.1). In fact, there is an isomorphism f such that $f = \lambda$ a.e.*

A proof of this lemma appears in Appendix B.

More generally, the same proof (but letting $\psi = \varphi$) shows the following.

4.2.11 Proposition: *Suppose $\alpha: S \times G \to G'$ is the cocycle coming from an orbit equivalence of the G-action on S and the G'-action on S'. If $\alpha \sim \alpha_\pi$ where $\pi: G \to G'$ is an isomorphism then identifying G and G' by this isomorphism, the actions of G on S and S' will be isomorphic.*

4.2.12 Example: Cocycles of Z-actions can be easily described. Namely, if S is a Z-space, and $\alpha: S \times Z \to H$ is a cocycle, let $f(s) = \alpha(s, 1)$. Then by repeated applications of the cocycle identity, $\alpha(s, n)$ can be expressed in terms of f for any n. Conversely, any function $f: S \to H$ can arise this way. Thus cocycles on Z-spaces can be identified with Borel functions $S \to H$.

We now consider cocycles on transitive G-spaces. Suppose $G_0 \subset G$ is a closed subgroup and that $\alpha: G/G_0 \times G \to H$ is a strict cocycle. Then the cocycle identity immediately implies that $\sigma_\alpha: G_0 \to H$ given by $\sigma_\alpha(g) = \alpha([e], g)$ for $g \in G_0$ is a Borel (and hence continuous) homomorphism. Furthermore, if α, β are strict cocycles and α and β are strictly equivalent, then σ_α and σ_β are conjugate homomorphisms, i.e., there is $h \in H$ such that $\sigma_\alpha(g) = h\sigma_\beta(g)h^{-1}$. We now make the following simple but important observation: Every (measurable) homomorphism $\sigma: G_0 \to H$ is of the form σ_α for some strict cocycle $\alpha: G/G_0 \times G \to H$. To see this, we first observe that it is true for the identity homomorphism $i: G_0 \to G_0$. Namely, choose a Borel section $\gamma: G/G_0 \to G$ of the natural projection with $\gamma([e]) = e$. Then for $x \in G/G_0$ and $g \in G$, $\gamma(x)g$ and $\gamma(xg)$ are equal when projected to G/G_0. Therefore, $\beta(x, g) = \gamma(x)g\gamma(xg)^{-1} \in G_0$, and it is immediate that this is a strict cocycle, and that $\sigma_\beta = i$. If $\sigma: G_0 \to H$ is any homomorphism, then $\alpha(x, g) = \sigma(\gamma(x)g\gamma(xg)^{-1})$ is a strict cocycle with $\sigma_\alpha = \sigma$. We summarize our discussion in the following.

4.2.13 Proposition: *The map*

$$\{\text{Strict cocycles } G/G_0 \times G \to H\} \to Hom(G_0, H),$$

$\alpha \to \sigma_\alpha$, *is a surjection which induces a bijection between strict equivalence classes of strict cocycles and conjugacy classes of homomorphisms. Furthermore, any* α *is strictly equivalent to some* α' *such that* $\alpha'(G/G_0 \times G) = \sigma_\alpha(G_0)$.

Proof: In light of the above discussion, all that remains to be verified is that σ_α and σ_β conjugate implies α and β are strictly equivalent. We first observe that from the cocycle identity, letting γ be a section as above, that

$$\alpha([e], \gamma(x))\alpha(x, g)\alpha([e], \gamma(xg))^{-1} = \alpha([e], \gamma(x)g\gamma(xg)^{-1}) = \sigma_\alpha(\gamma(x)g\gamma(xg)^{-1}),$$

and that we have a similar formula for β. If $\sigma_\alpha(\cdot) = h^{-1}\sigma_\beta(\cdot)h$, then we immediately have

$$\alpha(x, g) = \varphi(x)\beta(x, g)\varphi(xg)^{-1}$$

where

$$\varphi(x) = \alpha([e], \gamma(x))^{-1}h\beta([e], \gamma(x)).$$

As an illustration we present the following:

4.2.14 Example: Consider $O(n)$ (the orthogonal group) acting on $S^{n-1} \cong O(n)/O(n-1)$. As in Example 4.2.3, we have the derivative (strict) cocycle $\alpha: S^{n-1} \times O(n) \to GL(n-1)$. By the above proposition, this corresponds to a homomorphism $O(n-1) \to GL(n-1)$, and it is easy to verify from the construction of α that the homomorphism is just the usual inclusion map.

For transitive G-spaces, the study of arbitrary cocycles can be reduced to that of strict cocycles.

4.2.15 Proposition [*Mackey* 1]: *Consider cocycles on the G-space* G/G_0, *where* $G_0 \subset G$ *is closed. Then*

(*i*) *If* α *is any cocycle, there is a strict cocycle* α' *such that for each* g, $\alpha(s, g) = \alpha'(s, g)$ *a.e.*

(*ii*) *If two strict cocycles are equivalent, they are strictly equivalent.*

Proof: (i) If G_1 is a conjugate of G_0 in G, then we have an identification of G-spaces $G/G_0 \cong G/G_1$, and hence we are free to replace G_0 by a conjugate. If α is a cocycle, then by Fubini we have that for almost all $x \in G/G_0$, $\alpha(x, gh) = \alpha(x, g)\alpha(xg, h)$ for almost all $(g, h) \in G \times G$. Replacing G_0 by a conjugate if necessary, we can assume that this holds for $x = [e]$. Thus $\alpha([e], gh) = \alpha([e], g)\alpha([e]g, h)$ for almost all g, h, and so for any $a \in G_0$

$$\alpha([e], agh) = \alpha([e], ag)\alpha([e]g, h) \text{ a.e.}$$

Let $\sigma(g) = \alpha([e], g)$. Then from the above two equations, we have that for each $a \in G_0$,

$$\sigma(ag)^{-1}\sigma(agh) = \sigma(g)^{-1}\sigma(gh), \text{ i.e., } \sigma(ag)\sigma(g)^{-1} = \sigma(agh)\sigma(gh)^{-1},$$

for almost all g, h. Thus, for each a, the function $g \to \sigma(ag)\sigma(g)^{-1}$ considered as an element of $F(G, H)$ is invariant under translation by almost all elements $h \in G$, and since a conull subgroup cannot be proper (Appendix B) this is invariant under all G. By Lemma 2.2.16, for each a, there is a (unique) element $\pi(a) \in H$ such that for each $a \in G_0$ and almost all $g \in G$, $\sigma(ag)\sigma(g)^{-1} = \pi(a)$. The last equation clearly implies that π is a homomorphism. (To see that π is Borel, embed H as a Borel space in $[0, 1]$, and note that $\pi(a) = \int \sigma(ag)\sigma(g)^{-1}dg$, where dg is a probability measure in the class of Haar measure.) Let β be a strict cocycle corresponding to the homomorphism $\pi: G_0 \to H$, and let $\lambda(g) = \beta([e], g)$. Then for $a \in G_0$ $\lambda(ag)\lambda(g)^{-1} = \pi(a) = \sigma(ag)\sigma(g)^{-1}$ for almost all g, i.e., $\sigma(g)^{-1}\lambda(g) = \sigma(ag)^{-1}\lambda(ag)$. Once again by Lemma 2.2.16, we can find $\varphi: G \to H$ such that $\varphi(ag) = \varphi(g)$ for all a, g, and $\varphi(g) = \sigma(g)^{-1}\lambda(g)$ a.e. Now for almost all x, g we have

$$\alpha([x], g) = \sigma(x)^{-1}\sigma(xg) = \varphi(x)\lambda(x)^{-1}\lambda(xg)\varphi(xg)^{-1} = \varphi(x)\beta([x], g)\varphi(xg)^{-1},$$

the latter being a strict cocycle, say δ. We need to show that for all g, $\alpha([x], g) = \delta([x], g)$ for almost all $[x]$. Let $A = \{g \in G \mid \alpha([x], g) = \delta([x], g)$ for almost all $x\}$. By the above, A is conull, and it is easy to see it is closed under multiplication. Hence, $A = G$, and the proof of (i) is complete.

(ii) It clearly suffices to see that if α and β are strict cocycles, and $\alpha = \beta$ a.e., then α and β are strictly equivalent. By Fubini, there is a point $x \in G/G_0$ such that for almost all g, $\alpha(x, g) = \beta(x, g)$. As in (i), we can assume $x = [e]$. Let $\sigma(g) = \alpha([e], g)$, $\lambda(g) = \beta([e], g)$. Then for each $a \in G_0$, we also have $\sigma(ag) = \lambda(ag)$ for almost all g, and hence for each a, we can find g such that $\sigma(ag)\sigma(g)^{-1} = \lambda(ag)\lambda(g)^{-1}$. Therefore $\alpha([e], a) = \beta([e], a)$ for all $a \in A$, and (ii) then follows from Proposition 4.2.13.

In Proposition 4.2.11, we saw the consequences of a cocycle coming from an orbit equivalence to be equivalent to a cocycle of the form α_π where π is a homomorphism. We now examine the implications of this property for a cocycle on a transitive space.

4.2.16 Proposition: *Let $\alpha: G/G_0 \times G \to H$ be a strict cocycle corresponding (under Proposition 4.2.13) to the homomorphism $\sigma: G_0 \to H$. If $\alpha \sim \alpha_\pi$ for some homomorphism $\pi: G \to H$ then σ extends to a homomorphism $G \to H$ (which is a conjugate of π).*

Proof: By 4.2.15 we can assume α and α_π are strictly equivalent, i.e., $\alpha(x, g) = \varphi(x)\pi(g)\varphi(xg)^{-1}$ for all x, g. In particular, if $x = [e]$ and $g \in G_0$, $\sigma_\alpha(g) = \alpha([e], g) = \varphi([e])\pi(g)\varphi([e])^{-1}$, which obviously extends to a homomorphism $G \to H$.

Propositions 4.2.11 and 4.2.16 show that the problem of extending a homomorphism and the problem of showing that orbit equivalent actions are equivalent, problems which at first glance appear unrelated, are in fact closely related, both being special cases of the problem of showing that a cocycle is equivalent to one of the form α_π. We shall be exploring these problems for semisimple groups in Chapter 5.

We began our discussion of cocycles by considering twisted actions on spaces of functions, and we now return to these considerations. In particular, we consider the fixed points for such actions.

4.2.17 Definition: Suppose $\alpha: S \times G \to H$, and T is a left H-space. A function $f: S \to T$ will be called α-invariant if it is a fixed point for the α-twisted action of G on $F(S, T)$. Equivalently, for each g, $\alpha(s, g)f(sg) = f(s)$ a.e.

4.2.18 Example: (a) If $T = H$, then there is an α-invariant function $S \to H$ if and only if α is trivial, i.e., equivalent to the identity cocycle.

(b) More generally, if $T = H/H_0$, then there is an α-invariant function $S \to H/H_0$ if and only if $\alpha \sim \beta$ where $\beta(S \times G) \subset H_0$. For if φ is α-invariant, then for each g, $\beta(s, g) = \varphi(s)^{-1}\alpha(s, g)\varphi(sg) \in H_0$ a.e., and by changing β on a null set, we have $\beta(S \times G) \subset H_0$. Conversely, if $\alpha \sim \beta$ with β taking values in H_0, and φ implements the equivalence, then φ^{-1} will be an α-invariant function.

(c) If $\alpha \sim \beta$, then there is an α-invariant function $S \to T$ if and only if there is a β-invariant function.

If G is transitive, we can relate α-invariant functions to fixed points of homomorphisms corresponding to α.

4.2.19 Proposition: *Suppose $\alpha: G/G_0 \times G \to H$ is a strict cocycle and $\sigma: G_0 \to H$ the corresponding homomorphism. If T is an H-space, then there is an α-invariant function $G/G_0 \to T$ if and only if there is a fixed point in T for the action of $\sigma(G_0)$.*

Proof: By 4.2.13 we can replace α by a strict cocycle whose values lie in $\sigma(G_0)$. It is then clear that if $t \in T$ is $\sigma(G_0)$ invariant, $\varphi(s) = t$ for all s will be α-invariant. To see the converse, let us first extend σ to a map $G \to H$ by setting $\sigma(g) = \alpha([e], g)$. Then $\sigma | G_0$ is a homomorphism, but moreover $\sigma(hg) = \sigma(h)\sigma(g)$ for $h \in G_0$ and $g \in G$. We can recover α from σ via the cocycle identity. Namely, $\alpha([x], g) = \alpha([e]x, g) = \alpha([e], x)^{-1}\alpha([e], xg)$, so $\alpha([x], g) = \sigma(x)^{-1}\sigma(xg)$. Thus, if φ is α-invariant, we have that for each g and almost all x, $\sigma(xg)\varphi([x]g) = \sigma(x)\varphi([x])$. By Fubini, we can choose $[x] \in G/G_0$ such that this holds for almost all $g \in G$, and hence $\sigma(y)\varphi([y]) = \sigma(x)\varphi([x])$ for almost all $y \in G$. Set $\sigma(x)\varphi([x]) = a \in T$. Then for $h \in G_0$, $\sigma(hy)\varphi([hy]) = a$ for a.e. $y \in G$, i.e., $\sigma(h)\sigma(y)\varphi([y]) = a$ for a.e. $y \in G$. As $\sigma(y)\varphi([y]) = a$ a.e., we deduce $\sigma(h)a = a$ for all $h \in G_0$.

Remark: The above proof shows not only that there is a G_0-fixed point, but that the fixed point $a \in T$ can be chosen such that for almost all $g \in G$, $\sigma(g)\varphi([g]) = a$, where $\sigma(g) = \alpha([e], g)$.

We remark that if we are dealing with Borel functions themselves rather than function spaces $F(S, T)$ where functions are identified if they agree a.e., and the α-invariance held everywhere, not just a.e., then 4.2.19 would just be a matter of formal algebraic manipulation. However, function spaces like $F(S, T)$ are in general much better behaved than the space of Borel functions, and in particular, it is much easier to deduce the existence of fixed points from general functional analytic arguments. Thus the necessity for the a.e. argument in the above proof.

If the space T is a linear space, and H acts on T by linear transformations, then $F(S, T)$ will also be a linear space and the α-twisted action of G will be a linear action. In particular, if $S = G/G_0$, and we have a linear representation of G_0 on T, then we have an associated cocycle and hence an associated twisted linear action. Thus, we pass from a linear representation of G_0 on a space E to a linear representation of G on $F(G/G_0, E)$. This is just the classical notion of an induced representation. If the original representation has further structure, e.g. is an isometric representation on a Banach space, or a unitary representation on a Hilbert space, we would like the induced representation to have the same structure, and we achieve this by passing to a suitable subspace. More precisely, suppose E is a separable Banach space, $\mathrm{Iso}(E)$ the group of isometric automorphisms of E, and that $\sigma: G_0 \to \mathrm{Iso}(E)$ is a continuous representation (where $\mathrm{Iso}(E)$ has the strong operator topology). Assume for the moment that G/G_0 has a σ-finite invariant measure. Then we have an induced continuous representation of G by isometries on the Banach space $L^1(G/G_0,E) \subset F(G/G_0,E)$ given by $(\pi(g)f)(x) = \alpha(x, g)f(xg)$ where α is a cocycle corresponding to σ. The representation is by isometries because $\alpha(x, g) \in \mathrm{Iso}(E)$ and G preserves the measure on G/G_0. The representation is easily seen to be measurable, and hence it is continuous (Appendix B). If G does not have an invariant measure on G/G_0, let $\rho: G/G_0 \times G \to \mathbb{R}^+$ be the Radon-Nikodym derivative for a quasi-invariant measure. Then we define the induced representation of G on $L^1(G/G_0,E)$ by $(\pi(g)f)(x) = \rho(x, g)\alpha(x, g)f(xg)$. (We remark that this makes sense for any cocycle α taking values in $\mathrm{Iso}(E)$ defined on any G space.) We remark that whether or not there is an invariant measure, the formula for the adjoint representation π^* of G on $L^1(G/G_0, E)^* \cong L^\infty(G/G_0, E^*)$, namely $\pi^*(g) = (\pi(g)^{-1})^*$, is given by $(\pi^*(g)\varphi)(x) = \alpha^*(x, g)\varphi(xg)$, where α^* is the "adjoint cocycle" $\alpha^*(x, g) = (\alpha(x, g)^{-1})^*$.

Similarly, if E is a Hilbert space and $\sigma: G_0 \to U(E)$ is a representation, the induced representation $\pi: G \to U(L^2(G/G_0, E))$ is given by $(\pi(g)f)(x) = \rho(x, g)^{1/2}\alpha(x, g)f(xg)$ where α is a cocycle corresponding to σ. For finite groups, induced representations were studied by Frobenius, and the general theory for locally compact groups was developed by [Mackey 1].

We illustrate the ideas of this section by returning to an unfinished point in section 4.1., namely the proof of 4.1.6(a).

4.2.20 Proposition: *If G is amenable and $H \subset G$ is a closed subgroup, then H is amenable.*

Proof: Suppose $\sigma: H \to \mathrm{Iso}(E)$ is a representation where E is a separable Banach space and $A \subset E^*$ is an affine H-space. Form the adjoint of the induced representation, π^*, of G on $L^\infty(G/H, E^*)$. Let $\tilde{A} = \{\varphi \in L^\infty(G/H, E^*) \mid \varphi(x) \in A \text{ a.e.}\}$, which we can identify with $F(G/H, A)$. We can choose the cocycle corresponding to σ to have the same image as σ, and then $\tilde{A}$ becomes an affine G-space. Since G is amenable, there is a G-fixed point in $F(G/H, A)$, and by 4.2.19, there is an H-fixed point in A. This completes the proof.

Just as we can induce representations of a subgroup to a group, one can also induce actions. Namely, suppose (S, μ) is a G-space, (T, v) is a (right)-H-space and that $\alpha: S \times G \to H$ is a strict cocycle. On $(S \times T, \mu \times v)$, define a measure class preserving G-action by $(s, t)\cdot g = (sg, t\alpha(s, g))$. (The reader is invited to re-examine the picture following Definition 4.2.2, but now with the arrows reversed, since we are acting on the right.) This is called the skew product action defined by α, and when endowed with this action, we denote $S \times T$ by $S \times_\alpha T$. Of course, if $G = H$ and $\alpha = \alpha_i$, where i is the identity homomorphism, then this reduces to the usual product action. It is also straightforward to verify that if α and β are strictly equivalent, then $S \times_\alpha T$ and $S \times_\beta T$ are isomorphic G-spaces. (Once again, the picture following 4.2.2 makes this clear.) Suppose now that $S = G/H$ and that $\alpha: G/H \times G \to H$ corresponds to the identity $H \to H$ (e.g. $\alpha(x, g) = \gamma(x)g\gamma(xg)^{-1}$, where $\gamma: G/H \to G$ is a section). Then for any H-space T, we can form the G-space $G/H \times_\alpha T$.

4.2.21 Definition: The G-space $G/H \times_\alpha T$ is called the G-space induced from the H-space T.

Remarks: (1) The induced G-action is ergodic if and only if the action of H on T is ergodic. To see this, observe that a Borel set $A \subset G/H \times_\alpha T$ corresponds to a Borel map $G/H \to L^\infty(T)$, $x \to \chi_{A_x}$ where $A_x = \{t \in T | (x, t) \in A\}$. A is G-invariant if and only if this map is $\pi \circ \alpha$-invariant in the sense of Definition 4.2.17, where π is the representation of H on $L^\infty(T)$. Our assertion can then be deduced from 4.2.19. (Cf. 2.2.2.)

(2) There is another, more standard, and more natural (in that it does not involve the choice of a section) definition. Namely, let H act on $G \times T$ by the product action $(g, t)\cdot h = (gh, th)$, and let $X = (G \times T)/H$. We also have a G-action on $G \times T$ given by $(g, t)\cdot g_0 = (g_0^{-1}g, t)$, and as this commutes with the H-action, there is an induced action of G on X. Furthermore, if we project a probability measure on $G \times T$ in the class of the natural product measure to X, this will be quasi-invariant under G. We claim that X and $G/H \times_\alpha T$ are isomorphic G-spaces. Namely, as above, let $\gamma: G/H \to G$ be a Borel section of the natural projection $G \to G/H$ and define $\Phi: G/H \times_\alpha T \to X$ by $\Phi(y, t) = p(\gamma(y)^{-1}, t)$ where $p: G \times T \to (G \times T)/H = X$ is the natural map. One can readily check that Φ is a measure class preserving Borel isomorphism. To see it is a G-map, it suffices to see that $\Phi(yg, t\gamma(y)g\gamma(yg)^{-1}) = p(g^{-1}\gamma(y)^{-1}, t)$, i.e., that $(\gamma(yg)^{-1}, t\gamma(y)g\gamma(yg)^{-1})$ and $(g^{-1}\gamma(y)^{-1}, t)$ are in the same H-orbit. But acting upon the latter by $\gamma(y)g\gamma(yg)^{-1} \in H$, we obtain the former.

In Chapter 5 we will be able to deduce certain results about actions of a semisimple Lie group G using the rich structure of G. The inducing process will then enable us to extend some aspects of these results to actions of lattice subgroups. Here, we shall content ourselves with one further remark.

4.2.22 Proposition: *Suppose S is a G-space and $\Gamma \subset G$ a closed subgroup. Then the action of G induced by the (restricted) Γ-action on S is isomorphic to the product action of G on $G/\Gamma \times S$.*

Proof: The product action is the skew product $G/\Gamma \times_{\beta} S$ where $\beta: G/\Gamma \times G \to G$ is given by $\beta(x, g) = g$. Since $\beta([e], g) = \alpha([e], g)$ for $g \in \Gamma$ (both being equal to g), α and β are strictly equivalent by 4.2.13, so by our remarks above, the skew products are isomorphic.

We conclude this section by describing an invariant of a cocycle taking values in a locally compact group which, although we will not be using it in the sequel, is very important for a number of considerations. (For example, see the discussion at the end of section 4.3.) Let (S, μ) be an H-space, and for simplicity, suppose the action is ergodic. Let $\alpha: S \times H \to G$ be a cocycle, where G is locally compact. We shall define an ergodic G-space X, called the Mackey range of α, which will (up to isomorphism) be an invariant of the cohomology class of α. If $H \subset G$ is a closed subgroup, and $\alpha = \alpha_i$ where $i: H \to G$ is the inclusion map, the Mackey range will be the G-space induced from the H-space S, and in general the construction of the Mackey range will be a natural generalization of the construction of induced actions.

Form the skew product H-space $S \times_{\alpha} G$. If α is a strict cocycle, the formula $(s, g) \cdot h = (sh, g\alpha(s, h))$ actually defines an action, but for an arbitrary cocycle, what we have is a "near action", i.e., for each g, $s \to sg$ is a measure class preserving Borel isomorphism $S \to S$, but for $g, h \in G$, we only have $s \cdot gh = (sg) \cdot h$ a.e. (Actually, every near action is "equal almost everywhere" to an action. See Appendix B.) In the case of inducing, we formed the space of H-orbits $(S \times G)/H$ and let G act on this space. However, for a general strict cocycle α, the space of orbits may be badly behaved, and one may even have that H acts properly ergodically on $S \times_{\alpha} G$. We thus require a replacement for the space of H-orbits. Let $I = \{f \in L^{\infty}(S \times_{\alpha} G) \mid \text{for each } h, f(zh) = f(z) \text{ a.e.}\}$. Then I is easily seen to be a weak-*-closed subalgebra of $L^{\infty}(S \times G)$, and hence there is a standard measure space (X, ν) and a measure class preserving Borel map $\varphi: S \times G \to X$ such that $\varphi^*(L^{\infty}(X)) = I \subset L^{\infty}(S \times G)$. If $(S \times_{\alpha} G)/H$ is countably separated, then we can naturally identify $X \cong (S \times_{\alpha} G)/H$. (We remark that X and φ are unique up to null sets.) As in the inducing process, G acts on $S \times_{\alpha} G$ by $(s, g)g_0 = (s, g_0^{-1}g)$, and this commutes with the near action of H. Therefore, G leaves the algebra I invariant, and hence (see Appendix B) we may choose X to be a G-space and φ to be a G-map.

4.2.23 Definition: The G-space (X, ν) is called the Mackey range of the cocycle α.

The action was introduced in [Mackey 5]. We list some properties of the Mackey range. Once again, as we will not be using them in the present work, we omit the proofs, but invite the reader to supply them. (For non-discrete groups, the technical results of Appendix B should be useful. Some detailed proofs can be found in [Ramsay 1] and [Zimmer 1].)

4.2.24 Proposition [*Mackey* 5]
(*a*) *The Mackey range action is always ergodic* (*assuming the original H-action is ergodic*).
(*b*) *If the skew product action of H on $S \times_\alpha G$ is ergodic, the Mackey range is the action of G on a point.*
(*c*) *The Mackey range is* (*up to isomorphism*) *a cohomology invariant of α.*
(*d*) *The cocycle is trivial* (*i.e., equivalent to $i(s, g) = e$*) *if and only if the Mackey range is the action of G on itself by translation.*
(*e*) *α is equivalent to a cocycle taking values in a subgroup $G_0 \subset G$ if and only if the Mackey range is induced from an action of G_0.*
(*f*) *If $S = H/H_0$ and $\alpha: H/H_0 \times H \to G$ corresponds to a homomorphism $\sigma: H_0 \to G$, then the Mackey range is the action of G on $\overline{G/\sigma(H_0)}$.*

Finally, we mention one further example. Suppose S_i is an essentially free ergodic G_i space, $i = 1, 2$, and that the actions are orbit equivalent. Let $\alpha: S_1 \times G_1 \to G_2$ be a cocycle corresponding to an orbit equivalence. Then the Mackey range of α is the G_2 action on S_2.

4.3 Amenable actions

If G is an amenable group, any action of G will inherit certain properties from the group. On the other hand, it is possible that a given action of a non-amenable group may also have these properties. This leads us to the notion of an amenable action of a group, first introduced in [Zimmer 3]. This notion has proven useful in a variety of situations. In particular, we shall see that if $\Gamma \subset SL(2, \mathbb{R})$ is a lattice, then the action of Γ on the boundary $\bar{\mathbb{R}}$ (see introduction) is amenable, in spite of the fact that Γ is not. More generally, if G is a Lie group and $\Gamma \subset G$

is a lattice, then Γ acts amenably on G/H if $H \subset G$ is an amenable group. In particular, this will be the case if G is semisimple and H is a minimal parabolic subgroup, thus generalizing the example above of Γ acting on $\bar{\mathbb{R}}$. The results on amenable actions we present here follow [Zimmer 3, 4].

Amenability of G demands the existence of a fixed point in every affine G-space. Amenability of an action of G on a space (S, μ) will be the condition that G has a fixed point in certain affine G-spaces which are nicely related to S. To describe these affine actions, suppose E is a separable Banach space and $\alpha: S \times G \to \mathrm{Iso}(E)$ is a cocycle. We wish to describe certain G-invariant compact convex sets in $L^\infty(S, E^*)$, where G acts on the latter by the adjoint of the α-twisted action on $L^1(S, E)$, namely $(\pi^*(g)\varphi)(s) = \alpha^*(s, g)\varphi(sg)$ where, as in section 4.2, α^* is the adjoint cocycle $\alpha^*(s, g) = (\alpha(s, g)^{-1})^*$. One possibility is to choose A to be a compact convex subset of the unit ball $E_1^* \subset E^*$ such that $\alpha^*(s, g)A = A$. Then $F(S, A) \subset L^\infty(S, E^*)$ will be a compact convex G-invariant subset. However, it is also possible to vary the set A as we move from point to point. Thus, suppose that for each $s \in S$, we have a compact convex subset $A_s \subset E_1^*$, which varies in a Borel fashion, i.e., $\{(s, A_s)\} \subset S \times E_1^*$ is a Borel subset, and which satisfy the condition that for each g, $\alpha^*(s, g)A_{sg} = A_s$ for almost all s. Then $F(S, \{A_s\}) = \{\varphi: S \to E_1^* \mid \varphi(s) \in A_s \text{ for almost all } s\}$ is a compact convex subset of $L^\infty(S, E^*)$ (with the weak-*-topology as the dual space of $L^1(S, E)$), and it is clearly G-invariant under the α^*-twisted action. That $F(S, \{A_s\})$ is closed in the unit ball of $L^\infty(S, E^*)$ can be seen by standard arguments. See [Zimmer 3] for details.

4.3.1 Definition: An affine G-space of the form $F(S, \{A_s\})$ (with the α^*-twisted action) will be called an affine G-space over S. The action of G on S is called amenable if every affine G-space over S has a fixed point. (We recall once again that the condition that φ be a fixed point is that for each g, $\alpha^*(s, g)\varphi(sg) = \varphi(s)$ for almost all s.)

If G is amenable, then clearly every G action is amenable. It is not a priori clear that non-amenable groups should admit amenable actions. As a first example, we have the following.

4.3.2 Proposition: *If $H \subset G$ is a closed subgroup, then the action of G on G/H is amenable if and only if H is an amenable group.*

Proof: The proof that the G action on G/H amenable implies H amenable is exactly the same as the proof of Proposition 4.2.20. All that was used in that proof was amenability of G on G/H, not amenability of G itself.

Conversely, suppose H is amenable and $F(G/H, \{A_x\})$ is an affine G-space over G/H, for some cocycle $\alpha: G/H \times G \to \mathrm{Iso}(E)$. By 4.2.15 we can assume α is strict. Let $\sigma(g) = \alpha([e], g)$. Then by applying 4.2.19 (and the remark following its proof) to the action of Iso(E) on compact subsets of E_1^*, there is a $\sigma(H)$-invariant compact convex set $A \subset E_1^*$ such that for almost all $g \in G$, $\sigma(g)A_{[g]} = A$. Since H is amenable, we can choose an H-fixed point $a \in A$, and define $\varphi([g]) = \sigma(g)^{-1}(a)$. This is well defined since a is H-invariant, and we have $\varphi([g]) \in A_{[g]}$ a.e. Finally, for all $x \in G/H$, $g \in G$, we have $\alpha(x, g)\varphi(xg) = \sigma(x)^{-1}\sigma(xg)\sigma(xg)^{-1}(a) = \varphi(x)$, showing that φ is α-invariant.

One situation in which amenability of an action implies amenability of the acting group is if there is a finite invariant measure.

4.3.3 Proposition: *If G acts amenably on S, and acts with finite invariant measure, then G is amenable.*

Proof: Suppose A is an affine G-space for a representation $\pi: G \to \mathrm{Iso}(E)$. Then $F(S, A)$ is an affine G-space over S for the cocycle α_π (4.2.6). Thus, there is an $\alpha_\pi{}^*$-invariant function $\varphi: S \to A$, so that $\pi^*(g)\varphi(sg) = \varphi(s)$ (for each g and almost all s). Converting to a right action of G on A, we have for each g, $\varphi(sg) = \varphi(s)g$ a.e. If μ is the G-invariant measure on S, then $\varphi_*(\mu)$ will be a G-invariant measure on A, and the barycenter $b(\varphi_*(\mu))$ will be a G-fixed point.

We now prove amenability of some other actions. In general, if X and Y are G-spaces and there is a measure class preserving G map $X \to Y$, then amenability of G on Y implies amenability of G on X [Zimmer 3]. However, we shall only be using this in the case of product actions, and as the technical details of the proof are simpler for these actions, we shall only present the proof in that case.

4.3.4 Proposition: *Suppose S and Y are G-spaces, and that G acts amenably on S. Then G acts amenably on $S \times Y$.*

Proof: Suppose $\alpha: S \times Y \times G \to \mathrm{Iso}(E)$. Let $F = L^1(Y, E)$, so $F^* = L^\infty(Y, E^*)$. If $F(S \times Y, \{A_{(s,y)}\})$ is G-invariant, let $B_s \subset F^*$ be $B_s = F(Y, \{A_{(s,y)}\}_{y \in Y})$. We have a natural identification of $F(S \times Y, \{A_{(s,y)}\})$ with $F(S, \{B_s\})$. Further, the G-action on the latter is given by the β^*-twisted action where $\beta: S \times G \to \mathrm{Iso}(F)$ is given by $[\beta(s, g)f](y) = \alpha(s, y, g) \cdot f(yg)$. By amenability of the G-action on S, there is a G-invariant function in $F(S, \{B_s\})$, and hence one in $F(S \times Y, \{A_{(s,y)}\})$.

As a consequence, we can now deduce amenability of the actions mentioned at the beginning of this section.

4.3.5 Theorem: *Suppose S is an amenable G-space. If $\Gamma \subset G$ is a closed subgroup, then S is an amenable Γ-space.*

Proof: $G/\Gamma \times S$ is amenable by the preceding proposition, and by 4.2.22, $G/\Gamma \times S$ is isomorphic as a G-space to the action induced from the Γ-action on S. Therefore, it suffices to prove the following lemma.

4.3.6 Lemma: *Suppose $\Gamma \subset G$ is a closed subgroup, S a Γ-space. If the action induced to G is amenable, so is the Γ-action on S.*

Proof: Let $\alpha: S \times \Gamma \to \mathrm{Iso}(E)$, and $F(S, \{A_s\}) = B$ an associated affine Γ-space over S. To see that there is a Γ-fixed point in B, it suffices by 4.2.19 to see that there is a G-fixed point in $F(G/\Gamma, B)$, where the action is the induced action, i.e., given by the β-twisting where $\beta: G/\Gamma \times G \to \Gamma$ is a cocycle corresponding to the identity homomorphism $\Gamma \to \Gamma$. Since the induced action of G on $G/\Gamma \times_\beta S$ is amenable, it suffices to show that the action of G on $F(G/\Gamma, B)$ can be identified with that of an affine G-space over $G/\Gamma \times_\beta S$. Define a cocycle $\tilde{\alpha}: G/\Gamma \times_\beta S \times G \to \mathrm{Iso}(E)$ by $\tilde{\alpha}(x, s, g) = \alpha(s, \beta(x, g))$, and let $A_{(x,s)} = A_s$. Then $F(G/\Gamma \times_\beta S, \{A_{(x,s)}\})$ is an affine G-space over $G/\Gamma \times_\beta S$, and the map $\Phi: F(G/\Gamma, B) \to F(G/\Gamma \times_\beta S, \{A_{(x,s)}\})$, $((\Phi f)(x))(s) = f(x, s)$ is clearly a bijective G-map.

We remark that the converse of Lemma 4.3.6 is also true. (See [Zimmer 4]).

4.3.7 Corollary: *Suppose Γ, $H \subset G$ are closed subgroups. If H is amenable, then the Γ-action on G/H is amenable. If $\Gamma \subset G$ is a lattice, the Γ-action on G/H is amenable if and only if H is amenable.*

Proof: The first assertion follows immediately from Theorem 4.3.5 and Proposition 4.3.2. To see the second assertion, suppose $A = F(G/H, \{A_x\})$ is an affine G-space over G/H. Then by restriction it is clearly also an affine Γ-space over G/H. If Γ acts amenably on G/H, there is a Γ-fixed point in A, say $a \in A$. Then $g \to ag$ induces a G-map $f: G/\Gamma \to A$, and hence if μ is the G-invariant measure on G/Γ, $b(f_*(\mu))$ will be a G-fixed point in A (where b is the barycenter map). The result then follows from 4.3.2.

4.3.8 Example: Let $H \subset SL(n, \mathbb{R})$ be the subgroup leaving a flag (i.e., a chain of subspaces $\{0\} \underset{\neq}{\subset} V_1 \underset{\neq}{\subset} V_2 \underset{\neq}{\subset} \ldots \underset{\neq}{\subset} V_k = \mathbb{R}^n$) invariant, so that $SL(n, \mathbb{R})/H$ is the corresponding variety of all such flags. Then $SL(n, Z)$ acts amenably on this variety (with Lebesgue measure) if and only if the flag is full, i.e., $\dim V_k = k$, (and $k = n$). In particular, $SL(n, Z)$ acts amenably on $\mathbb{P}^{n-1}$ if and only if $n = 2$. More generally, if G is a non-compact semisimple Lie group, $P \subset G$ a parabolic subgroup, and $\Gamma \subset G$ a lattice, then Γ acts amenably on G/P if and only if P is a minimal parabolic subgroup.

A very useful consequence of the condition of amenability of an action is the following.

4.3.9 Proposition: *Suppose (S, μ) is an amenable G-space and that X is a compact metric G-space. Then, possibly after discarding an invariant Borel null set in S, there exists a measurable G-map $S \to M(X)$.*

Proof: Let π be the representation of G on $C(X)$ by isometries and let $\alpha: S \times G \to \mathrm{Iso}(C(X))$ be $\alpha = \alpha_\pi$. Then $F(S, M(X))$ is an affine G-space over S and hence there is a function $\varphi: S \to M(X)$ such that for each g, $\alpha^*(s, g)\varphi(sg) = \varphi(s)$ a.e., i.e., $\pi^*(g)\varphi(sg) = \varphi(s)$. Switching to a right action of G on $M(X)$, we obtain that for each g, $\varphi(sg) = \varphi(s)g$ for almost all s. But then discarding an invariant null set, we can change φ on a null set to obtain a G-map. (This last statement follows by the usual arguments if G is countable, and from Appendix B general.)

The existence of measurable G-maps from a G-space into $M(X)$ was first examined in some special cases by [Furstenberg 2].

Our main interest in amenability for actions in the present work will be the application to the discrete subgroups and ergodic actions of semisimple Lie groups (for instance, via the above examples). However, we will now indicate, without proof, some of the theory of orbit equivalence for amenable actions. We present this both because of the striking nature of the results, and because it will put into better perspective our results in Chapter 5 concerning the orbit equivalence of actions of semisimple groups. For some geometric applications of amenability of groups and actions (and foliations), see [Gromov 1], [Brooks 1], and [Zimmer 5].

We begin with the following observation asserting that for free actions, amenability is an invariant of orbit equivalence.

4.3.10 Proposition: *Suppose (S_i, μ_i) is a G_i-space, $i = 1, 2$, and suppose the actions are essentially free and orbit equivalent. Then the G_1 action on S_1 is amenable if and only if the G_2 action on S_2 is amenable.*

Proof: For technical simplicity, we present a proof for G_i discrete. The general case can be reduced to the discrete case via a general technique of [Feldman, Hahn, Moore 1]. Let $\theta: S_1 \to S_2$ be the orbit equivalence and $\alpha: S_1 \times G_1 \to G_2$ the corresponding cocycle [Example 4.2.8]. Since G_1 is discrete, we can assume, by passing to a conull subset, that α is strict. If $\beta: S_2 \times G_2 \to \mathrm{Iso}(E)$ is a cocycle, and $F(S_2, \{A_t\})$ an affine G_2 space over S_2, then $\tilde{\beta}: S_1 \times G_1 \to \mathrm{Iso}(E)$, $\tilde{\beta}(s, g) = \beta(\theta(s), \alpha(s, g))$ is a cocycle and $F(S_1, \{A_{\theta(s)}\})$ is an affine G_1-space over S_1. If $\varphi \in F(S_1, \{A_{\theta(s)}\})$ is a $\tilde{\beta}$-invariant function, define $\psi(t) = \varphi(\theta^{-1}(t))$. For each $h \in H$, and almost all $t \in S_2$, we can write $(t, h) = (\theta(s), \alpha(s, g))$ for some $g \in G_1$, and since G_i is countable, it follows that $\psi \in F(S_2, \{A_t\})$ is a β-invariant function. Thus, if the G_1 action is amenable, so is the G_2 action.

4.3.11 Corollary: *Let S_i be an essentially free G_i-space with finite invariant measure, for $i = 1, 2$. Suppose the two actions are orbit equivalent. Then G_1 is amenable if and only if G_2 is amenable.*

Proof: 4.3.3 and 4.3.10.

The first major result concerning orbit equivalence was the fundamental theorem of H. Dye. Let Z be the group of integers.

4.3.12 Theorem [*Dye* 1, 2]: *All finite measure preserving (properly) ergodic Z-actions are orbit equivalent.*

This was later extended to the σ-finite case by Krieger.

4.3.13 Theorem [*Krieger* 2]: *All σ-finite (but not finite) measure preserving Z-actions are orbit equivalent.*

Krieger also extended the theorem to the case of quasi-invariant measure without equivalent invariant measure. He showed that a measurement of the extent to which the action fails to be measure preserving is a complete invariant of orbit equivalence. Namely, let X be an ergodic G-space, and $r:X \times G \to \mathbb{R}^+$ be the Radon-Nikodym cocycle. Let $\Delta: G \to \mathbb{R}^+$ be the modular function of G, and let $m: X \times G \to \mathbb{R}^+$ be $m(x, g) = r(x, g)\Delta(g)^{-1}$. We call m the modular cocycle. The Mackey range of this cocycle (Definition 4.2.23) will be an ergodic $\mathbb{R}^+$-action, which we call the modular flow or the modular range. For a unimodular group G, the modular flow will be translation of $\mathbb{R}^+$ on $\mathbb{R}^+$ itself if and only if the Radon-Nikodym cocycle is trivial (Proposition 4.2.24), and hence if and only if there is an equivalent invariant measure for the action (4.2.5).

4.3.14 Theorem [*Krieger* 2]: *For Z-actions with quasi-invariant measure, and not possessing (an equivalent) finite invariant measure, the modular flow is a complete invariant of orbit equivalence.*

Accessible expositions of Krieger's work are [Sutherland 1] and [Hamachi and Osikawa 1].

4.3.15 Example: Let us see how to compute the modular flow in some examples. Let $\Gamma \subset SL(2, \mathbb{R})$ be a lattice, and let P be the upper triangular subgroup. As is well known from Lie theory (e.g. [Helgason 1]), the Radon-Nikodym cocycle for the action of $G = SL(2, \mathbb{R})$ on G/P is the cocycle $\alpha: G/P \times G \to \mathbb{R}^+$ corresponding to the homomorphism $P \to \mathbb{R}^+$ given by Δ_P, the modular function of P. Clearly $r: G/P \times \Gamma \to \mathbb{R}^+$ is just $\alpha | G/P \times \Gamma$. Now the Γ action on $G/P \times_r \mathbb{R}^+$ that appears in the construction of the Mackey range is just the restriction to Γ of the G action on $G/P \times_\alpha \mathbb{R}^+$. Since $\Delta: P \to \mathbb{R}^+$ is surjective, as is well known, this is the action of G on $G/P \times_\alpha P/\ker \Delta_P$ which is transitive with stabilizer $\ker \Delta_P$. Thus as a Γ-space, $G/P \times_r \mathbb{R}^+$ is just the action of Γ on $G/\mathrm{Ker}\,\Delta_P$. Since $\ker \Delta_P$ is not compact, by Moore's ergodicity theorem (Chapter 2) Γ is ergodic on this space. Hence the modular flow is the action of $\mathbb{R}^+$ on a point. This computation can clearly be carried out on any simple non-compact Lie group. If Γ is a lattice in such a group G, and $P \subset G$ is a minimal parabolic, then the modular flow of the action of Γ on G/P will be the action of $\mathbb{R}^+$ on a point.

The Dye-Krieger theorems were extended over the years by a number of persons to include within its framework actions of larger classes of groups. (In fact Dye did not restrict himself to the integers.) This work has recently culminated with the following theorems.

4.3.16 Theorem [*Connes-Feldman-Weiss* 1]:
(*i*) *A free properly ergodic action of a discrete group is amenable if and only if it is orbit equivalent to a Z-action.*
(*ii*) *The Dye-Krieger theorems* (4.3.12–4.3.14) *hold for the class of amenable properly ergodic actions of discrete groups.*

4.3.17 Theorem [*Connes-Feldman-Weiss* 1]:
(*i*) *A free properly ergodic action of a continuous group is amenable if and only if it is orbit equivalent to an* $\mathbb{R}$*-action.*
(*ii*) *For such actions, the modular flow is a complete invariant of orbit equivalence. In particular, any two free properly ergodic actions of continuous amenable unimodular groups with invariant measure are orbit equivalent.*

4.3.18 Example: If G is a simple noncompact Lie group, $\Gamma \subset G$ at lattice, $P \subset G$ a minimal parabolic, we saw in Example 4.3.15 that the modular flow of Γ on G/P is independent of G and Γ. Since these actions are amenable, Theorem 4.3.16 implies that they are all orbit equivalent.

For some further results on amenable actions see section 9.2.

5 Rigidity

5.1 Margulis' superrigidity theorem and the Mostow-Margulis rigidity theorem

In this section we prove the rigidity theorems for lattices in semisimple Lie groups, making basic use of the results we have established in the preceding three chapters. Most of the results on semisimple groups we have obtained so far hold for an arbitrary semisimple group. In this section, we shall often require the assumption that $\mathbb{R}$-rank$(G) \geqq 2$. We recall that if G is a semisimple algebraic group defined over k, then k-rank(G) (or sometimes, by abuse of notation, k-rank(G_k)) is defined to be the maximal dimension of an abelian k-subgroup of G which is k-split, i.e., which can be diagonalized over k. If G is a connected semisimple Lie group then we can realize $\mathrm{Ad}(G)$ as a subgroup of finite index in the $\mathbb{R}$-points of an $\mathbb{R}$-group (Proposition 3.1.6). We then define $\mathbb{R}$-rank(G) to be the $\mathbb{R}$-rank of this algebraic group. Thus $\mathbb{R}$-rank$(SL(n, \mathbb{R})) = n - 1$, the $\mathbb{R}$-split abelian subgroup of maximal dimension being the diagonal matrices of determinant one.

5.1.1 Theorem (*Mostow-Margulis Rigidity Theorem*): *Let G, G' be connected semisimple Lie groups with trivial center and no compact factors and suppose $\Gamma \subset G$ and $\Gamma' \subset G'$ are lattices. Assume Γ is irreducible and $\mathbb{R}$-rank$(G) \geqq 2$. Suppose $\pi: \Gamma \to \Gamma'$ is an isomorphism. Then π extends to a rational isomorphism $\pi: G \to G'$.*

This was first shown for cocompact lattices by [Mostow 2] and then for non-cocompact lattices by [Margulis 3]. The theorem is true for $\mathbb{R}$-rank 1 groups as well excluding $PSL(2, \mathbb{R})$. This is due to [Mostow 2] for Γ cocompact, and Mostow's argument was completed to the non-cocompact case by [Prasad 1]. However, here we shall only be proving the result under the hypothesis of 5.1.1.

Theorem 5.1.1 follows directly from a more general result of [Margulis 1] which deals with the question of extending an arbitrary homomorphism $\Gamma \to G'$ to G, without the assumption that the image be a lattice in G'. In fact, Margulis's results also give basic information about homomorphisms of Γ into groups of the form H_k, where k is any local field of characteristic 0 and H is connected, semisimple, and defined over k. This generalized rigidity result ("superrigidity") is the central part of the proof of Margulis' arithmeticity theorem, which we discuss in the next chapter.

5.1.2 Theorem (*Margulis' Superrigidity Theorem*): *Let G be a connected semisimple algebraic $\mathbb{R}$-group, $\mathbb{R}$-rank$(G) \geqq 2$, and assume $G^0_{\mathbb{R}}$ has no compact factors. Let $\Gamma \subset G^0_{\mathbb{R}}$ be an irreducible lattice. Suppose k is a local field of characteristic* 0, *and let H be a connected algebraic k-group, almost simple over k. Assume $\pi:\Gamma \to H_k$ is a homomorphism with $\pi(\Gamma)$ Zariski dense. Then*
(*i*) *If $k = \mathbb{R}$, H is $\mathbb{R}$-simple* (*equivalently, center free*), *and $H_{\mathbb{R}}$ is not compact, then π extends to a rational homomorphism $G \to H$ defined over $\mathbb{R}$* (*and hence defines a homomorphism $G_{\mathbb{R}} \to H_{\mathbb{R}}$.*)
(*ii*) *If $k = \mathbb{C}$ and H is simple* (*equivalently, center free*), *then either* (*a*) $\overline{\pi(\Gamma)}$ *is compact* (*Hausdorff topology*); *or* (*b*) *π extends to a rational homomorphism $G \to H$.*
(*iii*) *If k is totally disconnected, then $\overline{\pi(\Gamma)}$ is compact* (*Hausdorff topology*).

For Theorem 5.1.1, the relevant case of Theorem 5.1.2 is of course $k = \mathbb{R}$.

Proof of Theorem 5.1.1 from 5.1.2: Let G be as in 5.1.1 and $\underline{G}$ a connected semisimple algebraic $\mathbb{R}$-group with trivial center so that $G = \underline{G}^0_{\mathbb{R}}$ (Proposition 3.1.6). We can express each simple factor of G' in a similar fashion, and by applying (i) of 5.1.2 to each simple factor of G', we deduce that π extends to a rational homomorphism $\underline{G}_{\mathbb{R}} \to H_{\mathbb{R}}$ where $H^0_{\mathbb{R}} = G'$. (We can apply 5.1.2 since $\pi(\Gamma) = \Gamma'$ is Zariski dense in G', and hence in H, by the Borel density theorem (3.2.5).) By connectedness of G, $\pi(G) \subset G'$. Since $\pi(\Gamma)$, and hence $\pi(\underline{G})$ is Zariski dense in H, and $\pi(\underline{G})$ is an algebraic subgroup of H, it follows that $\pi(\underline{G}) = H$. For any rational homomorphism over $\mathbb{R}$, $\pi(\underline{G}_{\mathbb{R}})$ is of finite index in $\pi(\underline{G})_{\mathbb{R}}$, and so we deduce $\dim \pi(G) = \dim(\pi(\underline{G}_{\mathbb{R}})) = \dim \pi(\underline{G})_{\mathbb{R}} = \dim H_{\mathbb{R}} = \dim G'$. Since G' is also connected, it follows that $\pi(G) = G'$. If π is not an isomorphism, let $N = \ker(\pi)$. Since G is center free, N is of positive dimension, and by irreducibility of Γ, Γ is dense when projected to G/N. Since π factors to a map $G/N \to G'$, this would imply that $\pi(\Gamma)$ is topologically dense in G', which contradicts discreteness of $\Gamma' \subset G'$.

We begin the proof of Theorem 5.1.2 for the real field.

Proof of 5.1.2 for $k = \mathbb{R}$: We begin by showing that it suffices to find a rational Γ-map between homogeneous spaces of G and H. (We remark that if H acts on a space, so does Γ, via the homomorphism $\pi:\Gamma \to H_{\mathbb{R}}$.)

5.1.3 Lemma: *Suppose $P \subset G$ and $L \subset H$ are proper algebraic $\mathbb{R}$-subgroups, and that there is a rotational Γ-map $\varphi: G/P \to H/L$ defined over $\mathbb{R}$. Then π extends to a rational homomorphism $G \to H$ defined over $\mathbb{R}$.*

Proof: Let $gr(\pi) = \{(\gamma, \pi(\gamma)) \in G \times H\}$ be the graph of π, and $\overline{gr(\pi)}$ the Zariski closure. We claim that $\overline{gr(\pi)}$ is the graph of a homomorphism $G \to H$. We first note that the projection of $\overline{gr(\pi)}$ onto G must be all of G, since on one hand this projection contains Γ and hence is Zariski dense in G by the Borel density theorem (3.2.5), and on the other hand is the image under a regular homomorphism of an algebraic group hence must be an algebraic subgroup of G. Thus to see that $\overline{gr(\pi)}$ is the graph of a homomorphism, we suppose that $(g, h_1), (g, h_2) \in \overline{gr(\pi)}$. Let $R = R(G/P, H/L)$ be the space of rational maps $G/P \to H/L$. Then $G \times H$ acts on R by $[(g, h)\cdot\varphi](x) = \varphi(xg)\cdot h$. The condition that φ be a Γ-map is exactly the assertion that φ is invariant under the action of $gr(\pi)$. Since the set of points in $G \times H$ leaving φ invariant is an algebraic subgroup, this implies that $\overline{gr(\pi)}$ leaves φ invariant, and hence, with g, h_i as above, $\varphi(xg) = \varphi(x)h_1$ and $\varphi(xg) = \varphi(x)h_2$. Thus, $h_1h_2^{-1}$ leaves $\varphi(G/P)$ pointwise fixed. However, since $\varphi(x\gamma) = \varphi(x)\pi(\gamma)$, $\pi(\Gamma)$ leaves $\varphi(G/P)$ invariant, and hence it leaves the Zariski closure $\overline{\varphi(G/P)}$ invariant. Since $\pi(\Gamma)$ is Zariski dense in H, H leaves $\overline{\varphi(G/P)}$ invariant, and so $\varphi(G/P)$ must be Zariski dense in H/L. Thus, $h_1h_2^{-1}$ leaves all H/L pointwise fixed. Therefore, $h_1h_2^{-1} \in \bigcap_{h\in H} hLh^{-1}$, a normal subgroup. Since H is center free, $h_1h_2^{-1} = e$, showing that $\overline{gr(\pi)}$ is the graph of a homomorphism. Finally, since the projection $\overline{gr(\pi)} \to G$ is a regular bijective homomorphism and hence has a regular inverse, and the projection $\overline{gr(\pi)} \to H$ is clearly regular, the extended homomorphism $G \to H$ is regular. Since $\pi(\Gamma) \subset H_{\mathbb{R}}$ and Γ is Zariski dense in G, π is defined over $\mathbb{R}$.

We are therefore faced with the problem of constructing such a rational map φ. We will show that we can do this where P is a minimal parabolic subgroup defined over $\mathbb{R}$. For example, if $G = PSL(n, \mathbb{C})$, so $G_{\mathbb{R}} = PSL(n, \mathbb{R})$, then P will be the image in $PSL(n, \mathbb{C})$ of the subgroup of upper triangular matrices. In general, we have $G^0_{\mathbb{R}}/P_0$ identified as a Zariski dense subset of G/P where we set $P_0 = G^0_{\mathbb{R}} \cap P$. Therefore, if we have a rational map $\varphi: G/P \to H/L$ which is defined on a Zariski dense subset of $G^0_{\mathbb{R}}/P_0$ and as a map $G^0_{\mathbb{R}}/P_0 \to H/L$ is a Γ-map, then $\varphi: G/P \to H/L$ will also be a Γ-map. Therefore, it suffices to show that we can find a proper algebraic $\mathbb{R}$-group $L \subset H$ and a rational Γ-map $\varphi: G^0_{\mathbb{R}}/P_0 \to H_{\mathbb{R}}/L_{\mathbb{R}}$, defined on a Zariski dense subset of $G^0_{\mathbb{R}}/P_0$. The proof now breaks down into two major steps.

Step 1: There is a proper algebraic $\mathbb{R}$-subgroup $L \subset H$ and a measurable Γ-map $\varphi: G^0_{\mathbb{R}}/P_0 \to H_{\mathbb{R}}/L_{\mathbb{R}}$. (Actually, φ may only be defined on a Γ-invariant conull set.)

Step 2: Any such measurable Γ-map is essentially rational (i.e., agrees a.e. with a rational map (Definition 3.4.3).)

In light of our observations above, the proof of the above two statements completes the proof of the theorem for $k = \mathbb{R}$. The proof of both of these statements will involve the results of the preceding chapters.

Proof of Step 1: Let $Q \subset H$ be a (proper) parabolic subgroup defined over $\mathbb{R}$ (so that Q is $\mathbb{R}$-cocompact in H). Then $H_{\mathbb{R}}/Q_{\mathbb{R}}$ is a compact metrizable Γ-space. Since P is a minimal $\mathbb{R}$-parabolic subgroup, $P_{\mathbb{R}}$ is amenable, and hence the Γ-action on $G^0_{\mathbb{R}}/P_0$ is amenable by Corollary 4.3.7. Thus by Proposition 4.3.9, there is a measurable Γ-map $\varphi: G^0_{\mathbb{R}}/P_0 \to M(H_{\mathbb{R}}/Q_{\mathbb{R}})$. By Corollary 3.2.17, the action of $H_{\mathbb{R}}$ on $M(H_{\mathbb{R}}/Q_{\mathbb{R}})$ is smooth, so that the orbit space $M(H_{\mathbb{R}}/Q_{\mathbb{R}})/H_{\mathbb{R}}$ is countably separated. Let $\bar{\varphi}$ be the composition of φ with the natural map $M(H_{\mathbb{R}}/Q_{\mathbb{R}}) \to M(H_{\mathbb{R}}/Q_{\mathbb{R}})/H_{\mathbb{R}}$. Since φ is a Γ-map, we have (on a Γ-invariant conull set in $G^0_{\mathbb{R}}/P_0$), $\varphi(x\gamma) = \varphi(x)\pi(\gamma)$ so that $\bar{\varphi}(x\gamma) = \bar{\varphi}(x)$. Since $\bar{\varphi}$ is a map into a countably separated space, and Γ is ergodic on $G^0_{\mathbb{R}}/P_0$ by Moore's theorem (2.2.6), it follows that $\bar{\varphi}$ is essentially constant. In other words, on a conull set, φ takes values in a single $H_{\mathbb{R}}$ orbit in $M(H_{\mathbb{R}}/Q_{\mathbb{R}})$. By 3.2.18 and 3.2.19, the stabilizer of this orbit is of the form $L_{\mathbb{R}}$ where $L \subset H$ is a proper algebraic $\mathbb{R}$-group. Since φ takes values in this orbit, we can (possibly discarding a further Γ-invariant null set) view φ as a Γ-map $G^0_{\mathbb{R}}/P_0 \to H_{\mathbb{R}}/L_{\mathbb{R}}$ (once again, with the understanding that φ need only be defined on a Γ-invariant conull set). This completes the proof of Step 1.

Preliminaries to the proof of Step 2: We now want to show that our measurable map defined on $G^0_{\mathbb{R}}/P_0$ is essentially rational. Before proceeding to the central part of the argument, we shall make some preliminary reductions using the structure theory of G. We shall state those structural results for arbitrary G, but here as usual, we will verify them only for $G = SL(n, \mathbb{C})$.

In Chapter 3 we have already seen some situations in which we can deduce essential rationality of a measurable map. In particular, if φ is a measurable unipotent representation of the real points of a connected unipotent $\mathbb{R}$-group, then φ is rational. Our map is defined on $G^0_{\mathbb{R}}/P_0$ which of course is not a unipotent group. The first step then will be to replace $G^0_{\mathbb{R}}/P_0$ by such a group.

5.1.4 Lemma: *With G and P as above, there exists a connected unipotent $\mathbb{R}$-subgroup $U \subset G$ such that the product $m: P \times U \to G$ is injective, the image $m(P \times U)$ is a Zariski dense $\mathbb{R}$-open subset of G, and $m: P \times U \to m(P \times U)$ is an $\mathbb{R}$-biregular isomorphism of $\mathbb{R}$-varieties. Furthermore, the induced map $U_{\mathbb{R}} \to G^0_{\mathbb{R}}/P_0$ is a measure space isomorphism (modulo null sets).*

This image is called the "big cell" and a proof in general can be found in any of the standard references on algebraic groups. (In fact, U is the unipotent radical of the parabolic opposite to P.) Here, we shall only describe the proof for $G = SL(n, \mathbb{C})$. In this case, P is the subgroup of G consisting of upper triangular matrices. Let U be the subgroup of lower triangular matrices with all diagonal entries equal to 1. The assertion of Lemma 5.1.4 can be reduced to the following statement in linear algebra. If $A \in SL(n, \mathbb{C})$, let A_k be the principal (i.e., upper left-hand corner) $k \times k$ submatrix. Then A can be written $A = BC$ where $B \in U$ and $C \in P$ if and only if for all $1 \leq k \leq n$, $\det A_k \neq 0$. This representation of A is unique and the entries of B and C are polynomial functions (over $\mathbb{Q}$) of the entries of A, and $(\det A_k)^{-1}$, $1 \leq k \leq n$. This linear algebra statement can be readily verified by induction on n, for example, and deducing 5.1.4 from this is then simply a matter of terminology.

We remark that Lemma 5.1.4 implies that any rational function on U defines a P-invariant rational function on $P \times U$ and hence a rational function on G/P.

Returning to our map φ, we can lift φ back to a function $\varphi: G^0_{\mathbb{R}} \to H_{\mathbb{R}}/L_{\mathbb{R}}$. We have the commutative diagram

$$\begin{array}{ccccc} G^0_{\mathbb{R}} & \supset & U_{\mathbb{R}} & \longrightarrow & G^0_{\mathbb{R}}/P_0 \\ \downarrow & & \downarrow & & \downarrow \\ G & \supset & U & \longrightarrow & G/P \end{array}$$

and in light of Lemma 5.1.4, to see that φ is essentially rational, it suffices to see that $\varphi | U_{\mathbb{R}}$ is essentially rational (where φ is now defined on $G^0_{\mathbb{R}}$). However, for any $g \in G^0_{\mathbb{R}}$, we also have the diagram

$$\begin{array}{ccccc} G^0_{\mathbb{R}} & \supset & U_{\mathbb{R}} \cdot g & \longrightarrow & G^0_{\mathbb{R}}/P_0 \\ \downarrow & & \downarrow & & \downarrow \\ G & \supset & U \cdot g & \longrightarrow & G/P \end{array}$$

and hence it also suffices to see that $\varphi | U_{\mathbb{R}} \cdot g$ is essentially rational. Thus, we deduce the following.

5.1.5 Lemma: *To prove Step 2, it suffices to show that for some $g \in G^0_{\mathbb{R}}$, $u \to \varphi(ug)$ is an essentially rational function on $U_{\mathbb{R}}$.*

Thus, we have reduced the problem to showing rationality of a measurable map on a unipotent group, and thus we have some hope of applying our earlier results. Although our map is defined on a group, it is not a homomorphism, as the target space is not even a group. We do, however, have some sort of algebraic relation, namely that the map φ is a Γ-map. Thus, we might hope to be able to force this algebraic relation to show that $u \to \varphi(ug)$ only depends upon a unipotent homomorphism of u, and hence would be rational. This, however, is not possible. When we try to force the algebra, we shall need some commutativity of $U_{\mathbb{R}}$ with some elements of $A^0_{\mathbb{R}}$, where A is the $\mathbb{R}$-split abelian subgroup of maximal dimension in P. For $SL(n, \mathbb{C})$, this asks for commutativity of the lower triangular group with a (non-identity) element of the positive diagonal matrices (of determinant 1), which is of course impossible. This is in fact true for any G, namely, the centralizer of $U_{\mathbb{R}}$ in $A^0_{\mathbb{R}}$ is trivial. However, if $\mathbb{R}\text{-rank}(G) \geqq 2$, i.e., $\dim A_{\mathbb{R}} \geqq 2$ (as it is in the hypothesis of Theorem 5.12), we can find unipotent subgroups of $U_{\mathbb{R}}$ whose centralizer in $A^0_{\mathbb{R}}$ is not trivial. The idea of the proof is then to show (i) that this implies (roughly speaking) that φ is rational on these subgroups and (ii) that rationality on these subgroups implies rationality on $U_{\mathbb{R}}$. We therefore first wish to describe how $U_{\mathbb{R}}$ can be built from subgroups whose centralizers in $A^0_{\mathbb{R}}$ are nontrivial.

If $t \in A^0_{\mathbb{R}}$, let C_t be the centralizer of t in G and $C^u_t = C_t \cap U$. Then C_t and C^u_t are algebraic $\mathbb{R}$-subgroups. Let us consider the example $G = SL(3, \mathbb{C})$, so that $A^0_{\mathbb{R}}$ is the two dimensional group of positive diagonal unimodular matrices. If

$$t = \begin{bmatrix} \lambda & 0 & 0 \\ 0 & \lambda & 0 \\ 0 & 0 & \lambda^{-2} \end{bmatrix}, \text{ then } C_t = \left\{ \left(\begin{array}{c|c} M & \begin{matrix} 0 \\ 0 \end{matrix} \\ \hline \begin{matrix} 0 & 0 \end{matrix} & \alpha \end{array} \right) \quad \text{where } \alpha = (\det M)^{-1} \right\},$$

and

$$C^u_t = \left\{ \begin{bmatrix} 1 & 0 & 0 \\ a & 1 & 0 \\ 0 & 0 & 1 \end{bmatrix} \right\}.$$

Similarly, by varying $t \in A^0_{\mathbb{R}}$ we can also obtain the subgroups

$$\left\{ \begin{bmatrix} 1 & 0 & 0 \\ 0 & 1 & 0 \\ 0 & b & 1 \end{bmatrix} \right\} \text{ and } \left\{ \begin{bmatrix} 1 & 0 & 0 \\ 0 & 1 & 0 \\ c & 0 & 1 \end{bmatrix} \right\}$$

as groups of the form C^u_t. An analogous construction applies in $SL(n, \mathbb{C})$, and in fact for any G under consideration. More precisely, the structure theory of semisimple groups implies the following in general.

5.1.6 Lemma: *There exist $t_1, \ldots, t_n \in A^0_{\mathbb{R}}$, $t_i \neq e$, and connected algebraic (unipotent) $\mathbb{R}$-subgroups $U_i \subset C^u_{t_i}$ such that*

(i) *the product map* $\prod_{i=1}^{n} U_i \to U$ *is an* $\mathbb{R}$*-isomorphism of varieties;*

(ii) *for each r,* $\prod_{i=r}^{n} U_i$ *is a subgroup of U and* $\prod_{i=r}^{n} U_i$ *is normal in* $\prod_{i=r-1}^{n} U_i$.

We remark that Lemma 5.1.6 depends in an essential way upon the condition that $\mathbb{R}$-rank$(G) \geqq 2$. This lemma is in fact the only point in the proof of Theorem 5.1.2 for which we will need this hypothesis.

We can now reduce the rationality property required in Lemma 5.1.5 to rationality along the subgroups U_i.

5.1.7 Lemma: *To prove Step 2, it suffices to show that if* $t \in A^0_{\mathbb{R}}$, $t \neq e$, *and* $V \subset C^u_t$ *is a connected algebraic* $\mathbb{R}$*-subgroup, then for almost all* $g \in G^0_{\mathbb{R}}$, $u \to \varphi(ug)$ *is an essentially rational function on* $V_{\mathbb{R}}$.

Proof: Let U_i be as in Lemma 5.1.6. We shall verify that for almost all $g \in G^0_{\mathbb{R}}$, $u \to \varphi(ug)$ is an essentially rational function on $U_{\mathbb{R}}$, which suffices by Lemma 5.1.5. We claim that by induction on $n - r$, for almost all g, $u \to \varphi(ug)$ is essentially rational on $\prod_{i=r}^{n} (U_i)_{\mathbb{R}}$. By 5.1.6, this suffices. For $r = n$, the hypotheses of 5.1.7 immediately yield the required essential rationality on $(U_n)_{\mathbb{R}}$. Suppose we have, for almost all g, essential rationality on $\prod_{r}^{n}(U_i)_{\mathbb{R}}$. Let $u \in \prod_{r}^{n}(U_i)_{\mathbb{R}}$ and $c \in (U_{r-1})_{\mathbb{R}}$, and for each $g \in G^0_{\mathbb{R}}$, consider the function $\varphi_g(c, u) = \varphi(cug)$. By our hypothesis, for each u and almost all g, φ_g is rational in c. Thus by Fubini, for almost all g, φ_g is rational in c for almost all u. On the other hand, we also have $\varphi_g(c, u) = \varphi((cuc^{-1})cg)$. By (ii) of Lemma 5.1.6, $cuc^{-1} \in \prod_{r}^{n}(U_i)_{\mathbb{R}}$, and hence by our inductive assumption, we obtain that for each c, and almost all g, φ_g is essentially rational in u. Once again, by Fubini, we obtain that for almost all g, φ_g is rational in u for almost all c. We recall that any connected unipotent $\mathbb{R}$-group is $\mathbb{R}$-isomorphic as a variety to $\mathbb{C}^{\ell}$ for some ℓ. (For example, just use the expotential map as in the proof of 3.4.1.) Thus, we may apply Theorem 3.4.4, and conclude that for almost all g, φ_g is essentially rational, proving the lemma.

As we saw in section 3.4, to see that a function on a unipotent group is rational it suffices to see that it depends only on a measurable representation which extends to a larger semisimple group. (See 3.4.1 and 3.4.2.) Thus, we can make one further reduction.

5.1.8 Lemma: *To prove Step 2, it suffices to see that if $t \in A^0_{\mathbb{R}}$, $t \neq e$, then for almost all $g \in G^0_{\mathbb{R}}$ there exists:*

(a) an $\mathbb{R}$-subvariety $W_g \subset H/L$ such that the map $\varphi_g:(C_t)^0_{\mathbb{R}} \to H_{\mathbb{R}}/L_{\mathbb{R}}$, $\varphi_g(c) = \varphi(cg)$ satisfies $\varphi_g(c) \in W_g$ for almost all c;

(b) an $\mathbb{R}$-algebraic group Q_g which acts $\mathbb{R}$-regularly on W_g;

(c) a measurable homomorphism $h_g:(C_t)^0_{\mathbb{R}} \to (Q_g)_{\mathbb{R}}$;

(d) a point $x_g \in W_g \cap H_{\mathbb{R}}/L_{\mathbb{R}}$;

such that $\varphi_g(c) = x_g \cdot h_g(c)$ for almost all $c \in (C_t)^0_{\mathbb{R}}$.

Proof: We assume the conditions above, and verify that the conditions of Lemma 5.1.7 are satisfied. There is a connected semisimple subgroup of $(C_t)_{\mathbb{R}}$ that contains $V_{\mathbb{R}}$ for any unipotent $\mathbb{R}$-subgroup $V \subset C_t$. This is immediate for the example (i.e., $SL(n, \mathbb{C})$) we considered above, and follows in general from the structural result that C_t is reductive. Thus, if the equation $\varphi_g(c) = x_g \cdot h_g(c)$ were to hold for all $c \in (C_t)^0_{\mathbb{R}}$, $h_g | V_{\mathbb{R}}$ would be unipotent by 3.4.2, and therefore extend to a unipotent homomorphism $V \to Q_g$. Hence, by Proposition 3.4.1, for almost all g, $\varphi_g | V_{\mathbb{R}}$ would be rational, and the conditions of 5.1.7 would be satisfied. However, as the above equation holds only a.e. on $(C_t)^0_{\mathbb{R}}$, and $V_{\mathbb{R}}$ is of smaller dimension and hence of measure zero, a small further argument is necessary. Namely, for each $u \in V_{\mathbb{R}}$, and almost all $g \in G^0_{\mathbb{R}}$, $c \in (C_t)^0_{\mathbb{R}}$, we have $\varphi(ucg) = x_g \cdot h_g(uc) = x_g \cdot h_g(u) h_g(c)$. By Fubini, there is some fixed $c \in (C_t)^0_{\mathbb{R}}$ such that for almost all g, we have this equation for almost all $u \in V_{\mathbb{R}}$. Since $h_g | V_{\mathbb{R}}$ is unipotent, the argument above shows that for almost all $g \in G^0_{\mathbb{R}}$, $u \to x_g \cdot h_g(u) h_g(c)$ is rational. Thus, for almost all g, $u \to \varphi(ucg)$ is essentially rational on $V_{\mathbb{R}}$, and since c is fixed, we have $u \to \varphi(ug)$ is essentially rational for almost all g, verifying the conditions of Lemma 5.1.7.

Proof of Step 2: Fix $t \in A^0_{\mathbb{R}}$, $t \neq e$, and let $C = (C_t)^0_{\mathbb{R}}$. We shall verify the conditions of Lemma 5.1.8. By Proposition 3.5.2, it suffices to show that for almost all $g \in G^0_{\mathbb{R}}$, there is an $H_{\mathbb{R}}$-orbit in $F(C, H_{\mathbb{R}}/L_{\mathbb{R}})$ such that $(\varphi_g)_c$ lies in this orbit for almost all $c \in C$. We claim first that it suffices to see that there is an $H_{\mathbb{R}}$-orbit in $F(C, H_{\mathbb{R}}/L_{\mathbb{R}})$ such that for almost all $g \in G^0_{\mathbb{R}}$, φ_g lies in this orbit. For then, for each fixed $c \in C$, φ_{cg} lies in this orbit and hence by Fubini, for almost all g, φ_{cg} will lie in this orbit for almost all $c \in C$. Since $\varphi_{cg} = (\varphi_g)_c$, this verifies our assertion. We now show that almost all φ_g, $g \in G^0_{\mathbb{R}}$, lie in a single $H_{\mathbb{R}}$-orbit.

Define the map $\Phi: G^0_{\mathbb{R}} \to F(C, H_{\mathbb{R}}/L_{\mathbb{R}})$ by $\Phi(g) = \varphi_g$. Let $T = \{t^n\} \subset A^0_{\mathbb{R}}$, the cyclic subgroup generated by t. We remark that since $A^0_{\mathbb{R}}$ is connected and $\mathbb{R}$-split, it is isomorphic to $\mathbb{R}^\ell$ for some ℓ, and hence T is not compact. We also recall that for $p \in P_0$, $\varphi(pg) = \varphi(g)$. Thus

$$\begin{aligned}\varphi_{tg}(c) &= \varphi(ctg) \\ &= \varphi(tcg) \quad \text{(by definition of } C\text{)} \\ &= \varphi(cg) \quad \text{(since } T \subset P_0\text{)} \\ &= \varphi_g(c)\end{aligned}$$

Thus, $\Phi(tg) = \Phi(g)$ for all g, and hence we can consider Φ as a measurable map $\Phi: G^0_{\mathbb{R}}/T \to F(C, H_{\mathbb{R}}/L_{\mathbb{R}})$. Recall also that φ is a Γ-map (at least on a conull set). Thus for each c, γ, and almost all g,

$$\varphi_{g\gamma}(c) = \varphi(cg\gamma) = \varphi(cg)\cdot\pi(\gamma) = \varphi_g(c)\cdot\pi(\gamma).$$

By Fubini, for each γ and almost all g, this is true for almost all c. Thus, for each γ and almost all g, we have $\Phi(g\gamma) = \Phi(g)\cdot\pi(\gamma)$ where $\pi(\gamma) \in H_{\mathbb{R}}$ acts on $F(C, H_{\mathbb{R}}/L_{\mathbb{R}})$ pointwise. Let $\bar{\Phi}: G^0_{\mathbb{R}}/T \to F(C, H_{\mathbb{R}}/L_{\mathbb{R}}/H_{\mathbb{R}}$ be the composition of Φ with projection onto the orbit space. Then we have for each $\gamma \in \Gamma$, $\bar{\Phi}(g\gamma) = \bar{\Phi}(g)$ a.e., i.e., $\bar{\Phi}$ is essentially Γ-invariant. However, by Moore's theorem (2.2.7), Γ is ergodic on $G^0_{\mathbb{R}}/T$. By Proposition 3.3.1, the orbit space $F(C, H_{\mathbb{R}}/L_{\mathbb{R}}/H_{\mathbb{R}}$ is countably separated, and hence, by Corollary 2.2.8, $\bar{\Phi}$ is constant on a conull set. In other words, on a conull set in $G^0_{\mathbb{R}}$, all $\Phi(g)$ are in the same $H_{\mathbb{R}}$- orbit. This completes the proof of Step 2, and with it the proof of Theorem 5.1.2 for $k = \mathbb{R}$.

Proof of Theorem 5.1.2 for $k \neq \mathbb{R}$: In the proof of Step 1, the results of section 3.2. show that the local field k is immaterial through and including the point that we deduce the existence of a measurable Γ-map $G^0_{\mathbb{R}}/P_0 \to H_k/S$ where $S \subset H_k$ is the stabilizer of a measure on $M(H_k/Q_k)$. By Corollary 3.2.18, for $k = \mathbb{R}$, S is the set of real points of an algebraic group. As we saw in section 3.2, this is special to the real field. However, using Corollary 3.2.19, for $k \neq \mathbb{R}$, we deduce the following variant of Step 1.

Step 1′: There is a measurable Γ-map $\varphi: G^0_{\mathbb{R}}/P_0 \to H_k/S$ where either

(a) $S = L_k$ where $L \subset H$ is an algebraic k-group of strictly smaller dimension; or
(b) S is compact.

Thus, the essentially new situation that may arise when $k \neq \mathbb{R}$ is condition (b). In this case, we claim that $\overline{\pi(\Gamma)}$ is compact. This follows from the following general observation.

5.1.9 Proposition [*Zimmer* 12]: *Suppose S is an ergodic Γ-space with quasi-invariant measure (Γ any locally compact group) such that Γ is also ergodic on $S \times S$. Suppose $\pi : \Gamma \to H$ is a homomorphism where H is also locally compact, and that there is a measurable Γ-map $\varphi : S \to H/K$ where $K \subset H$ is a compact subgroup. Then $\overline{\pi(\Gamma)}$ is compact.*

Proof: Let μ be the quasi-invariant probability measure on S, and let $\nu = \varphi_*(\mu)$, a probability measure on H/K. The map $\varphi \times \varphi : S \times S \to H/K \times H/K$ is a Γ-map, $(\varphi \times \varphi)_*(\mu \times \mu) = \nu \times \nu$, and $\mu \times \mu$ is ergodic under Γ by hypothesis. Thus, $\nu \times \nu$ is also ergodic under Γ. We next observe that the H action on $H/K \times H/K$ is smooth. To see this, it suffices by Lemma 3.2.13 to see that K is smooth on H/K, which follows from Corollary 2.1.21 since K is compact. Since Γ acts by $\pi(\Gamma) \subset H$, the natural map $H/K \times H/K \to (H/K \times H/K)/H$ is Γ-invariant, and since $\nu \times \nu$ is ergodic under Γ and $(H/K \times H/K)/H$ is countably separated, this map is essentially constant. In other words, $\nu \times \nu$ is supported on an H-orbit, say A, in $H/K \times H/K$. By Fubini's theorem, there is a point $x \in H$ such that ν is supported on $\{[y] \in H/K \mid ([x], [y]) \in A\}$. But such a set must be a xKx^{-1}-orbit in H/K, which is compact. In other words, support$(\nu) \subset H/K$ is compact. Since $\pi(\Gamma)$ leaves ν quasi-invariant, $\pi(\Gamma)$ leaves support(ν) invariant, and since K is compact, $\pi(\Gamma)$ leaves a compact set $B \subset H$ invariant (under translation). But then $\pi(\Gamma) \subset BB^{-1}$, and so $\overline{\pi(\Gamma)}$ is compact.

To apply this proposition to case (b) in Step 1′, we need only show that the irreducible lattice Γ is ergodic on $G^0_{\mathbb{R}}/P_0 \times G^0_{\mathbb{R}}/P_0$. However, the action of P_0 on $G^0_{\mathbb{R}}/P_0$ has an orbit of full measure, and with stabilizer $A_{\mathbb{R}} \cap P_0$, where A is the $\mathbb{R}$-split abelian subgroup (contained in P) of maximal dimension. (For example, if $G = SL(n, \mathbb{C})$, and P_0 = upper triangular real matrices, let $\bar{P}_0$ = lower triangular real matrices. Then P_0 and $\bar{P}_0$ are conjugate subgroups of $G^0_{\mathbb{R}} = SL(n, \mathbb{R})$, and so we can identify $G^0_{\mathbb{R}}/P_0 \cong G^0_{\mathbb{R}}/\bar{P}_0$. Our assertion then follows from the discussion following lemma 5.1.4.) Thus, the $G^0_{\mathbb{R}}$ action on $G^0_{\mathbb{R}}/P_0 \times G^0_{\mathbb{R}}/P_0$ has an orbit of full measure and with stabilizer $A_{\mathbb{R}} \cap P_0$. Thus, as a measurable Γ-space, $G^0_{\mathbb{R}}/P_0 \times G^0_{\mathbb{R}}/P_0 \cong G^0_{\mathbb{R}}/A_{\mathbb{R}} \cap P_0$. However, since $A_{\mathbb{R}} \cap P_0$ is not compact (for $SL(n, \mathbb{C})$, this is just the real diagonal subgroup), Γ is ergodic on this space by Moore's theorem (2.2.7).

Thus, to complete the proof of Theorem 5.1.2 for $k \neq \mathbb{R}$, it remains only to consider case (a) in Step 1′. For $k = \mathbb{C}$, exactly the same argument as the proof for $k = \mathbb{R}$ now goes through and we thus deduce that π extends to a rational homomorphism $G \to H$. For $k \neq \mathbb{R}, \mathbb{C}$, i.e., k totally disconnected, the proof of Step 2 goes through if we replace "rational" by "constant". The point is that since each $(U_i)_{\mathbb{R}}$ is connected, any homomorphism of $(U_i)_{\mathbb{R}}$ into the k-points of a k-group (which is totally disconnected) must be constant. Therefore, in the case $k \neq \mathbb{R}, \mathbb{C}$, we deduce the existence of an essentially constant Γ-map $\varphi: G^0_{\mathbb{R}}/P_0 \to H_k/L_k$, where $L \subset H$ is a proper k-subgroup. This means that there is a fixed point for $\pi(\Gamma)$ acting on H_k/L_k, i.e., $\pi(\Gamma)$ is contained in a conjugate of L_k. But this contradicts the assumption that $\pi(\Gamma)$ is Zariski dense in H_k. Hence, for $k \neq \mathbb{R}, \mathbb{C}$, the only case which can occur in Step 1′ is case (b), and hence $\overline{\pi(\Gamma)}$ is always compact. This completes the proof of Theorem 5.1.2 in all cases.

Remarks: Margulis' first proof of Theorem 5.1.2 appeared in [Margulis 1]. The proof of Step 2 that we have presented is essentially the proof of that paper. Margulis' original proof of Step 1 was based on the "multiplicative ergodic theorem" of [Oseledec 1]. (See also [Raghunathan 2].) In [Margulis 2], a different approach to the proof of Step 1 was presented, using the existence of measurable Γ-maps $G^0_{\mathbb{R}}/P_0 \to M(X)$ where X is a compact metric Γ-space. Here, we have seen this property as a consequence of amenability of Γ acting on $G^0_{\mathbb{R}}/P_0$; this particular consequence of amenability for this action was first established in [Furstenberg 2]. The proof of Step 1 we have presented here based upon amenability and the smoothness of actions of algebraic groups acting on suitable spaces of measures, is from [Zimmer 8]. This latter proof was developed in order to be able to extend the super-rigidity theorem to cocycles of ergodic actions. (Cf. the remarks following the proof of Proposition 4.2.16.)

5.2 Rigidity and orbit equivalence of ergodic actions

In this section we shall prove the following rigidity theorem for ergodic actions.

5.2.1 Theorem [*Zimmer* 8]: (*Rigidity for ergodic actions of semisimple Lie groups*): *Suppose G, G' are connected semisimple Lie groups with finite center and no compact factors. Suppose S (resp. S') is an essentially free ergodic irreducible G (resp. G')-space with finite invariant measure, and assume that the actions are orbit equivalent. Assume $\mathbb{R}$-rank$(G) \geq 2$. Then*

(*i*) *G and G' are locally isomorphic.*
(*ii*) *In the center-free case, $G \cong G'$, and identifying G and G' via this isomorphism, the actions of G on S and S' are isomorphic.*

Remarks: (a) We saw in section 4.2 (see the paragraph following the proof of Proposition 4.2.16) that the problem of extending homomorphisms defined on

subgroups and the problem of showing that orbit equivalent actions are isomorphic are closely related, and in fact are both special cases of a problem concerning cocycles. We shall deduce Theorem 5.2.1 from a general superrigidity theorem for cocycles, which will then subsume both Theorem 5.2.1, and the Mostow–Margulis Theorem (5.1.1), and generalize Margulis' superrigidity Theorem (5.1.2). We indicated in section 5.1 that Margulis' superrigidity theorem can be applied to problems other than the classical rigidity problem in 5.1.1, e.g. to arithmeticity which will be discussed in Chapter 6. Similarly, the superrigidity theorem for cocycles we will prove (Theorem 5.2.5 below) can be applied to problems other than orbit equivalence. See for example, section 9.4, and in a more geometric direction [Zimmer 15, 18, 19].
(b) At the conclusion of section 4.3, we described the orbit equivalence theory for amenable actions. The salient feature of that theory is that orbit equivalence is a very weak condition for amenable actions. In particular, for actions of discrete amenable groups with finite invariant measure, orbit equivalence of actions in no way enables one to distinguish the acting groups, and even given the group, in no way enables one to distinguish the actions. Thus, Theorem 5.2.1 shows that the orbit equivalence theory for semisimple Lie groups is in a sense diametrically opposed to the theory in the amenable case.
(c) The rigidity of lattices in semisimple groups can be considered as a generalization of a type of rigidity of lattices in Euclidean spaces. (This is the point of view in the introduction.) On the other hand, the rigidity of ergodic actions of semisimple groups is not a generalization of a phenomenon for actions of Euclidean groups, as our discussion in (b) indicates.
(d) We indicated in section 5.1 that the rigidity theorem for lattices (5.1.1) is also true for simple groups of $\mathbb{R}$-rank 1 as long as we exclude $PSL(2, \mathbb{R})$. It is therefore natural to enquire as to whether or not Theorem 5.2.1 is also true for these groups. While some partial information has been obtained ([Zimmer 16]), this question is open as of this writing.

Before turning to the proof of 5.2.1, let us present an example.

5.2.2 Corollary [*Zimmer* 8]: *For* $i = 1, 2$, *let* G_i *be a connected simple Lie group with finite center,* $\Gamma_i \subset G_i$ *a lattice, and* S_i *an essentially free ergodic* Γ_i*-space with finite invariant measure. Suppose* $\mathbb{R}$*-rank*$(G_2) \geqq 2$, *and that the* Γ_1 *action on* S_1 *and the* Γ_2 *action on* S_2 *are orbit equivalent. Then* G_1 *and* G_2 *are locally isomorphic.*

Proof: Let X_i be the G_i action induced from the Γ_i-action on S_i (Definition 4.2.21). Since $\Gamma_i \subset G_i$ is a lattice, and Γ_i acts with finite invariant measure on S_i, G_i acts with finite invariant measure on X_i. It is a straightforward consequence of the definition that G_i will be essentially free on X_i if Γ_i is so on S_i. It is also straightforward to see that if $X_i = G_i/\Gamma_i \times_{\alpha_i} S_i$ (as in Definition 4.2.21), then $x, y \in X_i$ are in the same G_i orbit if and only if their projections to S_i are in the same Γ_i orbit. Thus, if $\theta: S_1 \to S_2$ is an orbit equivalence, and $f: G_1/\Gamma_1 \to G_2/\Gamma_2$ is a measure class preserving Borel isomorphism (Appendix A), then the map $\varphi: X_1 \to X_2$ given by $\varphi([g], s) = (f([g]), \theta(s))$ will be an orbit equivalence between the G_i actions. We can therefore apply Theorem 5.2.1.

5.2.3 Corollary: *As we vary n, $n \geqq 2$, the actions of $SL(n, Z)$ on $\mathbb{R}^n/Z^n$ are mutually non-orbit equivalent.*

Proof: We have verified ergodicity of $SL(n, Z)$ on $\mathbb{R}^n/Z^n$ in Example 2.1.5. Thus, it only remains to check essential freeness. This can be done directly in this case, or we can use the following observation.

5.2.4 Lemma: *Let Γ be a discrete group acting on a space S with quasi-invariant measure μ. If the action is not essentially free, then there is some $\gamma \in \Gamma$ and some $A \subset S$ with $\mu(A) > 0$ such that $s \cdot \gamma = s$ for all $s \in A$.*

Proof: If for all $\gamma \in \Gamma$, $S_\gamma = \{s \in S \mid s\gamma = s\}$ is null, then Γ acts freely on the conull set

$$\bigcap_{\lambda \in \Gamma} [(\bigcap_{\gamma \in \Gamma} (S - S_\gamma)) \cdot \lambda].$$

To see that this is impossible for an element of $SL(n, Z)$ acting on $\mathbb{R}^n/Z^n$, we need only observe that $SL(n, Z)$ acts by automorphisms on $\mathbb{R}^n/Z^n$, and hence the fixed point set of any element is a closed subgroup. If it had positive measure it would have to be all of $\mathbb{R}^n/Z^n$ since the latter is connected. But if an element of $SL(n, Z)$ acts trivially on $\mathbb{R}^n/Z^n$, it is clearly the identity. (For example, it then acts trivially on $L^2(\mathbb{R}^n/Z^n)$, and hence on the character group $(\mathbb{R}^n/Z^n)^\wedge \cong Z^n$. Cf. Example 2.1.5.)

Corollary 5.2.2 of course implies significantly more about the actions of $SL(n, Z)$ on $\mathbb{R}^n/Z^n$. For example, if we fix n, and choose $p \neq n$, then there is no measurable action of $SL(p, Z)$ on $\mathbb{R}^n/Z^n$ with (a.e.) the same orbits as the standard $SL(n, Z)$ action.

In Corollary 5.2.2, we restricted our attention to lattices in simple groups, because it is not necessarily true that if we induce an action of an irreducible lattice Γ to G we obtain an irreducible action. We shall consider these and other examples later in this section. (See in particular 5.2.12).

We now state the superrigidity theorem for cocycles, generalizing Margulis' superrigidity theorem (5.1.2).

5.2.5 Theorem [*Zimmer* 8] (*Superrigidity for cocycles*): *Suppose G is a connected semisimple algebraic $\mathbb{R}$-group, $\mathbb{R}$-rank$(G) \geqq 2$, and assume $G^0_{\mathbb{R}}$ has no compact factors. Suppose S is an irreducible ergodic $G^0_{\mathbb{R}}$-space with finite invariant measure. Let H be a connected algebraic k-group, almost simple over k, where k is a local field of characteristic 0. Suppose $\alpha: S \times G^0_{\mathbb{R}} \to H_k$ is a cocycle such that α is not equivalent to a cocycle taking values in a subgroup L_k where $L \subset H$ is a proper algebraic k-subgroup. Then*

(*i*) *If $k = \mathbb{R}$, H is $\mathbb{R}$-simple (equivalently, center free), and $H_{\mathbb{R}}$ is not compact, then there is a rational homomorphism $\pi: G \to H$ defined over $\mathbb{R}$ such that $\alpha \sim \alpha_{\pi | G^0_{\mathbb{R}}}$, i.e., α is equivalent to the cocycle $(s, g) \to \pi(g)$.*

(*ii*) *If $k = \mathbb{C}$ and H is simple (equivalently, center free), then either*

(*a*) *α is equivalent to a cocycle taking values in a compact subgroup of H; or*

(*b*) *there is a rational homomorphism $\pi: G \to H$ such that $\alpha \sim \alpha_{\pi | G^0_{\mathbb{R}}}$.*

(*iii*) *If k is totally disconnected, then α is equivalent to a cocycle taking values in a compact subgroup of H_k.*

As in the situation of lattices, the relevant case for the rigidity theorem for actions (5.2.1) is the case $k = \mathbb{R}$. Before we deduce 5.2.1 from 5.2.5, let us first observe that Margulis superrigidity (5.1.2) also follows from 5.2.5. To see this, suppose $\pi: \Gamma \to H_k$ is a homomorphism as in 5.1.2, and let $\alpha: G^0_{\mathbb{R}}/\Gamma \times G^0_{\mathbb{R}} \to H_k$ be a strict cocycle corresponding to π (Proposition 4.2.13). We can then obtain 5.1.2 from 5.2.5 by applying the latter and Proposition 4.2.16 to α once we observe the following general fact.

5.2.6 Lemma: *Suppose G is locally compact, $G_0 \subset G$ a closed subgroup, and $\alpha: G/G_0 \times G \to H$ a cocycle corresponding to a homomorphism $\pi: G_0 \to H$ (4.2.13, 4.2.15). If $L \subset H$ is a subgroup, then α is equivalent to a cocycle taking values in L if and only if $\pi(G_0)$ is contained in a conjugate of L.*

Proof: This follows from 4.2.13, 4.2.15.

Proof of Theorem 5.2.1 from 5.2.5: Let G, G' be as in Theorem 5.2.1. We first consider the case in which both groups have trivial center. We wish to apply Proposition 4.2.11. We can write $G \cong \underline{G}^0_{\mathbb{R}}$ where $\underline{G}$ is a connected semisimple algebraic $\mathbb{R}$-group with trivial center. Let H be a simple factor of G', and write $H = \underline{H}^0_{\mathbb{R}}$, where $\underline{H}$ is a connected simple adjoint $\mathbb{R}$-group. Let $\alpha: S \times G \to G'$ be the cocycle corresponding to an orbit equivalence $\theta: S \to S'$ (Example 4.2.8). We first claim that α is not equivalent to a cocycle into a proper algebraic subgroup of G'. If it were, we could write $\alpha(s, g) = \varphi(s)^{-1}\beta(s, g)\varphi(sg)$ where $\varphi: S \to G'$ and $\beta(s, g) \in L'$, $L' \subset G'$ proper algebraic, and for each g the identity holds a.e. Define a map $\lambda: S' \to G'$ by $\lambda(y) = \varphi(\theta^{-1}(y))$. The map $(s, g) \to (\theta(s), \alpha(s, g))$ is a measure class preserving bijection between conull subsets of $S \times G$ and $S' \times G'$. (This would be clear if G, G' were discrete. For the general case, see [Zimmer 9]). Since $\varphi(s)\alpha(s, g)\varphi(sg)^{-1} \in L'$ for almost all (s, g), we have $\lambda(y)g'\lambda(yg')^{-1} \in L'$ for almost all $(y, g') \in S' \times G'$. In other words, $\lambda(y)g' \equiv \lambda(yg')$ in G'/L' for almost all y, g'. If we let μ be the G'-invariant probability measure on S', then viewing λ as a map $S' \to G'/L'$, we then have that for almost all g', $\lambda_*(\mu)$ is g'-invariant. Since a conull subgroup of G' must be all of G' (Appendix B) this implies that $\lambda_*(\mu)$ is a G'-invariant probability measure on G'/L'. If L' is proper, this is impossible by the Borel density theorem (3.2.5). This verifies our first assertion about α.

We also observe that if A is any automorphism of G', then $A \circ \alpha$ is not equivalent to a cocycle into a proper algebraic subgroup. This actually follows from the above paragraph since $A \circ \alpha$ is the cocycle corresponding to the orbit equivalence $\theta: S \to S'$, where G' now acts on S' by $(y, g') \to yA(g')$.

Now let $p: G' \to H$ be the projection and consider the cocycle $p \circ \alpha: S \times G \to H$. If this were equivalent to a cocycle into a proper algebraic subgroup $L \subset H$, then α would clearly be equivalent to a cocycle into $p^{-1}(L)$. (Just write $G' = H \times H'$ and examine the definitions.) Similarly, if A is any automorphism of H, then $A \circ p \circ \alpha: S \times G \to H$ is not equivalent to a cocycle into a proper algebraic subgroup (since A defines an automorphism of G' by taking the identity automorphism on H'). We wish to apply Theorem 5.2.5 to the cocycle $\beta = p \circ \alpha$ (for each simple factor of G'). The hypotheses and conclusions of 5.2.5 concern cocycles into $\underline{H}_{\mathbb{R}}$, and here we have cocycles into $H = \underline{H}^0_{\mathbb{R}}$. To compare them, we use the following.

5.2.7 Lemma: *Suppose G, H, J are locally compact groups with $J \subset H$ normal of finite index. Suppose $\alpha: S \times G \to J$ is a cocycle on an ergodic G-space. Then (i) If $\alpha: S \times G \to H$ is equivalent (as a cocycle into H) to a cocycle taking values in a subgroup $H_0 \subset H$, then there is an element $h \in H$ such that, letting A_h be conjugation*

by h, $A_h \circ \alpha$ is equivalent as a cocycle into J, to a cocycle taking values in $J \cap H_0$.
(ii) If $\alpha, \beta: S \times G \to J$ are cocycles which are equivalent as cocycles into H, then for some $h \in H$, $A_h \circ \alpha$ and β are equivalent as cocycles into J.

Let us postpone the proof of this lemma for a moment and continue our previous argument. From (i) of the lemma, and our previous observations, it follows that as a cocycle into $\underline{H}_{\mathbb{R}}$, $p \circ \alpha$ cannot be equivalent to a cocycle into a proper algebraic subgroup $L_{\mathbb{R}} \subset \underline{H}_{\mathbb{R}}$. For if it were, by (i) of the Lemma, $L_{\mathbb{R}} \supset \underline{H}^0_{\mathbb{R}}$, so $\dim L = \dim \underline{H}$, and $L = \underline{H}$ by connectedness of $\underline{H}$. We can therefore apply the case $k = \mathbb{R}$ of Theorem 5.2.5, and deduce that $p \circ \alpha \sim \alpha_{\pi|G}$ as cocycles into $\underline{H}_{\mathbb{R}}$ for some $\mathbb{R}$-rational homomorphism $\pi: \underline{G} \to \underline{H}$. We observe that since G is connected, $\pi(G) \subset H = \underline{H}^0_{\mathbb{R}}$. Thus by (ii) of Lemma 5.2.7, replacing π by $A_h \circ \pi$ for some $h \in H$, we have an $\mathbb{R}$-rational homomorphism such that $p \circ \alpha \sim \alpha_{\pi|G}$ as cocycles into $H = \underline{H}^0_{\mathbb{R}}$. Since this is true for each simple factor of G', there is a rational homomorphism $\tilde{\pi}: G \to G'$ such that $\alpha \sim \alpha_{\tilde{\pi}}$ as cocycles into G'. By Proposition 4.2.11, to complete the argument (in the center free case), it suffices to show that $\tilde{\pi}$ is an isomorphism. Since α is not equivalent to a cocycle into a proper algebraic subgroup, $\tilde{\pi}(G)$ cannot be contained in a proper algebraic subgroup. Arguing as in the conclusion of the proof that Theorem 5.1.2 implies 5.1.1 in the preceding section, we deduce that $\tilde{\pi}(G) = G'$. Finally, suppose $N = \ker(\tilde{\pi})$. From $\alpha \sim \alpha_{\tilde{\pi}}$, we deduce that there is a function $\varphi: S \to G'$ such that for each $g \in N$ we have $\varphi(s)^{-1}\alpha(s, g)\varphi(sg) = e$ for almost all s. We recall that α corresponds to an orbit equivalence $\theta: S \to S'$, and so we have for each g, $\theta(s) \cdot \alpha(s, g) = \theta(sg)$ a.e. Thus, for each $g \in N$, we have $\theta(s) \cdot \varphi(s) = \theta(sg) \cdot \varphi(sg)$ a.e. Because the action of G is irreducible and G has trivial center, if $N \neq \{e\}$, then $\dim N > 0$ and N is ergodic on S. But then the preceding sentence implies that the map $s \to \theta(s) \cdot \varphi(s)$ is constant on a conull set. In particular, since $\varphi(s) \in G'$, this implies that there is a single G' orbit in S' such that $\theta(s)$ is in this orbit for almost all s. In other words, G' is essentially transitive on S', which is impossible (since the actions are essentially free with finite invariant measure.)

Let us now indicate the modifications necessary in case the centers $Z(G)$, $Z(G')$ are finite. Let $\alpha: S \times G \to G'$ be the cocycle corresponding to the orbit equivalence. $Z(G)$ is finite, and acts freely on S, so $S/Z(G)$ is a standard Borel $G/Z(G)$ space. Let $q: S \to S/Z(G)$ be the natural projection, and let $f: S/Z(G) \to S$ be a Borel section of q (defined a.e.) (See Appendix A). Then if $s \in S$, there is a unique $h(s) \in Z(G)$ such that $s = f(q(s)) \cdot h(s)$. We remark that if $g \in Z(G)$, then $h(sg) = h(s)g$. Define $\varphi: S \to G'$ by $\varphi(s) = \alpha(f(q(s)), h(s))$. Then one immediately calculates that $\beta(s, g) = \varphi(s)\alpha(s, g)\varphi(sg)^{-1}$ is a cocycle equivalent to α such that for almost all s, $\beta(s, h) = e$ for $h \in Z(G)$. Changing β on a null set, we can assume $\beta(s, h) = e$ for all s. We can then define a cocycle $\tilde{\beta}: S/Z(G) \times G/Z(G) \to G'$

by $\tilde{\beta}(x, [g]) = \beta(s, g)$ where $s \in S$ such that $q(s) = x$. One checks as usual that the condition $\beta(s, h) = e$ ensures that this is well defined (a.e.). We note that $\tilde{\beta}$ is not equivalent to a cocycle into a subgroup of G' of the form $r^{-1}(L)$ where $r:G' \to G'/Z(G')$ is the natural projection and L is a proper algebraic subgroup of $G'/Z(G')$. If it were, β and hence α would be as well, and this is impossible by the same use of the Borel density theorem as in the center free case. Now let H be any simple factor of $G'/Z(G')$. By applying Theorem 5.2.5 to each $p \circ \tilde{\beta}$, where $p:G' \to H$ is projection, we obtain as in the center free case a rational homomorphism $\tilde{\pi}:G/Z(G) \to G'/Z(G')$ such that $\tilde{\beta} \sim \alpha_{\tilde{\pi}}$. It follows that β, and hence α, is equivalent to α_π where $\pi:G \to G'/Z(G')$ is the composition of $\tilde{\pi}$ with the natural projection. One sees π is surjective exactly as in the center free case, and to show that $N = \ker(\pi)$ is actually $Z(G)$, we use the same argument as in the center free case applied to the equation

$$\tilde{\theta}(s) \cdot \alpha(s, g) = \tilde{\theta}(sg) \quad \text{where} \quad \tilde{\theta}:S \to S'/Z(G')$$

is the composition of θ with the natural projection. If $N \neq Z(G)$, then the argument would show that $G'/Z(G')$ is essentially transitive on $S'/Z(G')$ which would again imply essential transitivity of G' on S$'$ providing a contradiction. Thus, to complete the proof of Theorem 5.2.1 from 5.2.5, it remains only to verify the lemma.

Proof of Lemma 5.2.7: The cocycle $\alpha:S \times G \to H$ is equivalent to a cocycle taking values in J and equivalent to a cocycle taking values in H_0. By Example 4.2.18, there are α-invariant functions $f:S \to H/H_0$ and $f':S \to H/J$. Then $\varphi = (f, f'):S \to H/H_0 \times H/J$ is also α-invariant. Thus (taking H to act on the left on the coset spaces), for all g, $\alpha(s, g)\varphi(sg) = \varphi(s)$ a.e. This implies that $\bar{\varphi}(sg) = \bar{\varphi}(s)$, where $\bar{\varphi}:S \to (H/H_0 \times H/J)/H$. The action of H on $H/H_0 \times H/J$ is smooth (since by 3.2.13, it suffices to see that H_0 is smooth on H/J, which it clearly must be since H/J is finite), i.e., $(H/H_0 \times H/J)/H$ is countably separated. By the ergodicity of G on S, $\bar{\varphi}$ is constant on a conull set, or equivalently, there is a single H-orbit in $H/H_0 \times H/J$ such that $\varphi(s)$ lies in this orbit for almost all s. Since J is normal, we can choose a point in the orbit such that the stabilizer is $J \cap H_0$, and we can then view φ as an α-invariant map $S \to H/J \cap H_0$. By Example 4.2.18, α is equivalent (as a cocycle into H) to a cocycle β taking values in $J \cap H_0$. Thus, to complete the proof of assertion (i) of Lemma 5.2.7, it suffices to prove assertion (ii).

Choose a function $\varphi:S \to H$ such that for each g, $\beta(s, g)^{-1}\varphi(s)\alpha(s, g) = \varphi(sg)$ a.e. Thus, $\tilde{\varphi}(s) = \tilde{\varphi}(sg)$ where $\tilde{\varphi}$ is the composition of φ with the projection $H \to J\backslash H/J$. Since $J\backslash H/J$ is finite and G is ergodic on S, $\tilde{\varphi}$ is constant on a conull set, say $\tilde{\varphi}(s) = [h_0]$ for some $h_0 \in H$. This means that for almost all s, $\varphi(s)$ lies in the same $J{:}J$ double coset in H, so we can choose Borel maps φ_1, $\varphi_2{:}S \to J$ such that $\varphi(s) = \varphi_1(s)h_0\varphi_2(s)$ a.e. But then we have for each g, $\beta(s, g)^{-1}\varphi_1(s)h_0\varphi_2(s)\alpha(s, g) = \varphi_1(sg)h_0\varphi_2(sg)$ a.e. and so $h_0\alpha(s, g)h_0^{-1} = \psi(s)\beta(s, g)\psi(sg)^{-1}$ where $\psi{:}S \to J$ is given by $\psi(s) = (h_0\varphi_2(s)^{-1}h_0^{-1})\varphi_1(s)^{-1}$. This completes the proof of Lemma 5.2.7.

Having seen that the superrigidity theorem for cocycles for $k = \mathbb{R}$ implies the rigidity theorem for orbit equivalence (5.2.1), we now turn to the proof of the superrigidity theorem for cocycles itself. The proof will be a generalization of the proof we gave of Margulis' superrigidity theorem, and we will refer the reader back to the proof of Theorem 5.1.2 for some points of the proof.

Proof of Theorem 5.2.5: A basic tool in the proof of Theorem 5.1.2 was the action of Γ on the homogeneous space $G^0_{\mathbb{R}}/P_0$, where $P_0 = P \cap G^0_{\mathbb{R}}$ (and P is a minimal $\mathbb{R}$-parabolic subgroup). Here, we shall be using as a replacement for this action the product action of $G^0_{\mathbb{R}}$ on the product space $S \times G^0_{\mathbb{R}}/P_0$. We observe two important features of this action (which also hold for the action of a lattice on $G^0_{\mathbb{R}}/P_0$), namely ergodicity (Moore's theorem, 2.2.15), and amenability (which follows from amenability of P_0, and Propositions 4.3.2, 4.3.4). Once again, for simplicity, we shall first consider the case $k = \mathbb{R}$. As with the proof of Theorem 5.1.2, we begin by observing that it suffices to construct a suitable map between homogeneous spaces.

5.2.8 Lemma: *It suffices to show that there is a proper algebraic $\mathbb{R}$-subgroup $L \subset H$, a cocycle $\beta \sim \alpha$, and an $\mathbb{R}$-rational map $\varphi{:}G/P \to H/L$ such that for each $g \in G^0_{\mathbb{R}}$ and almost all s, $\varphi(xg) = \varphi(x)\cdot\beta(s, g)$.*

Proof: We first observe that $\varphi(G/P)$ must be Zariski dense in H/L. Otherwise, letting $Q = \{h \in H \mid h$ leaves $\varphi(G/P)$ invariant$\}$, we have that Q is a proper subgroup defined over $\mathbb{R}$, and that for each $g \in G^0_{\mathbb{R}}$ and almost all s, $\beta(s, g) \in Q_{\mathbb{R}}$. Changing β on a null set, we would have that $\beta(S \times G^0_{\mathbb{R}}) \subset Q_{\mathbb{R}}$ which contradicts our assumptions about α. This verifies Zariski density of $\varphi(G/P)$. Furthermore, we have for each g and all s, t in a conull set in S, $\varphi(xg) = \varphi(x)\beta(s, g)$ and $\varphi(xg) = \varphi(x)\beta(t, g)$, so that $\beta(s, g)\beta(t, g)^{-1}$ leaves $\varphi(G/P)$ pointwise fixed. Since $\varphi(G/P)$ is Zariski dense, $\beta(s, g)\beta(t, g)^{-1}$ fixes all of H/L, i.e., $\beta(s, g)\beta(t, g)^{-1} \in \bigcap_{h \in H} hLh^{-1}$, and since H is $\mathbb{R}$-simple, $\bigcap_{h \in H} hLh^{-1} = \{e\}$. Thus, $\beta(s, g)$ is essentially independent of s, i.e., defines a measurable homomorphism

$\pi: G^0_{\mathbb{R}} \to H_{\mathbb{R}}$. To see that π is rational, we take the Zariski closure of the graph of π in $G \times H$, and using the equation $\varphi(xg) = \varphi(x)\pi(g)$, argue as in the proof of Lemma 5.1.13 to show that this closure is the graph of an $\mathbb{R}$-rational homomorphism $G \to H$.

As in the proof of Theorem 5.1.2, we observe that because $G^0_{\mathbb{R}}/P_0$ is Zariski dense in G/P, it suffices to construct an $\mathbb{R}$-rational map $\varphi: G/P \to H/L$, defined on a Zariski dense subset of $G^0_{\mathbb{R}}/P_0$ which as a map $G^0_{\mathbb{R}}/P_0 \to H_{\mathbb{R}}/L_{\mathbb{R}}$ satisfies the condition that for all g, $\varphi(xg) = \varphi(x)\beta(s, g)$ for almost all $s \in S$. The proof for $k = \mathbb{R}$ now breaks up into three major steps, the first two corresponding to steps 1 and 2 in the proof of Theorem 5.1.2.

Step 1: There is a proper algebraic $\mathbb{R}$-subgroup $L \subset H$ and a measurable map $\varphi: S \times G^0_{\mathbb{R}}/P_0 \to H_{\mathbb{R}}/L_{\mathbb{R}}$ such that for each g and almost all $(s, x) \in S \times G^0_{\mathbb{R}}/P_0$, we have

$$(*) \qquad \varphi(sg, xg) = \varphi(s, x)\alpha(s, g).$$

(Equivalently, writing the H action on H/L on the left, φ is an $\tilde{\alpha}$-invariant function for the cocycle $\tilde{\alpha}(s, x, g) = \alpha(s, g)$.)

Step 2: For any such function φ, for almost all s, $\varphi_s: G^0_{\mathbb{R}}/P_0 \to H_{\mathbb{R}}/L_{\mathbb{R}}$ given by $\varphi_s(x) = \varphi(s, x)$ is essentially rational.

Step 3: By changing α to an equivalent cocycle we can obtain (*) and Step 2 with φ_s independent of s in a conull subset of S.

By Lemma 5.2.8 and the remarks preceding Step 1, the proof of Steps 1, 2, 3 completes the proof of the theorem for $k = \mathbb{R}$.

Proof of Step 1: Let $Q \subset H$ be a proper parabolic subgroup defined over $\mathbb{R}$. Then $H_{\mathbb{R}}/Q_{\mathbb{R}}$ is a compact metrizable $H_{\mathbb{R}}$-space. We have a cocycle $\alpha: S \times G^0_{\mathbb{R}} \to H_{\mathbb{R}}$, and we define $\tilde{\alpha}: S \times G^0_{\mathbb{R}}/P_0 \times G^0_{\mathbb{R}} \to H_{\mathbb{R}}$ by $\tilde{\alpha}(s, x, g) = \alpha(s, g)$, so that $\tilde{\alpha}$ is also a cocycle. Thus, the space of measurable functions from $S \times G^0_{\mathbb{R}}/P_0$ into $M(H_{\mathbb{R}}/Q_{\mathbb{R}})$ (the space of probability measures on $H_{\mathbb{R}}/Q_{\mathbb{R}}$) is an affine $G^0_{\mathbb{R}}$-space over $S \times G^0_{\mathbb{R}}/P_0$ (Definition 4.3.1). As we remarked above, Propositions 4.3.2, 4.3.4, and amenability of P_0 imply that the action of $G^0_{\mathbb{R}}$ on $S \times G^0_{\mathbb{R}}/P_0$ is amenable. It follows from the definition of amenability that there is an $\tilde{\alpha}$-invariant function $\varphi: S \times G^0_{\mathbb{R}}/P_0 \to M(H_{\mathbb{R}}/Q_{\mathbb{R}})$, i.e., for each $g \in G^0_{\mathbb{R}}$, and almost all $(s, x) \in S \times G^0_{\mathbb{R}}/P_0$, we have $\tilde{\alpha}(s, x, g)\varphi(sg, xg) = \varphi(s, x)$. Using the definition of $\tilde{\alpha}$, and transferring to a right action of $H_{\mathbb{R}}$ on $M(H_{\mathbb{R}}/Q_{\mathbb{R}})$, we obtain $\varphi(sg, xg) = \varphi(s, x)\alpha(s, g)$. Let $\bar{\varphi}$ be the composition of φ with the natural map $M(H_{\mathbb{R}}/Q_{\mathbb{R}}) \to M(H_{\mathbb{R}}/Q_{\mathbb{R}})/H_{\mathbb{R}}$. Then for each $g \in G_{\mathbb{R}}^0$, $\bar{\varphi}(sg, xg) = \bar{\varphi}(s, x)$ for almost all (s, x). By Corollary 3.2.17, the action of $H_{\mathbb{R}}$ on $M(H_{\mathbb{R}}/Q_{\mathbb{R}})$ is smooth, and hence $\bar{\varphi}$ is an essentially $G_{\mathbb{R}}^0$-invariant map into a countably separated space. However, by Moore's ergodicity theorem (2.2.15), P_0 is ergodic on S, and so by Proposition 2.2.2, $G_{\mathbb{R}}^0$ is ergodic on $S \times G_{\mathbb{R}}^0/P_0$. It follows (Lemma 2.2.16) that $\bar{\varphi}$ is constant on a conull set. In other words, there is single $H_{\mathbb{R}}$-orbit in $M(H_{\mathbb{R}}/Q_{\mathbb{R}})$ such that on a conull set φ takes values in this orbit. By 3.2.18, 3.2.19, the stabilizer of this orbit is of the form $L_{\mathbb{R}}$ where $L \subset H$ is a proper algebraic $\mathbb{R}$-group. Thus, we can consider φ as a map $\varphi: S \times G_{\mathbb{R}}^0/P_0 \to H_{\mathbb{R}}/L_{\mathbb{R}}$, and this completes the proof of Step 1.

Proof of Step 2: We can make the same reductions here as we did in the preliminaries for Step 2 in the proof of Theorem 5.1.2. More precisely, we obtain the following analogue of Lemma 5.1.8. We fix $t \in A_{\mathbb{R}}^0$, $t \neq e$, and let $C = (C_t)_{\mathbb{R}}^0$ be the identity component of the real points of the centralizer of t. For $s \in S$, and $g \in G_{\mathbb{R}}^0$, define the map $\varphi_{s,g}: C \to H_{\mathbb{R}}/L_{\mathbb{R}}$ by $\varphi_{s,g}(c) = \varphi(s, cg)$, where we have lifted φ to a map $S \times G_{\mathbb{R}}^0 \to H_{\mathbb{R}}/L_{\mathbb{R}}$.

5.2.9 Lemma: (*Cf. Lemma* 5.1.8). *To prove Step* 2, *it suffices to show that for almost all* (s, g), *there exist:*
(*a*) *an* $\mathbb{R}$-*subvariety* $W_{s,g} \subset H/L$ *such that* $\varphi_{s,g}(c) \in W_{s,g}$ *for almost all* c;
(*b*) *an* $\mathbb{R}$-*algebraic group* $Q_{s,g}$ *which acts* $\mathbb{R}$-*regularly on* $W_{s,g}$;
(*c*) *a measurable homomorphism* $h_{s,g}: C \to (Q_{s,g})_{\mathbb{R}}$;
(*d*) *a point* $x_{s,g} \in W_{s,g} \cap H_{\mathbb{R}}/L_{\mathbb{R}}$ *such that* $\varphi_{s,g}(c) = x_{s,g} \cdot h_{s,g}(c)$ *for almost all* $c \in C$.

The sufficiency of the conditions in Lemma 5.2.9 follow just as do the sufficiency of those in Lemma 5.1.8 for the proof of Step 2 in Theorem 5.1.2. We now proceed to verify that these conditions hold. Using Proposition 3.2.5 as in the proof of Step 2 of Theorem 5.1.2 for $k = \mathbb{R}$, it suffices to show that for almost all $(s, g) \in S \times G_{\mathbb{R}}^0$, $\varphi_{s,g}$ all lie in a single $H_{\mathbb{R}}$-orbit in $F(C, H_{\mathbb{R}}/L_{\mathbb{R}})$.

Define the map $\Phi: S \times G_{\mathbb{R}}^0 \to F(C, H_{\mathbb{R}}/L)$ by $\Phi(s, g) = \varphi_{s,g}$. Let $T \subset A_{\mathbb{R}}^0 \subset P_0$ be the cyclic subgroup generated by t, so that (as in the proof of 5.1.2) T is not compact. We recall that for $p \in P_0$, $\varphi(s, pg) = \varphi(s, g)$ and hence

$$\varphi_{s,tg}(c) = \varphi(s, ctg) = \varphi(s, tcg) = \varphi(s, cg) = \varphi_{s,g}(c).$$

Thus, $\Phi(s, tg) = \Phi(s, g)$ for all $g \in G^0_{\mathbb{R}}$, and so we can consider Φ as a map $\Phi: S \times G^0_{\mathbb{R}}/T \to F(C, H_{\mathbb{R}}/L_{\mathbb{R}})$. By the invariance property of φ established in Step 1, we have that for each $h \in G^0_{\mathbb{R}}$, $c \in C$, and almost all s, $g \in S \times G^0_{\mathbb{R}}$,

$$\varphi_{(s,g)\cdot h}(c) = \varphi(sh, cgh) = \varphi(s, cg)\cdot\alpha(s, h) = \varphi_{s,g}(c)\cdot\alpha(s, h).$$

By Fubini, for each $h \in G^0_{\mathbb{R}}$, and almost all (s, g), this holds for almost all $c \in C$. In other words, for each $h \in G^0_{\mathbb{R}}$, $\Phi((s, g)\cdot h) = \Phi(s, g)\cdot\alpha(s, h)$ for almost all s, g. Let $\tilde{\Phi}$ be the composition of Φ with the natural map $F(C, H_{\mathbb{R}}/L_{\mathbb{R}}) \to F(C, H_{\mathbb{R}}/L_{\mathbb{R}})/H_{\mathbb{R}}$. Then $\tilde{\Phi}$ is essentially G-invariant. Furthermore, $H_{\mathbb{R}}$ acts smoothly on $F(C, H_{\mathbb{R}}/L_{\mathbb{R}})$ by Proposition 3.3.1, and $G^0_{\mathbb{R}}$ is ergodic on $S \times G^0_{\mathbb{R}}/T$. This latter assertion follows from Moore's theorem (2.2.15) and Proposition 2.2.2. Thus $\tilde{\Phi}$ is constant on a conull set, i.e., there is a single $H_{\mathbb{R}}$-orbit in $F(C, H_{\mathbb{R}}/L_{\mathbb{R}})$ such that for almost all (s, g), $\Phi(s, g)$ lies in this $H_{\mathbb{R}}$-orbit. By our observations above, this completes the proof of Step 2.

Proof of Step 3: As in Proposition 3.3.2, let

$$\mathrm{R} = \mathrm{Rat}(G_{\mathbb{R}}/P_{\mathbb{R}}, H_{\mathbb{R}}/L_{\mathbb{R}}) \subset F(G_{\mathbb{R}}/P_{\mathbb{R}}, H_{\mathbb{R}}/P_{\mathbb{R}})$$

be the set of essentially $\mathbb{R}$-rational maps $G_{\mathbb{R}}/P_{\mathbb{R}} \to H_{\mathbb{R}}/L_{\mathbb{R}}$. By Step 2, for each $s \in S$, we have an element $\varphi_s \in R$, and the map $\Phi: S \to R$, $\Phi(s) = \varphi_s$ is measurable. (We recall that $G^0_{\mathbb{R}}/P_0$ is Zariski dense in $G_{\mathbb{R}}/P_{\mathbb{R}}$. Thus, an essentially $\mathbb{R}$-rational map on the former uniquely determines an essentially $\mathbb{R}$-rational map on the latter.) We have an action of $G_{\mathbb{R}} \times H_{\mathbb{R}}$ on R given by $[\varphi\cdot(g, h)](x) = \varphi(xg^{-1})h$. We also remark that by Chevalley's theorem (3.1.4) and the fact that G/P is a projective variety, Proposition 3.3.2 applies to this situation. The invariance condition in Step 1 can be expressed as the assertion that for each $g \in G^0_{\mathbb{R}}$,

(*) $\quad \Phi(sg) = \Phi(s)\cdot(g, \alpha(s, g))$ for almost all s.

(We remark that this holds on $G_{\mathbb{R}}/P_{\mathbb{R}}$ as well as on $G^0_{\mathbb{R}}/P_0$ by Zariski density of the latter in the former.) Let $\tilde{\Phi}$ be the composition of Φ with the natural projection $R \to R/H_{\mathbb{R}}$. We therefore have from (*) that for each $g \in G^0_{\mathbb{R}}$, $\tilde{\Phi}(sg) = \tilde{\Phi}(s)\cdot g$ for almost all $s \in S$. By Proposition 3.3.2, $H_{\mathbb{R}}$ acts smoothly on R and hence $R/H_{\mathbb{R}}$ is a countably separated $G_{\mathbb{R}}$-space. Since $(R/H_{\mathbb{R}})/G_{\mathbb{R}} \cong R/(H_{\mathbb{R}} \times G_{\mathbb{R}})$, $(R/H_{\mathbb{R}})/G_{\mathbb{R}}$ is countably separated by 3.3.2. The composition of $\tilde{\Phi}$ with the natural projection $R/H_{\mathbb{R}} \to (R/H_{\mathbb{R}})/G_{\mathbb{R}}$ is essentially $G^0_{\mathbb{R}}$-invariant, and by ergodicity of $G^0_{\mathbb{R}}$ on S, we deduce that it is essentially constant. In other words, there is a single $G_{\mathbb{R}}$-orbit in $R/H_{\mathbb{R}}$ such that for almost all $s \in S$, $\tilde{\Phi}(s)$ lies in this orbit. Let $G_0 \subset G_{\mathbb{R}}$ be the stabilizer of this orbit. Then on a conull set, we can view $\tilde{\Phi}$ as a map $S \to G_{\mathbb{R}}/G_0$, and for each $g \in G^0_{\mathbb{R}}$, $\tilde{\Phi}(sg) = \tilde{\Phi}(s)\cdot g$ a.e. Since $G^0_{\mathbb{R}}$ is of finite index in $G_{\mathbb{R}}$, it clearly acts smoothly on $G_{\mathbb{R}}/G_0$. Using ergodicity of $G^0_{\mathbb{R}}$ on S again, we deduce that on a conull set, $\tilde{\Phi}(s)$ lies in a single $G^0_{\mathbb{R}}$-orbit in $G_{\mathbb{R}}/G_0$ or equivalently, in a single $G^0_{\mathbb{R}}$-orbit in $R/H_{\mathbb{R}}$. Let $G_1 \subset G^0_{\mathbb{R}}$ be the stabilizer of a point in this $G^0_{\mathbb{R}}$-orbit in $R/H_{\mathbb{R}}$. We can also view $\tilde{\Phi}$ as a map $S \to G^0_{\mathbb{R}}/G_1$ and this is essentially a $G^0_{\mathbb{R}}$-map, i.e., for each g, $\tilde{\Phi}(sg) = \tilde{\Phi}(s)g$ a.e. If μ is the $G^0_{\mathbb{R}}$-invariant probability measure on S, then $\tilde{\Phi}_*(\mu)$ will be a $G^0_{\mathbb{R}}$-invariant probability measure on $G^0_{\mathbb{R}}/G_1$.

We now consider the subgroup G_1. Let $f \in R$ be an element whose image in $R/H_{\mathbb{R}}$ is in the distinguished $G^0_{\mathbb{R}}$-orbit with stabilizer G_1. Let $M = \{(g, h) \in G_{\mathbb{R}} \times H_{\mathbb{R}} | f\cdot g = f\cdot h\}$. Then by Proposition 3.3.2, M is the set of real points of an $\mathbb{R}$-subgroup of $G \times H$. The group G_1 is simply the intersection with $G^0_{\mathbb{R}}$ of the projection of M onto $G_{\mathbb{R}}$, and hence G_1 has only finitely many components. But this and the existence of a $G^0_{\mathbb{R}}$-invariant probability measure on $G^0_{\mathbb{R}}/G_1$ implies that $G_1 = G^0_{\mathbb{R}}$ by Theorem 3.2.5(ii). This means that the $G^0_{\mathbb{R}}$-orbit in $R/H_{\mathbb{R}}$ which we have identified with $G^0_{\mathbb{R}}/G_1$ actually consists of a single point in $R/H_{\mathbb{R}}$. Since $\tilde{\Phi}(s)$ equals this point for almost all s, we deduce that there is a single $H_{\mathbb{R}}$-orbit in R such that for almost all s, $\Phi(s)$ lies in this $H_{\mathbb{R}}$-orbit. Let $\Phi_0 \in \mathrm{R}$ be some element in this orbit. We can choose a Borel section $\theta: \Phi_0 \cdot H_{\mathbb{R}} \to H_{\mathbb{R}}$ of the map $H_{\mathbb{R}} \to \Phi_0 \cdot H_{\mathbb{R}}$, $h \to \Phi_0 \cdot h$, (Appendix A), i.e., for each $\Phi \in \Phi_0 \cdot H_{\mathbb{R}}$, $\theta(\Phi) \in H_{\mathbb{R}}$ such that $\Phi_0 \cdot \theta(\Phi) = \Phi$. Define $\lambda: S \to H_{\mathbb{R}}$ by $\lambda(s) = \theta(\Phi(s))$, so that for almost all s, $\Phi_0 \cdot \lambda(s) = \Phi(s)$. From equation (*), $\Phi(sg) = \Phi(s)\cdot(g, \alpha(s, g))$, we deduce that for each g and almost all s,

$$\Phi_0 \cdot (e, \lambda(sg)) = \Phi_0 \cdot (g, \lambda(s)\alpha(s, g)),$$

i.e., $\Phi_0 = \Phi_0 \cdot (g, \lambda(s)\alpha(s, g)\lambda(sg)^{-1})$. Let $\beta(s, g) = \lambda(s)\alpha(s, g)\lambda(sg)^{-1}$. Then for all $x \in G_{\mathbb{R}}/P_{\mathbb{R}}$, $g \in G^0_{\mathbb{R}}$, we have $\Phi_0(x) = \Phi_0(xg^{-1})\cdot \beta(s, g)$, for almost all s, i.e., $\Phi_0(xg) = \Phi_0(x)\beta(s, g)$. Then Φ_0 is the required function, β the required equivalent cocycle, and the proof of Step 3, and with it the proof of Theorem 5.2.5 for $k = R$, is complete.

Proof of Theorem 5.2.5 for $k \neq \mathbb{R}$. As with the proof of Theorem 5.1.2 for $k \neq \mathbb{R}$, the proof of Step 1 above, for $k \neq \mathbb{R}$, enables us to deduce:

Step 1′: There is a measurable map $\varphi: S \times G^0_{\mathbb{R}}/P_0 \to H_k/M$ such that for all g, $\varphi(sg, xg) = \varphi(s, x)\alpha(s, g)$ a.e., where either

(a) $M = L_k$ where $L \subset H$ is an algebraic k-group of strictly lower dimension; or (b) M is compact.

As in 5.1.2, if $k = \mathbb{C}$, and we are in case (a) in Step 1′, the proof for $k = \mathbb{R}$ goes through. If $k \neq \mathbb{R}, \mathbb{C}$, and we are in case (a), then as in 5.1.2, for almost all s, the map $\varphi_s: G_{\mathbb{R}}^0/P_0 \to H_k/M$ is constant. Thus, in this situation, we can consider φ as a map $\varphi: S \to H_k/M$ such that for all g, $\varphi(sg) = \varphi(s)\cdot\alpha(s, g)$, i.e., φ is an α-invariant function (after we switch to a left action of H on H_k/M). (Definition 4.2.17.) By Example 4.2.18(b), this implies that α is equivalent to a cocycle into $M = L_k$, which is impossible given our hypothesis about α. Thus, case (a) in Step 1′ is impossible for $k \neq \mathbb{R}, \mathbb{C}$. It remains only to consider case (b), and for this we use the following generalization of Proposition 5.1.9.

5.2.10 Lemma: *Let G be locally compact, S, Y ergodic G-spaces such that G still acts ergodically on $S \times Y \times Y$. Let $\alpha: S \times G \to H$ be a cocycle, where H is locally compact, and suppose there an $\tilde{\alpha}$-invariant function $\varphi: S \times Y \to H/K$ where $K \subset H$ is compact and $\tilde{\alpha}(s, y, g) = \alpha(s, g)$. Then α is equivalent to a cocycle into a compact subgroup of H.*

To see that this suffices to complete the proof in case (b) above, it suffices to check that $G_{\mathbb{R}}^0$ acts ergodically on $S \times G_{\mathbb{R}}^0/P_0 \times G_{\mathbb{R}}^0/P_0$, where S is an irreducible ergodic $G_{\mathbb{R}}^0$-space. However, as we observed following the proof of Proposition 5.1.9, as a measurable $G_{\mathbb{R}}^0$-space $G_{\mathbb{R}}^0/P_0 \times G_{\mathbb{R}}^0/P_0$ is isomorphic to $G_{\mathbb{R}}^0/A_{\mathbb{R}} \cap P_0$. Thus, by Proposition 2.2.2, it suffices to see that $A_{\mathbb{R}} \cap P_0$ is ergodic on S, which in turn follows from the noncompactness of $A_{\mathbb{R}} \cap P_0$ and Moore's theorem 2.2.15.

We now turn to the proof of 5.2.10.

Proof of Lemma 5.2.10: Let m be a quasi-invariant ergodic measure on S, and μ a quasi-invariant ergodic measure on Y. We let H act on the right on H/K, so that $\tilde{\alpha}$-invariance of φ is the assertion that for each g and almost all (s, y), $\varphi(sg, yg) = \varphi(s, y)\, \alpha\, (s, g)$. Consider the map $\tilde{\varphi}: S \times Y \times Y \to H/K \times H/K$, $\tilde{\varphi}(s, y_1, y_2) = (\varphi(s, y_1), \varphi(s, y_2))$. Let $p: H/K \times H/K \to (H/K \times H/K)/H$ be the natural map. We recall (see the proof of Proposition 5.1.9) that $(H/K \times H/K)/H$ is countably separated since K is compact. The $\tilde{\alpha}$-invariance of φ implies that $p \circ \tilde{\varphi}$ is essentially G-invariant. By the assumption that G acts ergodically on $S \times Y \times Y$, $p \circ \tilde{\varphi}$ is constant on a conull set, i.e., there is a single H-orbit in $H/K \times H/K$ such that $\tilde{\varphi}(s, y_1, y_2)$ lies in this orbit for almost all (s, y_1, y_2). For each $s \in S$, let $\tilde{\varphi}_s: Y \times Y \to H/K \times H/K$ be given by $\tilde{\varphi}_s(y_1, y_2) = \tilde{\varphi}(s, y_1, y_2)$. Define $\varphi_s: Y \to H/K$ similarly. Then $\tilde{\varphi}_s = (\varphi_s, \varphi_s)$. By Fubini's theorem, for almost all s, $\tilde{\varphi}_s(y_1, y_2)$ lies in the distinguished H-orbit in $H/K \times H/K$ for almost all y_1, y_2. In other words, $(\tilde{\varphi}_s)_*(\mu \times \mu) = (\varphi_s)_*\mu \times (\varphi_s)_*\mu$ is a measure supported on this orbit. Arguing as in the proof of Proposition 5.1.9, we deduce that for almost all s, the support of $(\varphi_s)_*(\mu)$ is a compact subset of H/K. Let $\mathscr{C}$ be the space of compact subsets of H/K with the Hausdorff metric. (See e.g. 3.2.6.) Then $\mathscr{C}$ is a separable metrizable H-space. The map $s \to A_s =$ support $((\varphi_s)_*\mu)$ is a measurable map $\Phi:S \to \mathscr{C}$, and since φ is $\tilde{\alpha}$-invariant, it is straightforward to see that Φ is α-invariant.

We now claim that the action of H on $\mathscr{C}$ is smooth. If A, $B \in \mathscr{C}$ and $h_n \in H$ with $A \cdot h_n \to B$, then since A, B, K are compact it follows that $\{h_n\}$ lies in a compact subset of H. Thus, we can assume $h_n \to h \in H$, and so $B = A \cdot h$. Thus, all H-orbits on $\mathscr{C}$ are closed, verifying smoothness. Since $\Phi:S \to \mathscr{C}$ is α-invariant, i.e., for all g, $\Phi(s)\cdot\alpha(s, g) = \Phi(sg)$ a.e. (using a right action), $\tilde{\Phi}:S \to \mathscr{C}/H$ is essentially G-invariant. By ergodicity of G on S, and the fact that $\mathscr{C}/H$ is countably separated, $\tilde{\Phi}$ is essentially constant, i.e., there is a single H-orbit in $\mathscr{C}$ such that for almost all s, $\Phi(s)$ lies in this H-orbit. Thus, we can consider Φ as a α-invariant map $\Phi:S \to H/M$ where M is the stabilizer of an element in $\mathscr{C}$. It is clear that M must be compact since K is compact. It then follows from Example 4.2.18(b) that α is equivalent to a cocycle into M. This completes the proof of Lemma 5.2.10, and hence, completes the proof of Theorem 5.2.5 in all cases.

Remark: The reader will observe throughout the proof of Theorems 5.1.2 and 5.2.5 the constant interplay between ergodicity of certain actions and smoothness of other actions. One way in which this was used was to show that a cocycle was equivalent to one that took values in a subgroup. This type of result is generally useful, and we can formalize the technique we have employed in the following.

5.2.11 Lemma: (*Cocycle Reduction Lemma*). *Suppose $\alpha:S \times G \to H$ is a cocycle into a locally compact group H where G acts ergodically on S. Suppose X is a continuous H-space on which H acts smoothly. If there is an α-invariant function $\varphi:S \to X$, then there is a point $x \in X$, (with stabilizer H_x) such that α is equivalent to a cocycle taking values in H_x.*

Proof: We have for each g, $\varphi(sg) = \varphi(s)\alpha(s, g)$ a.e. Thus $\bar{\varphi}:S \to X/H$ is essentially invariant. X/H is countably separated by the smoothness assumption and since G is ergodic on S, $\bar{\varphi}$ is essentially constant. Thus, there is a point $x \in X$ such that for almost all s, $\varphi(s) \in x \cdot H$. Thus, changing φ on a set of measure 0, we can assume we have an α-invariant map $\varphi:S \to H/H_x$. The result then follows from Example 4.2.18(b).

We now discuss some examples of ergodic actions of semisimple groups with finite invariant measure. We have already seen one source of examples in Example 2.2.14. Namely, if G is a connected senisimple Lie group with no compact factors, $G \subset H$ where H is also semisimple, then the action of G on H/Γ will be an irreducible ergodic action if $\Gamma \subset H$ is an irreducible lattice. Another source of examples we have seen is actions induced from actions of lattices (see, e.g. the proof of Corollary 5.2.2). We applied this in Corollary 5.2.3 to the action of $SL(n, Z)$ on $\mathbb{R}^n/Z^n$. Another important class of ergodic actions of lattices (and hence by inducing, actions of simple groups) is actions defined by embedding the lattice in a compact group.

5.2.12 Example: Let G be a semisimple algebraic $\mathbb{Q}$-group, and $\Gamma = G_Z \cap G^0_{\mathbb{R}}$, so that Γ is a lattice in $G^0_{\mathbb{R}}$. Then Γ has a number of homomorphisms into compact groups. For example if Z_p is the compact commutative ring of p-adic integers, then for each p we have a homomorphism $\Gamma \to SL(n, Z_p)$. Similarly, via the map $Z \to Z/pZ$, we obtain homomorphisms $\Gamma \to SL(n, Z/pZ)$ for each p, and hence homomorphisms into products of groups of this type. Another standard arithmetic construction yields homomorphisms into compact Lie groups. For example, fix positive integers m, n, $n < m$, and consider the quadratic form $\sum_{i=1}^{m-n} x_i^2 - \sqrt{2} \sum_{i=m-n+1}^{m} x_i^2$. Let $0(m, n)$ be the subgroup of $GL(m, \mathbb{C})$ leaving this form invariant. Then $0(m, n)$ is a simple algebraic group defined over the field $\mathbb{Q}(\sqrt{2})$. Let σ be a field automorphism of $\mathbb{C}$ such that $\sigma(\sqrt{2}) = -\sqrt{2}$. This induces a map $\sigma: GL(m, \mathbb{C}) \to GL(m, \mathbb{C})$, and $\sigma(0(m, n))$ is then clearly the group leaving invariant the quadratic form $\sum_{i=1}^{m-n} x_i^2 + \sqrt{2} \sum_{i=m-n+1}^{m} x_i^2$. Thus, $\sigma(0(m, n))_{\mathbb{R}}$ is a compact Lie group. Let $\Gamma = 0(m, n)_{Z[\sqrt{2}]}$. Then $\sigma: \Gamma \to (0(m, n))_{\mathbb{R}}$ is an injective homomorphism into a compact Lie group. Furthermore, as we shall discuss further in Chapter 6, one can show that Γ is a lattice in the Lie group $0(m, n)_{\mathbb{R}}$, and that $\mathbb{R}$-rank $(0(m, n)) = \min(n, m - n)$.

By taking the closure of the image of Γ, we thus see that there are a number of examples of homomorphisms $\varphi: \Gamma \to K$, K compact, and $\varphi(\Gamma)$ dense in K. It follows from Lemma 2.2.13 that Γ acts ergodically on K, and hence on K/K_0 where $K_0 \subset K$ is any closed subgroup. We remark that if G is a connected semisimple Lie group without compact factors, and $\Gamma \subset G$ is an irreducible lattice, one can show that the action of G obtained by inducing an ergodic action of Γ of the above type is actually an irreducible ergodic action of G [Zimmer 13]. That this is not true for arbitrary ergodic actions of Γ can be seen from the following example. Let $\Gamma \subset G_1 \times G_2$ be an irreducible lattice, G_i non-compact. Let S be an ergodic G_1 space with finite invariant measure. The projection of Γ onto G_1 is dense by the irreducibility assumption, and hence Γ acts ergodically on S (by Corollary 2.2.17, for example). By Proposition 4.2.22, the induced $G_1 \times G_2$ action is just the product $G_1 \times G_2$ action on $S \times (G_1 \times G_2)/\Gamma$. However, G_2 does not act ergodically on this space, so the $G_1 \times G_2$ action is ergodic but not irreducible. Of course, for actions of lattices in simple groups, this question does not arise.

5.2.13 Example: We shall now sketch another construction of finite measure preserving ergodic actions of semisimple Lie groups. For each such G to which the rigidity theorem (5.2.1) applies this will provide uncountably many mutually non-orbit equivalent, essentially free, finite measure preserving ergodic actions. Let $\mathscr{H}$ be a real (separable) infinite dimensional Hilbert space, and let $\{e_i\}$ be an orthonormal basis. Let ν be the Gaussian measure on $\mathbb{R}$, i.e., $d\nu = (2\pi)^{-1}e^{-x^2/2}dx$. Let $\Omega = \prod_1^\infty \mathbb{R}$ and μ the product measure $\mu = \prod_1^\infty \nu$. We have a linear map $T:\mathscr{H} \to L^2(\Omega)$ given by $T(e_i) = p_i$ where $p_i:\Omega \to \mathbb{R}$ is the projection onto the i-th factor. Then T is an orthogonal isometry of $\mathscr{H}$ and $T(\mathscr{H})$. By the theory of Gaussian random processes (see [Kuo 1] for example), for each orthogonal operator U on $\mathscr{H}$, there exists an (essentially unique) a.e. defined measure preserving transformation $f_U:\Omega \to \Omega$ such that the induced map $f_U^*: L^2(\Omega, \mathbb{R}) \to L^2(\Omega, \mathbb{R})$ makes the following diagram commute:

$$\begin{array}{ccc} \mathscr{H} & \xrightarrow{\quad T \quad} & L^2(\Omega) \\ U\downarrow & & \downarrow f_U^* \\ \mathscr{H} & \xrightarrow[\quad T \quad]{} & L^2(\Omega). \end{array}$$

If G is locally compact and π is an orthogonal representation of G on $\mathscr{H}$, then for each $\pi(g)$ we obtain an a.e. defined transformation $f_{\pi(g)}$, and by Appendix B, we can choose these so as to define an action of G on Ω preserving the measure μ. Furthermore, letting $T_\mathbb{C}$ be the complexified map $\mathscr{H}_\mathbb{C} \to L^2(\Omega, \mathbb{C})$ we see that the complexification of the original representation π is a subrepresentation of σ, the representation of G on $L^2(\Omega, \mathbb{C})$ defined by translation. In fact, one can show that $\sigma \cong \sum_{n=0}^{\infty}{}^{\oplus} S^n(\pi_\mathbb{C})$ where $S^n(\pi_\mathbb{C})$ is the n-th symmetric power of the complexification of π. From this, one readily deduces [Segal 1] that the action of G is ergodic if and only if π has no finite dimensional invariant subspaces.

Suppose now that π_1, π_2 are infinite dimensional irreducible unitary representations. Viewing them as orthogonal representations, we then obtain two probability measure preserving ergodic actions of G. If these actions are equivalent, then $(\pi_2)_{\mathbb{C}}$ is a direct summand of $S^n((\pi_1)_{\mathbb{C}})$ for some n. Thus, for any given π_1, there can be at most countably other irreducible representations which will define the same ergodic action via this construction. Thus, if G is any locally compact group with an uncountable number of inequivalent irreducible infinite dimensional unitary representations, G has uncountably many inequivalent measure preserving ergodic actions.

Suppose now that π_1, π_2 as above but that the actions defined by π_1 and π_2 are equivalent modulo an automorphism A of G. (That is, if G acts on Ω, define a new action by $(s, g) \to s \cdot A(g)$. Equivalence modulo A means equivalence after modifying one action in this fashion.) Then $(\pi_2)_{\mathbb{C}} \circ A$ is a direct summand of $S^n((\pi_1)_{\mathbb{C}})$ for some n. If A is an inner automorphism, then $\pi_2 \circ A \cong \pi_2$. Thus, if G has uncountably many inequivalent infinite dimensional unitary representations, and $\mathrm{Out}(G)$, the outer automorphism group, is finite, then G has uncountably many probability measure preserving transformations no two of which are equivalent modulo an automorphism of G. Since noncompact semisimple Lie groups have this property, one then deduces from the rigidity theorem (5.2.1) that if G is as in 5.2.1, G has uncountably many mutually non-orbit equivalent finite measure preserving ergodic actions.

The rigidity theorem for ergodic actions (5.2.1) and Corollary 5.2.2 give us a good deal of information about orbit equivalence for actions of lattices with finite invariant measure. However, some of the most natural actions of lattices do not have finite invariant measure, for example the action of $SL(n, Z)$ on $\mathbb{R}^n$, $\mathbb{P}^{n-1}$, and other Grassmann or flag varieties. For the action on the variety of full flags, the action will be amenable and hence the results of Section 4.3 apply. (See Example 4.3.18). Here we wish to indicate without proof that one can extend the techniques of this chapter to obtain information on orbit equivalence for many natural actions of lattices without invariant measure.

We recall that any locally compact group H has a unique maximal normal amenable subgroup N (Proposition 4.1.12). If H is a connected Lie group, then by Corollary 4.1.9, H/N will be a connected semisimple Lie group with trivial center and no compact factors. Hence, if H is almost connected (i.e., H/H^0 is finite) the connected component of the identity of H/N will be a semisimple group of this type as well.

5.2.14 Theorem [*Zimmer* 9]: *Let G_i, $i = 1, 2$, be connected semisimple Lie groups with finite center and no compact factors, and let $\Gamma_i \subset G_i$ be an irreducible lattice. Let $H_i \subset G_i$ be an almost connected non-compact subgroup and $N_i \subset H_i$ the maximal normal amenable subgroup. Assume $\mathbb{R}$-rank$(H_1/N_1) \geqq 2$.* If *Γ_i acts essentially* *freely on* G_i/H_i *and the* Γ_1 *action on* G_1/H_1 *is orbit equivalent to the* Γ_2 *action on* G_2/H_2, then H_1/N_1 *and* H_2/N_2 *are locally isomorphic.*

Remark: If G, H are (real points of) algebraic groups, $H \supset Z(G)$ (the center of G), and H does not contain a normal non-central subgroup of G, then $\Gamma/\Gamma \cap Z(G)$ acts essentially freely on G/H. To see this, observe that for $\gamma \in \Gamma$, $\{x \in G/H \mid x \cdot \gamma = x\}$ is a closed subvariety, and hence if it is of positive measure, γ fixes all of G/H, which implies $\gamma \in Z(G)$. Essential freeness then follows from 5.2.4.

5.2.15 Corollary [*Zimmer* 9]: (*a*) *As we vary* n, $n \geqq 2$, *the actions of* $SL(n, Z)$ *on* $\mathbb{P}^{n-1}(\mathbb{R})$ *are mutually non-orbit equivalent.*
(*b*) *As we vary* n, $n \geqq 2$, *the actions of* $SL(n, Z)$ *on* $\mathbb{R}^n$ *are mutually non-orbit equivalent.*
(*c*) *For a fixed* $n \geqq 4$, *let* $G_{n,k}$ *be the Grassman variety of k-planes in* $\mathbb{R}^n$. *Then as k varies,* $1 \leqq k \leqq [n/2]$, *the actions of* $SL(n, Z)$ *on* $G_{n,k}$ *are mutually non-orbit equivalent.*

Proof: Let $H_n \subset SL(n, \mathbb{R})$ be the stabilizer of a point in $\mathbb{P}^{n-1}$ and N_n its maximal normal amenable subgroup. Then H_n/N_n is locally isomorphic to $PSL(n-1, \mathbb{R})$. If n, $p \geqq 3$, and $n \neq p$, it follows from Theorem 5.2.14 (using the above remark for essential freeness) that the actions of $SL(n, Z)$ on $\mathbb{P}^{n-1}$ and $SL(p, Z)$ on $\mathbb{P}^{p-1}$ are non-orbit equivalent. However, for $n = 2$, the action of $SL(2, Z)$ on $\mathbb{P}^1$ is amenable, while for $n \geqq 3$, the action of $SL(n, Z)$ on $\mathbb{P}^{n-1}$ is not amenable (Example 4.3.8). From Proposition 4.3.10, we deduce that $SL(2, Z)$ acting on $\mathbb{P}^1$ is not orbit equivalent to $SL(n, Z)$ acting on $\mathbb{P}^{n-1}$, $n \geqq 3$, and this completes the proof of (a). Assertion (b), (c) are proved similarly.

Finally, in concluding this section, we mention without proof one further result on orbit equivalence which can be obtained from the techniques developed here. If H is a connected Lie group, N the maximal normal amenable subgroup, and S an ergodic H-space, the action is called irreducible if the restriction of the action to any normal subgroup of H which properly contains N is still ergodic. If H is semisimple, with no compact factors, this reduces to our previous definition (2.2.11), and if H/N is simple, then this is no further restriction than ergodicity of H.

5.2.16 Theorem [*Zimmer* 9]: *For* $i = 1, 2$, *let* H_i *be a connected Lie group with maximal normal amenable subgroup* N_i. *Suppose* $\mathbb{R}$-*rank* $(H_1/N_1) \geqq 2$. *Let* S_i *be an essentially free irreducible ergodic* H_i *space with finite invariant measure, and suppose the* H_1 *action on* S_1 *and the* H_2 *action on* S_2 *are orbit equivalent. Then* H_1/N_1 *and* H_2/N_2 *are isomorphic and* N_1 *is compact if and only if* N_2 *is compact.*

For implications of these results for foliations, see [Zimmer 8]. For some further results on orbit equivalence, see chapters 9 and 10.

6 Margulis' Arithmeticity Theorems

6.1 Arithmeticity in groups of $\mathbb{R}$-rank $\geqq 2$

We recall from the introduction the following construction of lattices.
(1) If G is a connected semisimple algebraic $\mathbb{Q}$-group, then G_Z is a lattice in $G_{\mathbb{R}}$. (Theorem of Borel–Harish–Chandra, 3.1.7).
(2) If G is locally compact, $\Gamma \subset G$ a lattice and $\Gamma' \subset G$ a subgroup commensurable with Γ, then Γ' is a lattice. (We recall that Γ, Γ' commensurable means $[\Gamma:\Gamma \cap \Gamma']$, $[\Gamma':\Gamma \cap \Gamma'] < \infty$.)
(3) If $\Gamma \subset H$ is a lattice and $\varphi: H \to G$ is a surjective homomorphism with compact kernel, then $\varphi(\Gamma)$ is a lattice in G. (To see this, we first observe that since kernel(φ) is compact, $\varphi(\Gamma)$ is discrete. Furthermore, there is a natural H-map of H-spaces $\tilde{\varphi}: H/\Gamma \to G/\varphi(\Gamma)$, and hence if μ is the H-invariant probability measure on H/Γ, $\tilde{\varphi}_*(\mu)$ will be a G-invariant probability measure on $G/\varphi(\Gamma)$.)

A lattice in a semisimple Lie group is called arithmetic if it can be obtained from the above three procedures. More precisely:

6.1.1 Definition: Let G be a connected semisimple Lie group with trivial center and no compact factors. Let $\Gamma \subset G$ be a lattice. Then Γ is called arithmetic if there exist (i) a semisimple algebraic $\mathbb{Q}$-group H and (ii) a surjective homomorphism

$$\varphi: H^0_{\mathbb{R}} \to G \text{ with compact kernel,}$$

such that

$$\varphi(H_Z \cap H^0_{\mathbb{R}}) \text{ and } \Gamma \text{ are commensurable.}$$

We remark that we may assume that the center of $H^0_{\mathbb{R}}$ is trivial (simply by dividing it out if it were present) and by semisimplicity, $H^0_{\mathbb{R}}$ is isomorphic to $G \times K$ where K is a compact group, and φ is simply projection onto G.

Our main aim in this section is to prove Margulis' arithmeticity theorem.

6.1.2 Theorem [*Margulis* 1]: *Let G be a connected semisimple Lie group with trivial center and no compact factors. Let $\Gamma \subset G$ be an irreducible lattice. Assume $\mathbb{R}$-rank$(G) \geqq 2$. Then Γ is arithmetic.*

Before beginning the proof, we shall discuss a very useful operation on algebraic groups, namely "restriction of scalars." We do this both because it will be used in the proof of Theorem 6.1.2 and because it provides a basic construction of algebraic $\mathbb{Q}$-groups (and hence of arithmetic groups).

Suppose k is an algebraic number field, i.e., $\mathbb{Q} \subset k \subset \mathbb{C}$ and $d = [k, \mathbb{Q}] < \infty$. Suppose G is an algebraic k-group (or more generally, an affine k-variety). We wish to describe the construction of an algebraic $\mathbb{Q}$-group H such that G_k and $H_{\mathbb{Q}}$ are naturally isomorphic. For example, we can view $\mathbb{C}^N$ as a group defined over k, and the k-points are of course k^N. Choosing a basis of k over $\mathbb{Q}$, we can identify k^N and $\mathbb{Q}^{Nd}$, and $\mathbb{Q}^{Nd}$ can of course be considered as the $\mathbb{Q}$-points of the $\mathbb{Q}$-group $\mathbb{C}^{Nd}$. Thus, by expanding $\mathbb{C}^N$ to $\mathbb{C}^{Nd}$, the k-points become $\mathbb{Q}$-points. We will now describe this in a more formal fashion which will make clear the extension of this notion to arbitrary groups (or varieties).

Let $d = [k:\mathbb{Q}] < \infty$, $k \subset \mathbb{C}$, and choose a basis $\alpha_1, \ldots, \alpha_d$ of k over $\mathbb{Q}$. We recall that there are d distinct field embeddings $\sigma_1, \ldots, \sigma_d$, $\sigma_i: k \to \mathbb{C}$, and we choose $\sigma_1 =$ identity. It is well known that:
(i) $\{\sigma_i\}$ are linearly independent in the space of functions $k \to \mathbb{C}$ (which is a vector space over $\mathbb{C}$), and hence the $d \times d$ matrix $(\sigma_i(\alpha_j))$ is non singular; and
(ii) If $\alpha \in k$, then $\sigma_i(\alpha) = \alpha$ for all i if and only if $\alpha \in \mathbb{Q}$.

Now fix a positive integer N. We wish to describe the identification $k^N \cong \mathbb{Q}^{Nd}$ as an identification of k-points in $\mathbb{C}^N$ with $\mathbb{Q}$-points in $\mathbb{C}^{Nd}$. Each σ_i defines a map $\sigma_i^N: k^N \to \mathbb{C}^N$; we shall abuse notation and still denote this map by $\sigma_i: k^N \to \mathbb{C}^N$. For each $w \in k^N$, let $w' \in (\mathbb{C}^N)^d$ be given by $w' = (\sigma_1(w), \ldots, \sigma_d(w))$. Then $w \to w'$ is a bijection of k^N with its image, which we denote by $(k^N)' \subset (\mathbb{C}^N)^d$. If we let $p:(\mathbb{C}^N)^d \to \mathbb{C}^N$ be projection onto the first factor, then $p|(k^N)':(k^N)' \to k^N$ is the inverse map to $w \to w'$. These maps establish an isomorphism of k^N and $(k^N)'$ as $\mathbb{Q}$-vector spaces.

We identify $(\mathbb{C}^N)^d$ with $d \times N$ complex matrices. The group $GL(d, \mathbb{C})$ acts on the set of $d \times N$ matrices by multiplication on the left, and thus we have an action of $GL(d, \mathbb{C})$ on $(\mathbb{C}^N)^d$. Let $T \in GL(d, \mathbb{C})$ be the matrix $T = (\sigma_i(\alpha_j))$ described above. Then we consider T as a map $T:(\mathbb{C}^N)^d \to (\mathbb{C}^N)^d$. It is immediate from the definitions that $T|(\mathbb{Q}^N)^d$ is a bijection $T:(\mathbb{Q}^N)^d \to (k^N)'$. (Formally, then T provides an isomorphism of $(\mathbb{C}^N)^d$ with a $\mathbb{Q}$-variety such that identifying $(\mathbb{C}^N)^d$ with this $\mathbb{Q}$-variety, $[(\mathbb{C}^N)^d]_{\mathbb{Q}} = (k^N)'$.) If we let $\mathcal{O}$ be the ring of algebraic integers in k, then we can choose the basis $\alpha_1, \ldots, \alpha_d$ of k over $\mathbb{Q}$ such that $\alpha_i \in \mathcal{O}$ and $\mathcal{O} = \Sigma Z\alpha_i$. In this case T will satisfy the additional condition that $T((Z^N)^d) = (\mathcal{O}^N)'$. We also observe that if K is a field with $\sigma_i(k) \subset K \subset \mathbb{C}$ for all i, then $T((K^N)^d) = (K^N)^d$.

Suppose that $p \in k[X_1, \ldots, X_N]$. If $\sigma: k \to \mathbb{C}$ is a field embedding, we let p^σ be the corresponding polynomial with coefficients in $\sigma(k)$. We recall that σ can be extended to a field automorphism $\sigma \in \mathrm{Aut}(\mathbb{C})$. Thus, if $I \subset \mathbb{C}[X_1, \ldots, X_n]$ is an ideal, we let $I^\sigma = \{p^\sigma \mid p \in I\}$. Then I^σ is an ideal, and if I is generated by elements of $k[X_1, \ldots, X_n]$, I^σ is generated by elements of $\sigma(k)[X_1, \ldots, X_N]$. Thus, if $V \subset \mathbb{C}^N$ is a k-variety with annihilator I, then $V^\sigma \subset \mathbb{C}^N$, $V^\sigma = \{\sigma(x) \mid x \in V\}$ (where $\sigma: \mathbb{C}^N \to \mathbb{C}^N$) is a $\sigma(k)$-variety with annihilator I^σ. If $G \subset \mathbb{C}^{N \times N}$ is an algebraic k-group, it is clear that G^σ is an algebraic $\sigma(k)$-group.

If $V \subset \mathbb{C}^N$ is a k-variety, then $V^{\sigma_1} \times \ldots \times V^{\sigma_d} \subset (\mathbb{C}^N)^d$ is a variety. Let $q_r: (\mathbb{C}^N)^d \to \mathbb{C}^N$ be the r-th coordinate projection. Since V is defined over k, ΠV^{σ_i} can be described as the set of zeros in $(\mathbb{C}^N)^d$ of all polynomials of the form $p^{\sigma_i} \circ q_i$ where $p \in k[X_1, \ldots, X_N] \cap I$, where I is the ideal of V. Let $S: \mathbb{C}^d \to (\mathbb{C}^N)^d$ be the embedding of $\mathbb{C}^d$ as the first column (identifying, as above, $(\mathbb{C}^N)^d$ with $d \times N$ matrices). Then we can also describe ΠV^{σ_i} as the set of simultaneous zeroes of all coordinate entries of the maps $\tilde{p}: (\mathbb{C}^N)^d \to (\mathbb{C}^N)^d$ given by $\tilde{p} = T^{-1}(S(p^{\sigma_1} \circ q_1, \ldots, p^{\sigma_d} \circ q_d))$, for $p \in k[X_1, \ldots, X_N] \cap I$. The point of this construction is that $\tilde{p} \circ T$ is actually a polynomial (into $(\mathbb{C}^N)^d$) with coefficients in $\mathbb{Q}$. To see this, it suffices to see that $\tilde{p}(T((\mathbb{Q}^N)^d)) \subset (\mathbb{Q}^N)^d$, i.e. $\tilde{p}((k^N)') \subset (\mathbb{Q}^N)^d$. But if $x \in k^N$, we have

$$\tilde{p}(\sigma_1(x), \ldots, \sigma_d(x)) = T^{-1}(S(\sigma_1(p(x)), \ldots, \sigma_d(p(x)))).$$

We have $S(\sigma_1(p(x)), \ldots, \sigma_d(p(x))) \in (k^N)'$, and hence upon applying T^{-1} we obtain an element of $(\mathbb{Q}^N)^d$. Therefore, under the identification of $(\mathbb{C}^N)^d$ with $(\mathbb{C}^N)^d$ via T, this variety ΠV^{σ_i} is actually defined over $\mathbb{Q}$. Furthermore, $(\Pi V^{\sigma_i})_{\mathbb{Q}} = (V_k)'$ where V_k' is the image of V_k under the isomorphism $k^N \to (k^N)'$.

We summarize our discussion in the following. (We shall only be using this construction for groups, and hence just summarize for this case.)

6.1.3 Proposition (*Restriction of Scalars*) [*Weil* 1] *Let k be an algebraic number field ($[k:\mathbb{Q}] = d$), and suppose G is an algebraic k-group. Let $\sigma_1, \ldots, \sigma_d: k \to \mathbb{C}$ be the distinct field embeddings of k into $\mathbb{C}$ with $\sigma_1 =$ identity. Let $R_{k/\mathbb{Q}}(G) = \prod_{i=1}^{d} G^{\sigma_i}$, and for $g \in G_k$ let $g' = (\sigma_1(g), \ldots, \sigma_d(g))$. Let $(G_k)' = \{g' \mid g \in G_k\}$. Then $R_{k/\mathbb{Q}}(G)$ is (isomorphic to) an algebraic $\mathbb{Q}$-group such that $(R_{k/\mathbb{Q}}(G))_{\mathbb{Q}} = (G_k)'$, and $(R_{k/\mathbb{Q}}(G))_Z = (G_{\mathcal{O}})'$ where $\mathcal{O} \subset k$ is the subring of algebraic integers in k. The projection map $p: R_{k/\mathbb{Q}}(G) \to G$ onto the first factor is defined over k, and defines bijections $(R_{k/\mathbb{Q}}(G))_{\mathbb{Q}} \to G_k$, $(R_{k/\mathbb{Q}}(G))_Z \to G_{\mathcal{O}}$. Furthermore, if for all i, $\sigma_i(k) \subset K$ where K is a field $k \subset K \subset \mathbb{C}$, then each G^{σ_i} is defined over K and $(R_{k/\mathbb{Q}}(G))_K = \Pi (G^{\sigma_i})_K$.*

The group $R_{k/\mathbb{Q}}(G)$ is called the restriction of G to $\mathbb{Q}$. Theorem 3.1.7 implies:

6.1.4 Corollary: *If k is an algebraic number field, and G is a semisimple algebraic k-group, then $G_{\mathcal{O}}$ is isomorphic to a lattice in $[R_{k/\mathbb{Q}}(G)]_{\mathbb{R}}$.*

6.1.5 Example: We consider in this light Example 5.2.12. Namely, let $G = 0(m, n) \subset GL(m, \mathbb{C})$ be the subgroup leaving the form $\sum_{i=1}^{m-n} x_i^2 - \sqrt{2} \sum_{i=m-n+1}^{m} x_i^2$ invariant. Thus, G is an algebraic group defined over $k = \mathbb{Q}(\sqrt{2})$. Here $[k:\mathbb{Q}] = 2$, $\sigma_1 =$ identity, $\sigma_2 = \sigma$ where $\sigma(a + b\sqrt{2}) = a - b\sqrt{2}$, and $\mathcal{O} = Z[\sqrt{2}]$. Thus G^σ is the group leaving the form $\sum_{i=1}^{m-n} x_i^2 + \sqrt{2} \sum_{i=m-n+1}^{m} x_i^2$ invariant and $R_{k/\mathbb{Q}}(G) = G \times G^\sigma$. Since $k, \sigma(k) \subset \mathbb{R}$, $(G \times G^\sigma)_{\mathbb{R}} = G_{\mathbb{R}} \times (G^\sigma)_{\mathbb{R}}$. By Corollary 6.1.4 $(G_{\mathcal{O}})'$ is a lattice in $(G \times G^\sigma)_{\mathbb{R}}$, and since $(G^\sigma)_{\mathbb{R}}$ is compact, the projection of $G'_{\mathcal{O}}$ to $G_{\mathbb{R}}$ is a lattice. Thus, $G_{Z[\sqrt{2}]} \cap G^0_{\mathbb{R}}$ is an arithmetic lattice in $G^0_{\mathbb{R}}$.

With these preliminaries, we now turn to the proof of the arithmeticity theorem. As the reader will see, the basic part of the argument is Margulis' superrigidity theorem which we proved in section 5.1. During the course of the proof, we will need the fact that Γ is finitely generated. If all simple factors of G have $\mathbb{R}$-rank at least 2, this will be proven in Chapter 7. (See Theorems 7.1.4, 7.1.5.) For the result in general, see [Raghunathan 1].

Proof of Theorem 6.1.2. We shall change our notation and let G be a connected semisimple algebraic $\mathbb{Q}$-group with trivial center and let $\Gamma \subset G^0_{\mathbb{R}}$ be an irreducible lattice. We assume $\mathbb{R}\text{-rank}(G) \geqq 2$ and $G^0_{\mathbb{R}}$ has no compact factors. (We can assume we are in this situation by Proposition 3.1.6.) If $G \subset GL(n, \mathbb{C})$, let $L(G)$ be the Lie algebra of G, i.e., $L(G)$ is the tangent space at the identity, which we can consider as a linear subspace of $M(n \times n, \mathbb{C})$, the space of all $n \times n$ complex matrices. Since G is a $\mathbb{Q}$-group $L(G)$ will also be defined over $\mathbb{Q}$, i.e., it has a basis contained in $M(n \times n, \mathbb{Q})$. We recall that G acts on $M(n \times n, \mathbb{C})$ by conjugation, that $L(G)$ is invariant under this G-action, and that the representation $G \to GL(L(G))$ so obtained is the adjoint representation, denoted by Ad. Then Ad is defined over $\mathbb{Q}$, and since G has trivial center, Ad is faithful. The first step of the proof is the following

6.1.6 Lemma: *For all $\gamma \in \Gamma$, $Tr(Ad(\gamma))$ is a (real) algebraic number. (Here Tr denotes trace.)*

Proof: Since the group of field automorphisms $\text{Aut}(\mathbb{C})$ acts transitively on the transcendental numbers, it suffices to show that for each $\gamma\in\Gamma$, $\{\sigma(Tr(Ad(\gamma)))| \sigma\in\mathrm{Aut}(\mathbb{C})\}$ is a bounded set of complex numbers. Each $\sigma\in\mathrm{Aut}(\mathbb{C})$ acts on $M(n\times n,\mathbb{C})$ by $(a_{ij})\to(\sigma(a_{ij}))$, and since G and $L(G)$ are defined over $\mathbb{Q}$, $\sigma(G)=G$, $\sigma(L(G))=L(G)$. We first claim that for any γ, σ, $\sigma(Tr(Ad(\gamma)))=Tr(Ad(\sigma(\gamma)))$. Let $\{A_i\}$ be a basis of $L(G)$ contained in $M(n\times n,\mathbb{Q})$, and let $(c_{ij}(g))$ be the matrix of $Ad(g)$ with respect to this basis. Then $\sigma(\gamma)A_i\sigma(\gamma)^{-1}=\Sigma\, c_{ij}(\sigma(\gamma))A_j$, and since $A_i\in M(n\times n,\mathbb{Q})$, $\sigma(\gamma A_i\gamma^{-1})=\Sigma\, c_{ij}(\sigma(\gamma))A_j$ and hence $\gamma A_i\gamma^{-1}=\Sigma\,\sigma^{-1}(c_{ij}(\sigma(\gamma)))A_j$. Thus $\sigma^{-1}(c_{ij}(\sigma(\gamma)))=c_{ij}(\gamma)$, and hence $Tr(Ad(\sigma(\gamma)))=\sigma(Tr\,Ad(\gamma))$ as asserted. Thus, it suffices to show that $\{Tr(Ad(\sigma(\gamma)))|\sigma\in\mathrm{Aut}(\mathbb{C})\}$ is bounded for each $\gamma\in\Gamma$.

Let $G=\Pi H_i$ be the decomposition of G into simple factors, so $L(G)=\Sigma L(H_i)$, and $L(H_i)$ is $Ad(G)$ invariant. Thus, if we let $p_i{:}G\to H_i$ be projection and Ad_{H_i} the adjoint representation of H_i, we can write $Tr(Ad\,\sigma(\gamma))=\sum_i Tr(Ad_{H_i}(p_i(\sigma(\gamma))))$. Since Γ is Zariski dense in G by the Borel density theorem (3.2.5), $\sigma(\Gamma)$ clearly is as well, and hence $(p_i\circ\sigma)(\Gamma)$ is Zariski dense in H_i. Thus, by Margulis superrigidity (5.1.2), either $(p_i\circ\sigma)(\Gamma)$ is contained in a compact subgroup of H_i, or $(p_i\circ\sigma)|\Gamma$ extends to a rational surjective homomorphism $\pi{:}G\to H_i$. In the former case, all eigenvalues of $Ad_{H_i}(p_i(\sigma(\gamma)))$ have absolute value 1, and hence $|Tr(Ad_{H_i}(p_i(\sigma(\gamma))))|\leqq\dim H_i$. If $p_i\circ\sigma$ extends to π, then letting $c_g{:}G\to G$ and $c_h{:}H_i\to H_i$ be conjugation by $g\in G$ and $h\in H_i$, $\pi\circ c_g=c_{\pi(g)}\circ\pi$, and hence upon differentiating, we obtain for each $g\in G$ the commutative diagram

$$\begin{array}{ccc} L(G) & \xrightarrow{\;Ad(g)\;} & L(G) \\ {\scriptstyle d\pi}\big\downarrow & & \big\downarrow{\scriptstyle d\pi} \\ L(H_i) & \xrightarrow{\;Ad_{H_i}(\pi(g))\;} & L(H_i) \end{array}$$

Since $d\pi$ is surjective, any eigenvalue of $Ad_{H_i}(\pi(g))$ is an eigenvalue of $Ad(g)$, and in particular, for $\gamma\in\Gamma$, any eigenvalue of $Ad_{H_i}(p_i(\sigma(\gamma)))$ is an eigenvalue of $Ad(\gamma)$. Thus, if we let $e(\gamma)$ be the maximal absolute value of an eigenvalue of $Ad(\gamma)$, we deduce that for any $\sigma\in\mathrm{Aut}(\mathbb{C})$, $|Tr(Ad(\sigma(\gamma)))|\leqq(1+e(\gamma))\dim G$. This proves the lemma.

The next step is to pass from the conclusion that the traces are algebraic to the assertion that we can assume that the matrix entries themselves are algebraic.

6.1.7 Lemma: *Let K be the field of real algebraic numbers. Then for some m, there exists a faithful $\mathbb{R}$-rational representation $\pi{:}G\to GL(m,\mathbb{C})$ such that*

$\pi(\Gamma) \subset GL(m, K)$. Thus, identifying G with $\pi(G)$, we can assume that G is defined over K and $\Gamma \subset G_K$.

Proof: Let $T: G \to \mathbb{C}$ be given by $T(g) = Tr(Ad(g))$. This is a polynomial function, and hence the linear span V of $\{g \cdot T | g \in G\}$ is finite dimensional, where g acts by translation on polynomial functions, i.e., $(g \cdot T)(h) = T(hg)$. Let π be this representation of G on V. Assume for the moment that π is faithful. Since Γ is Zariski dense in G, V is also spanned by $\{\pi(\gamma) T | \gamma \in \Gamma\}$, (for otherwise we would have a Γ-invariant subspace under a rational representation of G which was not G-invariant.) Choose $\gamma_1, \ldots, \gamma_m \in \Gamma$ such that $\{\pi(\gamma_i) T\}$ is a basis for V. It suffices to show that with respect to this basis the matrices of $\pi(\gamma)$ have all entries in K, for all $\gamma \in \Gamma$. Since elements of V are polynomial functions, and $\{\pi(\gamma_i) T\}$ are linearly independent as functions on G, the Zariski density of Γ in G implies that they are linearly independent as functions on Γ. Therefore, we can find $s_1, \ldots, s_m \in \Gamma$ such that the matrix $A = (\pi(\gamma_i) T)(s_j)) = (T(s_j \gamma_i))$ is non-singular. By Lemma 6.1.6, $A \in M(m \times m, K)$. Let $C(\gamma) = (c_{ij}(\gamma))$ be the matrix of $\pi(\gamma)$ with respect to $\{\pi(\gamma_i) T\}$. Thus, $\pi(\gamma)(\pi(\gamma_i) T) = \sum_j c_{ij}(\gamma)(\pi(\gamma_j) T)$. Evaluating this equation at s_k, we obtain $T(s_k \gamma \gamma_i) = \sum_j c_{ij}(\gamma) T(s_k \gamma_j)$. Lemma 6.1.6 implies that $T(s_k \gamma \gamma_i) \in K$, so for some matrix $B \in M(m \times m, K)$, we have $B = C(\gamma) A$. Since $A \in M(m \times m, K)$ as well, and A is invertible, we deduce $C(\gamma)$ has entries in K.

Thus, to verify lemma 6.1.7 it suffices to show that the representation π is faithful. Suppose $g \in \ker(\pi)$. Then $g \cdot T = T$, i.e. $Tr(Ad(hg)) = Tr(Ad(h))$ for all $h \in G$. Let W be the linear span of $\{Ad(h) | h \in G\}$ in the space of all endomorphisms of the linear space $L(G)$. We then have $Tr(M Ad(g)) = Tr(M)$ for all $M \in W$. Since G is semisimple, Ad is a direct sum of irreducible representations. Thus we can write the representation $(Ad, L(G)) \cong \Sigma^{\oplus}(\sigma_{ij}, V_{ij})$ where each σ_{ij} is irreducible, $(\sigma_{ij}, V_{ij}) = (\sigma_{ik}, V_{ik})$, and for $i \neq r$, σ_{ik} and σ_{rk} are inequivalent. Any endomorphism of $\Sigma^{\oplus} V_{ij}$ leaving each V_{ij} invariant is given by a family $M_{ij} \in \mathrm{End}(V_{ij})$. From the irreducibility of σ_{ij} it follows (by Wedderburn's theorem) that under the above identification $L(G) \cong \Sigma^{\oplus} V_{ij}$, we have

$$W = \{M \in \mathrm{End}(\Sigma^{\oplus} V_{ij}) | M(V_{ij}) \subset V_{ij} \text{ all } i, j, \text{ and } M_{ij} = M_{ik} \text{ for all i, j, k}\}.$$

From the fact that $Tr(M Ad(g)) = Tr(M)$ for all $M \in W$, we then deduce that $Tr(S \sigma_{ij}(g)) = Tr(S)$ for all $S \in \mathrm{End}(V_{ij})$. It follows that $\sigma_{ij}(g) = I$ for all i, j, and hence that $Ad(g) = I$. Since G has trivial center, $g = e$ completing the proof.

We will now use the fact that the group Γ is finitely generated. We will prove this in the case in which all simple factors of G have $\mathbb{R}$-rank at least two in chapter 7. (See 7.1.5.)

6.1.8 Lemma: *With G as in the conclusion of 6.1.7, there is a real algebraic number field k* ($[k:\mathbb{Q}] < \infty$) *such that G is defined over k and* $\Gamma \subset G_k$.

Proof: If $\{\gamma_i\}$ is a finite set of generators of Γ, let k be the field generated by the matrix entries of γ_i. Then by 6.1.7, $[k:\mathbb{Q}] < \infty$, and clearly $\Gamma \subset GL(m, k)$. Since Γ is Zariski dense in G, it follows from Proposition 3.1.8 that G is defined over k.

We now complete the proof of the theorem. Let $R_{k/\mathbb{Q}}(G) = \prod_i G^{\sigma_i}$ be as in Proposition 6.1.3. Let $\alpha: G_k \to R_{k/\mathbb{Q}}(G)$ be the map $\alpha(g) = (\sigma_1(g), \ldots, \sigma_d(g))$. Since $\Gamma \subset G_k$, we have $\alpha(\Gamma) \subset [R_{k/\mathbb{Q}}(G)]_{\mathbb{Q}}$. Let H be the Zariski closure of $\alpha(\Gamma)$ in $R_{k/\mathbb{Q}}(G)$, so that H is an algebraic $\mathbb{Q}$-group (again by Proposition 3.1.8). The projection map $p: R_{k/\mathbb{Q}}(G) \to G$ (as in 6.1.3) satisfies $(p \circ \alpha)(\gamma) = \gamma$, and since Γ is Zariski dense in G, $p(H) = G$. Since G is semisimple the radical of H is sent to the identity by p, and since G has trivial center, the same is true for the center of H. Thus, dividing H by its radical and then the center of the resulting group, we deduce the following: There is a semisimple $\mathbb{Q}$-group H with trivial center, a surjective $\mathbb{R}$-map $p: H \to G$, and a homomorphism $\alpha: \Gamma \to H_{\mathbb{Q}}$ such that $p \circ \alpha$ is the identity and $\alpha(\Gamma)$ is Zariski dense in H. Since G is connected, $p(H^0) = G$, and hence by replacing Γ by the finite index subgroup $\Gamma \cap \alpha^{-1}(H^0)$, we can also assume that H is connected.

We now claim that $(\ker p)_{\mathbb{R}}$ is compact. We have an $\mathbb{R}$-isomorphism of algebriac $\mathbb{R}$-groups $H \cong G \times (\ker p)$ in such a way that p corresponds to projection onto G. Let F be an $\mathbb{R}$-simple factor of ker p, so we can write $H \cong G \times F \times F'$ as algebraic $\mathbb{R}$-groups, where F' is the product of the remaining $\mathbb{R}$-simple factors of ker p. Let $q: H \to F$ be projection. Since $\alpha(\Gamma)$ is Zariski dense in H, $(q \circ \alpha)(\Gamma)$ is Zariski dense in F, and since q is an $\mathbb{R}$-map, $(q \circ \alpha)(\Gamma) \subset F_{\mathbb{R}}$. We claim $F_{\mathbb{R}}$ is compact. If not, then by the case $k = \mathbb{R}$ of Margulis' superrigidity theorem (5.1.2), $q \circ \alpha$ extends to a rational homomorphism $h: G \to F$. But then $\{(g, h(g), f') \mid g \in G, f' \in F'\}$ would be a proper algebraic subgroup of H containing $\alpha(\Gamma)$, contradicting Zariski density of $\alpha(\Gamma)$ in H. This shows that $F_{\mathbb{R}}$ is compact for any $\mathbb{R}$-simple factor F of ker p, and hence $(\ker p)_{\mathbb{R}}$ is also compact.

For any prime number a, let $\mathbb{Q}_a$ be the field of a-adic numbers. We have a natural embedding $H_{\mathbb{Q}} \to H_{\mathbb{Q}_a}$, and hence we can also view α as a map $\alpha: \Gamma \to H_{\mathbb{Q}_a}$. By the totally disconnected case of Margulis superrigidity (5.1.2), $\alpha(\Gamma)$ is bounded in $H_{\mathbb{Q}_a}$. This means that the powers of each prime appearing in the denominators of the matrix entries of $\alpha(\gamma) \in H_{\mathbb{Q}}$ are uniformly bounded over $\gamma \in \Gamma$. More formally, if we let $Z_a \subset \mathbb{Q}_a$ be the a-adic integers, then $H_{Z_a} \subset H_{\mathbb{Q}_a}$ is open, and hence boundedness of $\alpha(\Gamma)$ implies that the image of $\alpha(\Gamma)$ in $H_{\mathbb{Q}_a}/H_{Z_a}$ is finite. Γ is finitely generated, and hence there are finitely many primes $a_1, \ldots, a_n$ such that any denominator of any matrix entry of $\alpha(\Gamma) \in H_{\mathbb{Q}}$ is a product of powers of the a_i. In other words, the map

$$\alpha(\Gamma)/\alpha(\Gamma) \cap H_Z \to \prod_{i=1}^{n} H_{\mathbb{Q}_{a_i}}/H_{Z_{a_i}}$$

is injective. Thus, from the finiteness of $\alpha(\Gamma)$ in $H_{\mathbb{Q}_a}/H_{Z_a}$, we deduce that $\alpha(\Gamma) \cap H_Z$ is of finite index in $\alpha(\Gamma)$. Applying p, we deduce that $\Gamma \cap p(H_Z)$ is of finite index in Γ. Since Γ is a lattice in $G_{\mathbb{R}}$, this implies that $\Gamma \cap p(H_Z)$ is also a lattice. By assertion 3 at the beginning of this chapter, $p(H_Z)$ is a lattice in $G_{\mathbb{R}}$ since $(\ker p)_{\mathbb{R}}$ is compact. Since $(\Gamma \cap p(H_Z)) \subset p(H_Z)$ is an inclusion of lattices, this implies that $\Gamma \cap p(H_Z)$ is of finite index in $p(H_Z)$. Thus, Γ and $p(H_Z)$ are commensurable (and hence so are Γ and $p(H_Z \cap H^0_{\mathbb{R}})$). This completes the proof of Theorem 6.1.2.

With notation as in Theorem 6.1.2, and Definition 6.1.1, if Γ is a non-co-compact lattice, the map $p: H^0_{\mathbb{R}} \to G$ can in fact be taken to have not only compact kernel, but trivial kernel, i.e., p can be chosen to be an isomorphism. The proof of this from Theorem 6.1.7 depends on the following compactness criterion, which should be considered as a companion result to Theorem 3.1.7.

6.1.9 Theorem [*Borel–Harish-Chandra* 1], [*Mostow–Tamagawa* 1]: *Let G be a semisimple algebraic $\mathbb{Q}$-group. Then $G_{\mathbb{R}}/G_Z$ (which is of finite volume by* 3.1.7) *is compact if and only if the only unipotent element in $G_{\mathbb{Q}}$ is the identity.*

We then deduce

6.1.10 Corollary [*Margulis* 1]: *If G is a connected semisimple algebraic $\mathbb{R}$-group with trivial center and $\mathbb{R}$-rank$(G) \geqq 2$, and $\Gamma \subset G^0_{\mathbb{R}}$ is an irreducible non-cocompact lattice, then there is a connected semisimple algebraic $\mathbb{Q}$-group H with trivial center and an isomorphism $p: H_{\mathbb{R}} \to G_{\mathbb{R}}$ such that $p(H_Z)$ and Γ are commensurable.*

Proof: By Theorem 6.1.2, we can find a connected semisimple algebraic $\mathbb{Q}$-group H with trivial center and an $\mathbb{R}$-rational surjective homomorphism $p: H \to G$ such that $p(H_Z)$ and Γ are commensurable and $(\ker p)_{\mathbb{R}}$ is compact. If Γ is not cocompact in $G^0_{\mathbb{R}}$, then H_Z is not cocompact in $H_{\mathbb{R}}$. Thus, by

Theorem 6.1.9, $H_{\mathbb{Q}}$ possesses non-trivial unipotent elements. Since ker p is a normal $\mathbb{R}$-subgroup, there is a normal $\mathbb{R}$-subgroup $L \subset H$ such that the product map $L \times \ker p \to H$ is an isomorphism of $\mathbb{R}$-groups. If we let $q: H \to \ker p$ be the projection map then q is $\mathbb{R}$-rational, and hence if $x \in H_{\mathbb{Q}}$ is unipotent, so is $q(x) \in (\ker p)_{\mathbb{R}}$. Since $(\ker p)_{\mathbb{R}}$ is compact, $q(x) = e$, and hence $H_{\mathbb{Q}} \cap L \neq \{e\}$. Since $L \subset H$ is normal, $H_{\mathbb{Q}} \cap L$ is normal in $H_{\mathbb{Q}}$, and since $H_{\mathbb{Q}} \subset H$ is Zariski dense (by 3.1.9) the Zariski closure of $H_{\mathbb{Q}} \cap L$, say L_1, is a normal subgroup of H. Since L_1 contains (by definition) a Zariski dense set of $\mathbb{Q}$-points, L_1 is a $\mathbb{Q}$-group by Proposition 3.1.8. Thus, $(L_1)_Z$ is a lattice in $(L_1)_{\mathbb{R}}$ by 3.1.7, and since $p: L \to G$ is an isomorphism, $p((L_1)_Z)$ is a lattice in $p(L_1)_{\mathbb{R}}$ and in particular is infinite. Since $p(H_Z)$ and Γ are commensurable, this implies $\Gamma \cap p(L_1)_{\mathbb{R}}$ is infinite. Because Γ is an irreducible lattice in $G^0_{\mathbb{R}}$, this implies that $p(L_1) = G$. (This follows from the general fact that if $G = \prod_{i=1}^{n} G_i$ is a product of connected simple Lie groups with trivial center and $\Gamma \subset G$ is an irreducible lattice, then $\Gamma_0 = \Gamma \cap \prod_{i=1}^{m} G_i = \{e\}$ as long as $m < n$. This is so because Γ_0 is normalized by Γ and $\prod_{i>m} G_i$, and hence Γ_0 is normal in G since $\Gamma \cdot \prod_{i>m} G_i$, is dense by the irreducibility assumption. Since G_i are center free and Γ_0 is discrete, this is impossible unless Γ_0 is trivial.) Thus, $L_1 = L$, and hence L is actually defined over $\mathbb{Q}$. Thus L_Z is a lattice in $L_{\mathbb{R}}$, and since $(\ker p)_{\mathbb{R}}$ is compact, L_Z is a lattice in $H_{\mathbb{R}}$. Thus $L_Z \subset H_Z$ is a subgroup of finite index, and $p(L_Z)$ and Γ are commensurable. Thus, replacing H by L, we obtain the required $\mathbb{Q}$-group.

6.2 The commensurability criterion

In this section we present a result of Margulis which gives a necessary and sufficient condition for a lattice to be arithmetic without the assumption that the $\mathbb{R}$-rank of the ambient Lie group is at least two.

6.2.1 Definition: Let G be a locally compact group and $\Gamma \subset G$ a closed subgroup. Let $\mathrm{Comm}_G(\Gamma) = \{g \in G \mid g\Gamma g^{-1}$ and Γ are commensurable$\}$. Then $\mathrm{Comm}_G(\Gamma)$ is called the commensurability subgroup of Γ in G. If G is understood, we shall sometimes denote this simply by $\mathrm{Comm}(\Gamma)$.

We remark that $\mathrm{Comm}(\Gamma)$ is a subgroup of G and $\Gamma \subset \mathrm{Comm}(\Gamma) \subset G$. If $\Gamma, \Gamma' \subset G$ with Γ and Γ' commensurable, then $\mathrm{Comm}(\Gamma) = \mathrm{Comm}(\Gamma')$.

As an example, we have the following.

6.2.2 Proposition [*Borel* 5]: *Let G be a connected semisimple algebraic $\mathbb{Q}$-group with trivial center. Then $G_{\mathbb{Q}} \subset Comm(G_Z)$ and if $G_{\mathbb{R}}$ has no compact factors, then $G_{\mathbb{Q}} = Comm(G_Z)$.*

Proof: Let $g \in G_{\mathbb{Q}}$, and let m be the least common multiple of all denominators of the entries of the matrices g, g^{-1}. Let $\Gamma = \{\gamma \in G_Z \mid \gamma \equiv I \bmod m^2\}$, so that $\Gamma \subset G_Z$ is clearly of finite index. If $\gamma \in \Gamma$, then $\gamma = I + m^2 B$ where B is an integral matrix, and hence $g\gamma g^{-1} = I + m^2 gBg^{-1}$ which is also an integral matrix. Thus $g\Gamma g^{-1} \subset G_Z$. Furthermore, since Γ is of finite index in G_Z, Γ is a lattice in $G_{\mathbb{R}}$ and hence so is $g\Gamma g^{-1}$. Therefore, $g\Gamma g^{-1}$ must be of finite index in G_Z. Since $g\Gamma g^{-1}$ is clearly of finite index in $gG_Z g^{-1}$, it follows that $g \in \text{Comm}(G_Z)$.

To see the converse, let $\mathbb{C}[G]$ and $\mathbb{Q}[G]$ denote as usual the space of regular functions on G and the space of $\mathbb{Q}$-regular functions on G respectively. Let V be the subspace of $\mathbb{C}[G]$ spanned by the matrix coefficient functions, i.e., $g \to g_{ij}$, and $V_{\mathbb{Q}}$ the subspace of $\mathbb{Q}[G]$ spanned by these functions. Then for some m, we can identify $V \cong \mathbb{C}^m$ in such a way that $V_{\mathbb{Q}} \cong \mathbb{Q}^m$. Define a representation of G on $\mathbb{C}[G]$ by $(\pi(g)f)(x) = f(g^{-1}xg)$. Then V is G-invariant, and $\pi: G \to GL(V)$ is a rational representation defined over $\mathbb{Q}$. Since G has trivial center, π is faithful, and thus π is a $\mathbb{Q}$-isomorphism of G with $\pi(G)$. Thus, if $g \in G$, to see that $g \in G_{\mathbb{Q}}$ it suffices to see that $\pi(g)f \in V_{\mathbb{Q}}$ for all $f \in V_{\mathbb{Q}}$.

Suppose that $g \in \text{Comm}(G_Z)$ and $f \in V_{\mathbb{Q}}$. Then we can write $\pi(g)f = f_0 + \sum_{i=1}^{n} c_i f_i$, where $f_i \in V_{\mathbb{Q}}$, $c_i \in \mathbb{C}$ and $(1, c_1, \ldots, c_n)$ are linearly independent over $\mathbb{Q}$. Since $g \in \text{Comm}(G_Z)$, $gG_Z g^{-1} \cap G_Z$ is a lattice in $G_{\mathbb{R}}$, and hence by the Borel density theorem (3.2.5), $gG_Z g^{-1} \cap G_Z$ is Zariski dense in G. If $x \in gG_Z g^{-1} \cap G_Z$, then $f(g^{-1}xg) = f_0(x) + \sum c_i f_i(x)$ where $f(g^{-1}xg)$, $f_0(x) \in \mathbb{Q}$. By linear independence of $\{1, c_i\}$ over $\mathbb{Q}$, we deduce that $f_i(x) = 0$, $f(g^{-1}xg) = f_0(x)$ for all x in a Zariski dense set. Thus, this holds for all $x \in G$, and hence $\pi(g)f \in V_{\mathbb{Q}}$. By the conclusion of the preceding paragraph, $g \in G_{\mathbb{Q}}$, and this completes the proof.

For an irreducible lattice, we have the following dichotomy.

6.2.3 Proposition: *Let G be a connected semisimple Lie group with trivial center and no compact factors, and let $\Gamma \subset G$ be an irreducible lattice. Then either $\Gamma \subset Comm(\Gamma)$ is a subgroup of finite index or $Comm(\Gamma)$ is dense in G (with the Hausdorff topology).*

Proof: Let $H = \overline{\text{Comm}(\Gamma)}$ and H^0 the (topological) connected component of the identity. Since $\Gamma \subset H$, Γ normalizes H^0, and hence $L(H^0)$, the Lie algebra of

H^0, is an invariant subspace of $L(G)$ under $Ad(\Gamma)$. Since Γ is Zariski dense in G, $L(H^0)$ is invariant under $Ad(G)$, and hence G normalizes H^0. If $H^0 = \{e\}$, then H and hence Comm(Γ) are discrete, and since $\Gamma \subset \text{Comm}(\Gamma) \subset G$ and Γ is a lattice in G, $[\text{Comm}(\Gamma)\colon \Gamma] < \infty$. On the other hand, if H^0 is a non-trivial normal subgroup then ΓH^0 is dense in G by irreducibility of Γ, and since $\Gamma H^0 \subset H$, $G = H$, completing the proof.

For $G = SL(n, \mathbb{R})$, and $\Gamma = SL(n, Z)$, it is clear from Proposition 6.2.2 that Comm(Γ) is dense in G with the Hausdorff topology.

The next proposition asserts that this is true in a more general setting.

6.2.4 Proposition: *Suppose G, Γ are as in Proposition* 6.2.3. *If Γ is arithmetic, then Comm(Γ) is dense in G.*

Proof: Let H be a connected semisimple $\mathbb{Q}$-group and $p\colon H^0_{\mathbb{R}} \to G$ a surjective homomorphism with compact kernel such that $p(H_Z \cap H^0_{\mathbb{R}})$ is commensurable with Γ. By 6.2.3 it suffices to show that Comm(Γ) is not discrete. It suffices to see that $p(\text{Comm}_{H_{\mathbb{R}}}(H_Z))$ is not discrete, and hence by 6.2.2 that $p(H_{\mathbb{Q}})$ is not discrete.

To establish this, we call on a result of [Borel 6] described in Chapter 10. Namely, if $a \in Z$ is a prime number, let $Z_{(a)} \subset \mathbb{Q}$ be the subring consisting of all rational numbers whose denominator is a power of a. We have natural non-discrete embeddings of $Z_{(a)}$ into $\mathbb{R}$ and $\mathbb{Q}_a$, where $\mathbb{Q}_a$ is the field of a-adic numbers. However, the image of $Z_{(a)}$ in $\mathbb{R} \times \mathbb{Q}_a$ will be discrete and in fact will be a lattice in $\mathbb{R} \times \mathbb{Q}_a$. [Borel 6] establishes that the analogous result is true for H. Namely, under the natural embedding $\lambda\colon H_{Z_{(a)}} \to H_{\mathbb{R}} \times H_{\mathbb{Q}_a}$ we have that $\lambda(H_{Z_{(a)}})$ is a lattice in $H_{\mathbb{R}} \times H_{\mathbb{Q}_a}$.

Returning to the proof of the proposition, by 3.1.12 we can choose the prime a so that $H_{\mathbb{Q}_a}$ is not compact. Since $H_Z \subset H_{Z_{(a)}} \subset H_{\mathbb{Q}}$ and H_Z is a lattice in $H_{\mathbb{R}}$, to see that $p(H_{\mathbb{Q}})$ is not discrete it suffices (by 6.2.2, 6.2.3) to see that if $K \subset H_{\mathbb{R}}$ is a compact, normal subgroup, then $KH_{Z_{(a)}}/KH_Z$ is not finite. The map

$$K\lambda(H_{Z_{(a)}})/K\lambda(H_Z) \to KH_{Z_{(a)}}/KH_Z$$

induced by projection onto $H_{\mathbb{R}}$ is easily seen to be injective, and hence it suffices to see that $K\lambda(H_{Z_{(a)}})/K\lambda(H_Z)$ is not finite. By the preceding paragraph $H_{\mathbb{R}} \times H_{\mathbb{Q}_a}/K\lambda(H_{Z_{(a)}})$ has a finite $H_{\mathbb{R}} \times H_{\mathbb{Q}_a}$-invariant measure. If $K\lambda(H_{Z_{(a)}})/K\lambda(H_Z)$ were finite, $H_{\mathbb{R}} \times H_{\mathbb{Q}_a}/K\lambda(H_Z)$ would have a finite invariant measure as well. But $\lambda(H_Z) \subset H_{\mathbb{R}} \times H_{Z_a}$ where $Z_a \subset \mathbb{Q}_a$ is the compact subring of a-adic integers. Thus we have an $H_{\mathbb{R}} \times H_{\mathbb{Q}_a}$-map $(H_{\mathbb{R}} \times H_{\mathbb{Q}_a})/K\lambda(H_Z) \to H_{\mathbb{Q}_a}/H_{Z_a}$. Therefore, $H_{\mathbb{Q}_a}/H_{Z_a}$ would have a finite $H_{\mathbb{Q}_a}$-invariant measure. Since H_{Z_a} is compact and $H_{\mathbb{Q}_a}$ is not compact, this is impossible.

The main result of this section is the converse to Proposition 6.2.4.

6.2.5 Theorem [*Margulis* 1]: *Let G be a connected semisimple Lie group with trivial center, no compact factors, and let* $\Gamma \subset G$ *be an irreducible lattice. If Comm*(Γ) *is dense in G then* Γ *is arithmetic.*

Thus, the dichotomy in Proposition 6.2.3 is exactly the dichotomy of non-arithmeticity and arithmeticity. To begin the proof, we first prove a replacement for the superrigidity theorem (5.1.2). The conclusions of the following theorem are exactly those of 5.1.2, but the assumption in 5.1.2 that the $\mathbb{R}$-rank be at least two is replaced by the assumption that the homomorphism of Γ extends to a homomorphism of Comm(Γ).

6.2.6 Theorem [*Margulis* 1]: *Let G be a connected semisimple algebraic* $\mathbb{R}$*-group such that* $G^0_{\mathbb{R}}$ *has no compact factors. Let* $\Gamma \subset G^0_{\mathbb{R}}$ *be an irreducible lattice, and assume that Comm*$_{G^0_{\mathbb{R}}}(\Gamma)$ *is dense in* $G^0_{\mathbb{R}}$. *Suppose k is a local field of characteristic 0, and let H be a connected algebraic k-group, almost simple over k. Let* $\pi : \mathrm{Comm}(\Gamma) \to H_k$ *be a homomorphism with* $\pi(\mathit{Comm}\ \Gamma)$ *Zariski dense. Then:*

(*i*) *If* $k = \mathbb{R}$, $H_{\mathbb{R}}$ *is* $\mathbb{R}$*-simple and non-compact, then* π *extends to an* $\mathbb{R}$*-rational homomorphism* $G \to H$.

(*ii*) *If* $k = \mathbb{C}$ *and H is simple (i.e. center free), then either*

  (*a*) $\overline{\pi(\Gamma)}$ *is compact*

  or (*b*) π *extends to a rational homomorphism* $G \to H$.

(*iii*) *If k is totally disconnected,* $\overline{\pi(\Gamma)}$ *is compact.*

Proof: Lemma 5.1.3 and Proposition 5.1.9 show that it suffices to prove the following two assertions (cf. Steps 1, 1′, 2 in the proof of 5.1.2).

(1) One of the following must be satisfied (and for $k = \mathbb{R}$, it must be (a)):

  (a) There is a proper algebraic k-subgroup $L \subset H$ and a measurable Comm(Γ)-map $\varphi : G^0_{\mathbb{R}}/P_0 \to H_k/L_k$;

  or (b) There is a compact subgroup $S \subset H_k$ and a measurable Γ-map $\varphi : G^0_{\mathbb{R}}/P_0 \to H_k/S$.

(2) In case (1a), for $k = \mathbb{R}$ or $\mathbb{C}$, φ is essentially rational, and for k totally disconnected, φ is essentially constant.

To deduce the theorem from assertions (1), (2), we observe that the proof of Lemma 5.1.3 and Proposition 5.1.9 immediately take care of $k = \mathbb{R}, \mathbb{C}$. For $k \neq \mathbb{R}, \mathbb{C}$, Proposition 5.1.9 shows that it suffices to consider case (1a). Just as in the conclusion of the proof of Theorem 5.1.2, if we had (1a), by (2) φ would be essentially constant, and this would imply that $\pi(\mathrm{Comm}(\Gamma)) \subset M_k$ where M is a proper algebraic k-subgroup of H (in fact a conjugate of L). This contradicts the assumption that $\pi(\mathrm{Comm}(\Gamma))$ is Zariski dense in H. This verifies that it suffices to prove assertions (1) and (2).

Assertion (1) is of course the analogue in the present context of Step 1 (for $k = \mathbb{R}$) and Step 1′ (for $k \neq \mathbb{R}$) of the proof of 5.1.2. We remind the reader that the assumption that $\mathbb{R}$-rank$(G) \geqq 2$ is not used in the proof of Steps 1, 1′ in 5.1.2. Therefore, we still have the assertions of Steps 1, 1′, yielding the existence of suitable Γ-maps. In our present case (1a), we need to obtain a Comm(Γ)-map. Therefore, assertion (1) follows from Steps 1, 1′ of the proof of Theorem 5.1.2 and the following result.

6.2.7 Lemma: *Suppose $\varphi : G^0_{\mathbb{R}}/P_0 \to H_k/L_k$ is a measurable Γ-map where $L \subset H$ is a proper algebraic k-subgroup. Assume further that L is of minimal dimension among all k-subgroups for which such a measurable Γ-map exists. Let $N(L^0)$ be the normalizer of L^0, and $q : H_k/L_k \to H_k/N(L^0)_k$ be the natural projection. Then $q \circ \varphi : G^0_{\mathbb{R}}/P_0 \to H_k/N(L^0)_k$ is a Comm(Γ)-map.*

To prove Lemma 6.2.7, we first show the following.

6.2.8 Lemma: *Assume the notation of* 6.2.7. *Let $\Gamma_0 \subset \Gamma$ be a subgroup of finite index. Suppose $\psi : G^0_{\mathbb{R}}/P_0 \to H_k/L_k$ is a measurable Γ_0-map. Then $q \circ \psi = q \circ \varphi$ (a.e.).*

Proof: Recall the Γ-action on the space of measurable maps $F(G^0_{\mathbb{R}}/P_0, H_k/L_k)$ given by $(\gamma \cdot f)(x) = f(x\gamma)\gamma^{-1}$. Thus $f \in F(G^0_{\mathbb{R}}/P_0, H_k/L_k)$ is (essentially) a Γ-map if and only if f is a fixed point under Γ. Since ψ is fixed under Γ_0, ψ has a finite orbit under the Γ-action, say $\Gamma \cdot \psi = \{\psi, \gamma_1 \cdot \psi, \ldots, \gamma_n \cdot \psi\}$. Let $S(n+1)$ be the symmetric group on $n+1$ letters, which acts naturally on $\prod_{i=0}^{n} H_k/L_k$. Define

$$\lambda : G^0_{\mathbb{R}}/P_0 \to \prod_{i=0}^{n} H_k/L_k \text{ by } \lambda(x) = (\psi(x), (\gamma_1 \cdot \psi)(x), \ldots, (\gamma_n \cdot \psi)(x)),$$

and let $\tilde{\lambda}$ be the composition of λ with the projection

$$\prod_{i=0}^{n} H_k/L_k \to \left(\prod_{i=0}^{n} H_k/L_k\right)/S(n+1).$$

Thus, $\tilde{\lambda}$ is a Γ-map. We therefore also have a Γ-map

$$(\varphi, \tilde{\lambda}): G^0_{\mathbb{R}}/P_0 \to H_k/L_k \times \left(\prod_{i=0}^{n} H_k/L_k\right)/S(n+1).$$

The action of H_k on $H_k/L_k \times \left(\prod_{i=0}^{n} H_k/L_k\right)/S(n+1)$ is smooth. This follows from 3.1.3. (More precisely, smoothness of this H_k action is equivalent to smoothness of the $H_k \times S(n+1)$ action on $H_k/L_k \times \prod_{i=0}^{n} H_k/L_k$. H_k is smooth on this space by 3.1.3, and hence so is $H_k \times S(n+1)$, by Lemma 3.2.14.) Thus, the space of H_k-orbits is countably separated, and $(\varphi, \tilde{\lambda})'$, the composition of $(\varphi, \tilde{\lambda})$ with projection to the space of H_k-orbits will be essentially constant, i.e., there is a single H_k-orbit such that $(\varphi, \tilde{\lambda})(x)$ lies in this orbit for almost all x. We can choose a point in this orbit whose stabilizer will be $L_k \cap M_k$, where M is the subgroup of H leaving some finite subset of H_k/L_k invariant. We can consider $(\varphi, \tilde{\lambda})$ as a Γ-map $(\varphi, \tilde{\lambda}): G^0_{\mathbb{R}}/P_0 \to H_k/L_k \cap M_k$. However, a subgroup of finite index in M_k leaves a finite set pointwise fixed, and so is of the form $\bigcap_{i=0}^{n} a_i L_k a_i^{-1}$ for some $a_i \in H_k$, where $([e], \{[a_0], \ldots, [a_n]\}) \in H_k/L_k \times \left(\prod_{i=0}^{n} H_k/L_k\right)/S(n+1)$ lies in the orbit containing almost all $(\varphi, \tilde{\lambda})(G^0_{\mathbb{R}}/P_0)$. Since $(\varphi, \tilde{\lambda})$ is a Γ-map, the minimality of dimension hypotheses on the group L implies $\dim(M \cap L) = \dim L$, and hence $\dim((\cap a_j L a_j^{-1}) \cap L) = \dim L$. This implies that for each i, $a_i L^0 a_i^{-1} = L^0$, i.e., $a_i \in N(L^0)_k$. Thus the image of $([e], \{[a_0], \ldots, [a_n]\})$ in $H_k/N(L^0)_k \times (\Pi H_k/N(L^0)_k)/S(n+1)$ lies in the diagonal and (since the diagonal is H_k-invariant), so does the entire orbit of $([e], \{[a_0], \ldots, [a_n]\})$. However, this contains almost all $(q(\varphi(x)), (\Pi q)(\tilde{\lambda}(x)))$, and hence for almost all $x \in G^0_{\mathbb{R}}/P_0$, $(q(\varphi(x)), q(\psi(x)))$ lies in the diagonal in $H_k/N(L^0)_k \times H_k/N(L^0)_k$. This completes the proof of Lemma 6.2.8.

Proof of Lemma 6.2.7. The action we defined at the beginning of the proof of Lemma 6.2.8 of course extends to an action of $\mathrm{Comm}(\Gamma)$. We have that φ is fixed under Γ. Let $g \in \mathrm{Comm}(\Gamma)$. Then there is a subgroup $\Gamma_0 \subset \Gamma$ of finite index such that $g^{-1}\Gamma_0 g \subset \Gamma$. It follows that $g \cdot \varphi$ is fixed under Γ_0. By Lemma 6.2.8, $q \circ (g \cdot \varphi) = q \circ \varphi$. But $q \circ (g \cdot \varphi) = g \cdot (q \circ \varphi)$, where the $\mathrm{Comm}(\Gamma)$ action on $F(G^0_{\mathbb{R}}/P_0, H_k/N(L^0)_k)$ is defined similarly. Thus, for all $g \in \mathrm{Comm}(\Gamma)$, $g \cdot (q \circ \varphi) =$

$q \circ \varphi$, i.e., $q \circ \varphi$ is a Comm(Γ)-map. (More precisely $q \circ \varphi$ is essentially a Comm(Γ)-map, and hence it agrees almost everywhere with a Comm(Γ)-map.)

This completes the proof of assertion (1) above, and hence to prove theorem 6.2.6 it suffices to verify assertion (2). The argument will be similar to that of the proof of Step 2 of theorem 5.1.2, although it is in fact a bit simpler.

We lift φ to a map $\varphi : G^0_{\mathbb{R}} \to H_k/L_k$, and φ is a Comm(Γ)-map. As in Lemma 5.1.5, it suffices to show that for some $g \in G^0_{\mathbb{R}}$, $u \to \varphi(ug)$ is an essentially rational function of $u \in U_{\mathbb{R}}$ if $k = \mathbb{R}$, $\mathbb{C}$, or an essentially constant function for $k \neq \mathbb{R}$, $\mathbb{C}$. For $g \in G^0_{\mathbb{R}}$, let $\varphi_g : G^0_{\mathbb{R}} \to H_k/L_k$ be given by $\varphi_g(a) = \varphi(ag)$. Then the proof of Lemma 5.1.8 shows that it suffices to show that for almost all $g \in G^0_{\mathbb{R}}$ there exist:

(a) a k-subvariety $W_g \subset H/L$ such that $\varphi_g(a) \in W_g$ for almost all $a \in G^0_{\mathbb{R}}$;
(b) an algebraic k-group Q_g which acts k-regularly on W_g:
(c) a measurable homomorphism $h_g : G^0_{\mathbb{R}} \to (Q_g)_k$;
(d) a point $x_g \in W_g \cap H_k/L_k$ such that for almost all $a \in G^0_{\mathbb{R}}$, $\varphi_g(a) = x_g \cdot h_g(a)$.

To see the existence of (a)–(d), it suffices by Proposition 3.5.2 to see that for almost all $g \in G^0_{\mathbb{R}}$, φ_g all lie in a single H_k-orbit in $F(G^0_{\mathbb{R}}, H_k/L_k)$. Let $\Phi : G^0_{\mathbb{R}} \to F(G^0_{\mathbb{R}}, H_k/L_k)$ be $\Phi(g) = \varphi_g$. If $h \in$ Comm(Γ), then for each c and almost all g, we have $\varphi_{gh}(c) = \varphi(cgh) = \varphi(cg) \cdot \pi(h) = \varphi_g(c)\pi(h)$. By Fubini, for each h and almost all g, this is true for almost all c. Thus, for each $h \in$ Comm(Γ), we have $\Phi(gh) = \Phi(g) \cdot \pi(h)$ for almost all $g \in G^0_{\mathbb{R}}$. Now Comm(Γ) acts ergodically on $G^0_{\mathbb{R}}$ by Lemma 2.2.13, and the action of H_k on $F(G^0_{\mathbb{R}}, H_k/L_k)$ is smooth by Proposition 3.3.1. The map $\overline{\Phi} : G^0_{\mathbb{R}} \to F(G^0_{\mathbb{R}}, H_k/L_k)/H_k$ is therefore an essentially Comm(Γ)-invariant map from an ergodic Comm(Γ)-space into a countably separated space, and therefore it is essentially constant. In other words, there is single H_k-orbit in $F(G^0_{\mathbb{R}}, H_k/L_k)$ such that $\Phi(g)$ lies in this orbit for almost all g. By our observations above, this completes the proof of Theorem 6.2.6.

The same proof shows the following.

6.2.9 Corollary: *With notation as in Theorem* 6.2.6, *assume* $\tilde{\Gamma} \subset Comm_{G^0_{\mathbb{R}}}(\Gamma)$ *is a subgroup of finite index with* $\Gamma \subset \tilde{\Gamma}$. *Assume* $\pi : \tilde{\Gamma} \to H_k$ *is a homomorphism with* $\pi(\tilde{\Gamma})$ *Zariski dense. Then the conclusion of Theorem* 6.2.6 *is still true.*

Proof: The proof of Theorem 6.2.6 will apply verbatim with Comm(Γ) replaced by $\tilde{\Gamma}$ once we know $\tilde{\Gamma}$ is dense in $G^0_{\mathbb{R}}$. But if A is the closure of $\tilde{\Gamma}$ in $G^0_{\mathbb{R}}$, since $\mathrm{Comm}(\Gamma)/\tilde{\Gamma}$ is finite, finitely many cosets of A will contain $\mathrm{Comm}(\Gamma)$ and hence $G^0_{\mathbb{R}}$, since $\mathrm{Comm}(\Gamma)$ is dense in $G^0_{\mathbb{R}}$. Since $G^0_{\mathbb{R}}$ is connected, $A = G^0_{\mathbb{R}}$.

We now return to the proof of the arithmeticity criterion.

Proof of Theorem 6.2.5. Once again, we change our notation from the statement of the theorem, and assume Γ is an irreducible lattice in $G^0_{\mathbb{R}}$ where G is a connected semisimple algebraic $\mathbb{Q}$-group with trivial center and $G^0_{\mathbb{R}}$ has no compact factors. In the proof of Theorem 6.1.2 (arithmeticity for irreducible lattices in $\mathbb{R}$-rank at least 2), the superrigidity theorem (5.1.2) was applied only to certain homomorphisms of Γ. The proof of Theorem 6.2.5 will be essentially the same as that of 6.1.2, except that we must verify that the homomorphisms of Γ in question actually extend to $\mathrm{Comm}(\Gamma)$ in order to be able to apply our substitute for superrigidity, namely Theorem 6.2.6.

If σ is an automorphism of $\mathbb{C}$, then we obtain an induced map $\sigma: G \to G$ (since G is defined over $\mathbb{Q}$). The homomorphism $\sigma|\Gamma$ obviously extends to $\mathrm{Comm}(\Gamma)$, and hence the proof of Lemma 6.1.6 shows that $Tr(Ad(\gamma))$ is a real algebraic number for all $\gamma \in \Gamma$. Lemmas 6.1.7 and 6.1.8 continue to be valid, so we may assume G is a k-group, $\Gamma \subset G_k$, where k is a real algebraic number field ($[k:\mathbb{Q}] < \infty$).

We now claim that $\Gamma \subset G_k$ implies that $\mathrm{Comm}(\Gamma) \subset G_k$. Let $g \in \mathrm{Comm}(\Gamma)$. Then there is a subgroup $\Gamma_0 \subset \Gamma$ of finite index with $g\Gamma_0 g^{-1} \subset \Gamma$. Since Γ_0 is also a lattice in $G^0_{\mathbb{R}}$, Γ_0 is Zariski dense in G, and hence (Proposition 3.1.10) the inner automorphism $\mathrm{Int}(g): G \to G$ is defined over k. Thus $Ad(g): L(G) \to L(G)$ is defined over k, i.e., $Ad(g) \in Ad(G)_k$. However $Ad: G \to Ad(G)$ is an isomorphism defined over k, hence $g \in G_k$.

It follows that the homomorphism $\alpha: \Gamma \to (R_{k/\mathbb{Q}}(G))_{\mathbb{Q}}$, $\alpha(\gamma) = (\sigma_1(\gamma), \ldots, \sigma_d(\gamma))$, as in the proof of 6.1.2, extends to a homomorphism $\alpha: \mathrm{Comm}(\Gamma) \to (R_{k/\mathbb{Q}}(G))_{\mathbb{Q}}$. Thus, letting H be the Zariski closure of $\alpha(\mathrm{Comm}(\Gamma))$, and dividing by the radical and then the center as in 6.1.2, we obtain a semisimple algebraic $\mathbb{Q}$-group H, an $\mathbb{R}$-rational epimorphism $p: H \to G$ and a homomorphism $\alpha: \mathrm{Comm}(\Gamma) \to H_{\mathbb{Q}}$ such that $\alpha(\mathrm{Comm}\ \Gamma)$ is Zariski dense and $p \circ \alpha =$ identity. By replacing Γ by a subgroup of finite index and $\mathrm{Comm}(\Gamma)$ by $\tilde{\Gamma}$, a subgroup of $\mathrm{Comm}(\Gamma)$ of finite index, $\Gamma \subset \tilde{\Gamma} \subset \mathrm{Comm}(\Gamma)$, we can assume H is connected. Using Corollary 6.2.9 of (the proof of) Theorem 6.2.6, we can complete the proof of Theorem 6.2.5 exactly as in the completion of the proof of Theorem 6.1.2.

7 Kazhdan's Property (T)

7.1 Kazhdan's property and some consequences

Let G be a locally compact (second countable) group and $\pi: G \to U(\mathscr{H})$ a unitary representation of G on the (separable) Hilbert space $\mathscr{H}$. Of course, $\mathscr{H}$ may or may not contain non-trivial vectors invariant under $\pi(G)$. We wish to define a notion of π almost containing invariant vectors.

7.1.1 Definition: Let $\varepsilon > 0$ and $K \subset G$ a compact subset. A unit vector $v \in \mathscr{H}$ is called (ε, K)-invariant if $\| \pi(g)v - v \| < \varepsilon$ for all $g \in K$. We say that π almost has invariant vectors if for all (ε, K), there exists an (ε, K)-invariant unit vector. (The vector may depend on (ε, K).)

7.1.2 Example: Let π be the regular representation of $\mathbb{R}$ on $L^2(\mathbb{R})$. Given $\varepsilon > 0$ and an interval $[a, b] \subset \mathbb{R}$, by choosing $[c, d] \subset \mathbb{R}$ to be a large enough interval, and letting $f_{[c,d]} \in L^2(\mathbb{R})$ be the normalized characteristic function of $[c, d]$, we will clearly have $\| \pi(t) f_{[c,d]} - f_{[c,d]} \| < \varepsilon$ for all $t \in [a, b]$. Thus, π almost has invariant vectors. However, π does not have (non-trivial) invariant vectors. The same argument shows that this is true for the regular representation of Z^n or $\mathbb{R}^n$.

A basic discovery of [Kazhdan 1] is that many semisimple groups and their lattice subgroups do not have representations with this type of behavior.

7.1.3 Definition: G has the Kazhdan property (or "property (T)") if any unitary representation of G which almost has invariant vectors actually has non-trivial invariant vectors.

Thus, Example 7.1.2 shows that $\mathbb{R}^n$ and Z^n do not have the Kazhdan property. However:

7.1.4 Theorem [*Kazhdan* 1]: *Let G be a connected semisimple Lie group with finite center each of whose factors has $\mathbb{R}$-rank at least 2. Then G, as well as any lattice subgroup of G, has the Kazhdan property.*

Actually, Kazhdan assumed that all simple factors had $\mathbb{R}$-rank at least 3.

The fact that $\mathbb{R}$-rank at least 2 is sufficient was observed independently by [Delaroche-Kirillov 1], [Vaserstein 1], and [Wang 1]. The proof of Theorem 7.1.4 will be given in section 7.4. Kazhdan's property has proved useful in a large variety of situations, only a few of which we will be considering. A basic consequence is the following.

7.1.5 Theorem [*Kazhdan* 1]: *Suppose* Γ *is a discrete group with the Kazhdan property. Then* Γ *is finitely generated.*

Proof: Let $\Gamma = \{\gamma_i\}$ be a listing of the elements of Γ, and Γ_n be the subgroup generated by $\{\gamma_1, \ldots, \gamma_n\}$. Let π_n be the representation of Γ on $L^2(\Gamma/\Gamma_n)$ given by translation, i.e., the representation induced from the identity representation of Γ_n. Define $\pi = \Sigma^{\oplus}\pi_n$. The representation π_n clearly contains a unit vector invariant under $\{\gamma_1, \ldots, \gamma_n\}$, namely $\chi_{\{[e]\}}$. Thus, π almost has invariant vectors. Since Γ has the Kazhdan property, π has non-trivial invariant vectors. The projection of an invariant vector in $\Sigma^{\oplus}L^2(\Gamma/\Gamma_n)$ onto each $L^2(\Gamma/\Gamma_n)$ will be invariant, and for some n, will be non-trivial. Hence for some n, π_n has non-trivial invariant vectors, which clearly implies Γ/Γ_n is finite. Since Γ_n is finitely generated, it follows that Γ is as well.

Another useful property is the following

7.1.6 Proposition: *If* $\varphi: G \to H$ *is a (continuous) homomorphism with* $\varphi(G)$ *dense in* H *and* G *has the Kazhdan property, so does* H.

Proof: If π is a representation of H which almost has invariant vectors, then $\pi \circ \varphi$ almost has invariant vectors. If v is $(\pi \circ \varphi)(G)$ invariant, it is also $\pi(H)$-invariant.

7.1.7 Corollary: *If* G *has the Kazhdan property, then every homomorphism into* $\mathbb{R}^n$ *is trivial.*

This follows immediately from 7.1.6 and the sentence preceding Theorem 7.1.4. In fact, Corollary 7.1.7 can be generalized considerably by using the following theorem.

7.1.8 Theorem [*Hulanicki* 1]: *If* G *is a locally compact (second countable) group, then* G *is amenable if and only if the regular representation of* G *on* $L^2(G)$ *almost has invariant vectors.*

The proof of this theorem will be given in section 7.2. We observe that the fact that amenability of G implies that the regular representation almost has invariant vectors is a generalization of Example 7.1.2. As a consequence of 7.1.8, we have:

7.1.9 Corollary: *If G is amenable, then G has the Kazhdan property if and only if G is compact.*

Proof: If G is amenable and has the Kazhdan property, then by Theorem 7.1.8 (and Definition 7.1.3), the regular representation has non-trivial invariant vectors. This implies that the Haar measure on G is finite, and hence G is compact.

Conversely, suppose G is compact and π almost has invariant vectors. If v is any unit vector, then $\tilde{v} = \int_G (\pi(g)v)dg$ will be a G-invariant vector. Further, if $\| \pi(g)v - v \| < \frac{1}{2}$ for all $g \in G$, then $\| \tilde{v} - v \| < \frac{1}{2}$, and $\tilde{v} \neq 0$. Since π almost has invariant vectors and G is compact, we can choose such a vector v.

We can then generalize Corollary 7.1.7 as follows:

7.1.10 Corollary: *Suppose G has the Kazhdan property and that H is amenable. Let $\varphi : G \to H$ be a homomorphism. Then $\overline{\varphi(G)}$ is compact.*

Proof: $\overline{\varphi(G)}$ has the Kazhdan property by 7.1.6, is amenable by 4.1.6, and hence is compact by 7.1.9.

7.1.11 Corollary: *Suppose Γ is a discrete group with the Kazhdan property. Then $\Gamma/[\Gamma, \Gamma]$ is finite.*

This follows from 7.1.10 and amenability of abelian groups (4.1.2). As another particular case of 7.1.10, we have (using 7.1.4):

7.1.12 Corollary [*Kazhdan* 1]: *Let G be a connected semisimple Lie group of finite center such that the $\mathbb{R}$-rank of every simple factor of G is at least two. Let $\Gamma \subset G$ be a lattice and $N \subset \Gamma$ a normal subgroup such that Γ/N is amenable. Then Γ/N is finite.*

[Margulis 7] has shown that this last result is true if we only assume $\mathbb{R}\text{-rank}(G) \geqq 2$ provided that Γ is an irreducible. More precisely:

7.1.13 Theorem [*Margulis* 7]: *Let G be a connected semisimple Lie group with no compact factors, finite center, and $\mathbb{R}$-rank$(G) \geqq 2$. Let $\Gamma \subset G$ be an irreducible lattice and $N \subset \Gamma$ a normal infinite subgroup. Then Γ/N has the Kazhdan property, and in particular if Γ/N is amenable then Γ/N is finite.*

We will not be proving 7.1.13 here, and instead refer the reader to Margulis' paper.

The remainder of this chapter is largely devoted to proving the assertions in this section which we have not yet verified: Theorems 7.1.4 (in section 7.4) and 7.1.8 (in section 7.2).

7.2 Amenability and unitary representations

In this section our aim is to prove that G is amenable if and only if the regular representation of G almost has invariant vectors. To this end, it will be convenient to introduce the notion of a mean on a function space.

7.2.1 Definition: Let (X, μ) be a standard measure space and $F \subset L^\infty(X)$ a (norm) closed subspace containing the constant functions, which is also closed under complex conjugation. A mean on F is a (norm) continuous linear functional $m: F \to \mathbb{C}$ such that

(a) $m(1) = 1$, where $\mathbb{C}$ is identified with the constant functions and hence with a subspace of $L^\infty(X)$.
(b) $f \geqq 0$ implies $m(f) \geqq 0$.
(c) $m(\bar{f}) = \overline{m(f)}$, or equivalently $m(Re\, f) = Re(m(f))$.

7.2.2 Example: Let $P(X) = \{\varphi \in L^1(X) \mid \varphi \geqq 0, \|\varphi\|_1 = 1\}$. Then $m_\varphi(f) = \int f\varphi\, d\mu$ defines a mean m_φ on $L^\infty(X)$, whenever $\varphi \in P(X)$.

7.2.3 Proposition: (*a*) *$\mathcal{M}_F$, the set of means on F, is a closed convex subspace of the unit ball in F^*, where the latter has the weak-*-topology.*
(*b*) *$\{m_\varphi | \varphi \in P(X)\}$ is weak-* dense in $\mathcal{M}_F$.*

Proof: (a) is clear. To see (b), suppose the contrary. Then by Hahn-Banach, for some $m \in \mathcal{M}_F$ there exists $\varepsilon > 0$ and $f \in F$ (recall that $F \cong (F^*)^*$, where F^* has the weak-*-topology), such that $Re(m(f)) \geqq Re(m_\varphi(f)) + \varepsilon$ for all $\varphi \in P(X)$. But this implies $m(Re\, f) > \sup_{\varphi \in P(X)} \int (Re\, f)\varphi = \text{ess sup}\,(\text{Re}(f))$. But this contradicts (b) in Definition 7.2.1.

If (X, μ) is a G-space, then G acts on $L^\infty(X)$ by isometries and hence, if $F \subset L^\infty(X)$ is a G-invariant subspace, G acts on $\mathcal{M}_F$. However, while the action of G is continuous in G if $L^\infty(X)$ has the weak-*-topology as the dual of $L^1(X)$, the action is not in general continuous on $L^\infty(G)$ if the latter has the norm topology, and hence we will not have continuity in general for the action on $\mathcal{M}_F$. (Of course, if G is discrete, this is not an issue.) In any case, a fixed point in $\mathcal{M}_F$ will be called a G-invariant mean on F. As an example of the usefulness of this notion, we have one direction of the equivalence we want.

7.2.4 Theorem: *Suppose the regular representation of G almost has invariant vectors. Then G is amenable.*

Proof: We first claim that there is a G-invariant mean on $L^\infty(G)$. Let π be the regular representation. By hypothesis, there is a sequence of unit vectors ψ_j in $L^2(G)$ such that as $j \to \infty$, $\|\pi(g)\psi_j - \psi_j\|_2 \to 0$ uniformly on compact subsets of G. (This is essentially just a reformulation of Definition 7.1.1.) Let $\varphi_j = |\psi_j|^2$, so that $\varphi_j \in P(G)$. An elementary argument then shows that $\|\pi(g)\varphi_j - \varphi_j\|_1 \to 0$. (Namely, $\|\pi(g)\psi_j - \psi_j\|_2 \to 0$ is equivalent to $Re(\int (\pi(g)\psi_j)\overline{\psi_j}) \to 1$. Since $|\int (\pi(g)\psi_j)\overline{\psi_j}| \leqq \int |\pi(g)\psi_j||\psi_j| \leqq 1$ by the Schwartz inequality, it follows that $\int |\pi(g)\psi_j||\psi_j| \to 1$ also. Thus $\|\pi(g)|\psi_j| - |\psi_j|\|_2 \to 0$. Hence

$$\begin{aligned}\|\pi(g)\varphi_j - \varphi_j\|_1 &= \int (|\pi(g)\psi_j|^2 - |\psi_j|^2)\\ &\leqq \int (|\pi(g)|\psi_j| - |\psi_j||)(|\pi(g)|\psi_j| + |\psi_j||)\\ &\leqq \|\pi(g)|\psi_j| - |\psi_j|\|_2 \, \|\pi(g)|\psi_j| + |\psi_j|\|_2\end{aligned}$$

We have seen that the first term approaches 0, and the second is bounded by 2.) By Proposition 7.2.3(a), $\mathcal{M}_{L^\infty(G)}$ is compact, and hence there is an accumulation point $m \in \mathcal{M}_{L^\infty(G)}$ of $\{m_{\varphi_j}\}$. We claim that m is G-invariant. For any $f \in L^\infty(G)$, by passing to a subsequence, we can assume $m_{\varphi_j}(f) \to m(f)$. Therefore, for any $g \in G$,

$$\begin{aligned}m(\pi(g)f - f) &= \lim m_{\varphi_j}(\pi(g)f - f)\\ &= \lim \int f(xg)\varphi_j(x)dx - \int f\varphi_j\\ &= \lim \int f(x)\varphi_j(xg^{-1})dx - \int f\varphi_j\\ &= \lim \int f[\pi(g^{-1})\varphi_j - \varphi_j].\end{aligned}$$

Thus, $|m(\pi(g)f - f)| \leqq \|f\|_\infty \overline{\lim_j} \|\pi(g^{-1})\varphi_j - \varphi_j\|_1 = 0$. Hence, m is a G-invariant mean.

To show amenability of G, let G act continuously on a compact metric space

X. Fix a point $x \in X$. For $f \in C(X)$, let $\tilde{f} \in L^\infty(G)$ be given by $\tilde{f}(g) = f(xg)$. Define $\mu(f) = m(\tilde{f})$. Since m is a G-invariant mean, it follows that $\mu \in C(X)^*$ is a G-invariant positive linear functional, $\mu(1) = 1$, and $\mu(\overline{f}) = \overline{\mu(f)}$. Thus, μ is identified with a G-invariant probability measure on X, verifying amenability of G.

To prove the converse to Theorem 7.2.4 (i.e. Theorem 7.1.8), we would simply like to reverse the steps of the above argument. This can be done in a straightforward fashion if G is discrete, but for a general locally compact group the argument will require some modification, although the basic idea is the same. We shall first present the proof for G discrete.

Proof of Theorem 7.1.8 for G discrete: Suppose G is discrete and amenable. We first claim that there is a G-invariant mean on $L^\infty(G)$. Since $\mathcal{M}_{L^\infty(G)}$ is a compact convex G-invariant subset of $(L^\infty(G))^*$, Proposition 4.1.4 would guarantee the existence of a G-fixed point in $\mathcal{M}_{L^\infty(G)}$ if $L^\infty(G)$ was separable in norm (which it is not in general). However, since G is countable, we can write $L^\infty(G) = \overline{\cup F_\alpha}$ where $F_\alpha \subset L^\infty(G)$ is a G-invariant separable closed subspace, closed under complex conjugation. For each α, we have the restriction map $(L^\infty(G))^* \to (F_\alpha)^*$, which yields a map $r_\alpha : \mathcal{M}_{L^\infty(G)} \to \mathcal{M}_{F_\alpha}$. By Proposition 4.1.4, there are fixed points in any $\mathcal{M}_{F_\alpha}$, as well as any $\mathcal{M}_{F_{\alpha_1}+\cdots+F_{\alpha_n}}$. Let $I_\alpha = \{m \in \mathcal{M}_{L^\infty(G)} | r_\alpha(m) \text{ is } G\text{-invariant}\}$. Since $\mathcal{M}_{F_{\alpha_1}+\cdots+F_{\alpha_n}}$ has fixed points, $\{I_\alpha\}$ is a family of compact subsets of $\mathcal{M}_{L^\infty(G)}$, any finite subcollection having a nonempty intersection. Since $\mathcal{M}_{L^\infty(G)}$ is compact, there exists a point $m \in \cap I_\alpha$, and since $\cup F_\alpha$ is dense in $L^\infty(G)$, m is a G-invariant mean on $L^\infty(G)$.

By Proposition 7.2.3(b), there is a net $\varphi_j \in P(X)$ such that $m_{\varphi_j} \to m$ in the weak-*-topology. We claim that G-invariance of m implies that for any fixed $g \in G$, $(\pi(g)\varphi_j - \varphi_j) \to 0$ weakly in $L^1(G)$, or in other words, for any $f \in L^\infty(G)$, $\int(\pi(g)\varphi_j - \varphi_j)f \to 0$. We have

$$\begin{aligned}\int(\pi(g)\varphi_j - \varphi_j)f &= \int(\pi(g)\varphi_j)f - \int\varphi_j f = \int\varphi_j[\pi(g^{-1})f] - \int\varphi_j f \\ &= m_{\varphi_j}(\pi(g^{-1})f - f).\end{aligned}$$

Thus, in the limit, we have $\int(\pi(g)\varphi_j - \varphi_j)f \to m(\pi(g^{-1})f - f) = 0$, since m is G-invariant.

The next step is to pass from weak convergence in $L^1(G)$ to strong convergence. We shall make use of a consequence of the Hahn-Banach theorem, namely that in a locally convex topological vector space, the weak and strong closure of a convex set are identical. Let $E = \prod_{g \in G} L^1(G)$ with the product topology, where each $L^1(G)$ has the norm topology. Then the weak topology on E is the product topology for the weak topology on $L^1(G)$. Define a linear map $T: L^1(G) \to E$ by $(T\varphi)_g = \pi(g)\varphi - \varphi$. The assertion of the preceding paragraph, that $(\pi(g)\varphi_j - \varphi_j) \to 0$ weakly for all g, implies that 0 is in the weak closure of $T(P(G))$. Since $T(P(G))$ is clearly convex, it follows that 0 is in the strong closure of $T(P(G))$, or in other words, there is a net $\psi_j \in P(G)$ such that $\|\pi(g)\psi_j - \psi_j\|_1 \to 0$ for each $g \in G$. (This argument is due to [Namioka 1].)

Finally, we wish to pass from L^1 to L^2. Suppose that for $g \in G$, and $\varepsilon > 0$, there is $\psi \in P(G) \subset L^1(G)$ such that $\|\pi(g)\psi - \psi\|_1 < \varepsilon$. Let $\varphi = \psi^{1/2}$. Then φ is a unit vector in $L^2(G)$, and $\int |\varphi(xg) - \varphi(x)|^2 \leqq \int |\psi(xh) - \psi(x)|$. (This last assertion just follows from the fact that for a, $b \geqq 0$, $|a - b|^2 \leqq |a^2 - b^2|$.) Thus $\|\pi(g)\varphi - \varphi\|_2 \leqq \varepsilon^{1/2}$. From this and the preceding paragraph it follows that the regular representation almost has invariant vectors. This completes the proof of 7.1.8 for G discrete.

We now turn to the proof of 7.1.8 for general locally compact (separable) G. If we examine the proof for G discrete, we see two problems in extending it to the continuous case. The first problem is that it is not immediately clear that there is a G-invariant mean on $L^\infty(G)$, because the action of G on $L^\infty(G)$ is not continuous (in G), and hence the argument in the discrete case will not apply directly. The second problem is that to establish that the regular representation almost has invariant vectors, we need to control $\|\pi(g)\varphi - \varphi\|_2$ uniformly for g in a compact set. In the discrete case, it therefore sufficed to do this pointwise in G. Thus, this part of the argument also requires modification.

It will be convenient now to switch to the left action of G on itself. Thus, regular representation will now mean left regular representation (i.e., $(\pi(g)f)(x) = f(g^{-1}x)$) with respect to left Haar measure, and similarly the action on spaces of means will be derived from the left translation on spaces of functions. We recall that $f: G \to \mathbb{C}$ is called (left) uniformly continuous if for all $\varepsilon > 0$, there is a neighborhood U of $e \in G$ such that $|f(yx) - f(x)| < \varepsilon$ for all $x \in G$, $y \in U$. Let $UCB(G)$ denote the space of (left) uniformly continuous bounded functions on G. Then we have $UCB(G) \subset L^\infty(G)$, and the action of G on $UCB(G)$ is continuous. $UCB(G)$ is not separable, but arguing as in the first paragraph of the proof of 7.1.8 for G discrete we see, using separability of G, the following:

7.2.5 Lemma: *If G is amenable, then there is a G-invariant mean on $UCB(G)$.*

We now wish to pass from a mean on $UCB(G)$ to one on $L^\infty(G)$. We shall do this by smoothing an L^∞-function to obtain a uniformly continuous one. If φ, f are measurable functions on G, we recall that $\varphi * f$ is defined to be the function $(\varphi * g)(x) = \int \varphi(g) f(g^{-1}x) dg$, as long as this exists for almost all $x \in G$. We recall some basic facts concerning convolution.

7.2.6 Proposition: *(a) If $f \in L^\infty(G)$, $\varphi \in P(G)$, then $\varphi * f \in UCB(G)$.*
(b) If $\varphi_n, \varphi \in L^1(G)$ and $\varphi_n \to \varphi$ in norm, then $\varphi_n * f \to \varphi * f$ in $L^\infty(G)$ for all $f \in L^\infty(G)$.
(c) For any G, there exists an approximate identity in $L^1(G)$, i.e. a sequence $e_n \in P(G)$ such that for all $\varphi \in L^1(G)$, $e_n * \varphi \to \varphi$ and $\varphi * e_n \to \varphi$ in L^1.
(d) For $\varphi \in L^1(G)$, $f \in L^\infty(G)$, and $g \in G$, $\varphi * (\pi(g)f) = [\Delta(g^{-1})(\rho(g^{-1})\varphi)] * f$, where Δ is the modular function of G, ρ is the representation of G on functions given by $(\rho(g)\varphi)(y) = \varphi(yg)$ (and π, is as above, the left regular representation).
(e) $(\pi(g)\varphi) * f = \pi(g)(\varphi * f)$.

If G is discrete, $\varphi \in P(G)$ and $f \in L^\infty(G)$, then $\varphi * f = \sum_{g \in G} \varphi(g)(\pi(g)f)$. Since $\varphi(g) \geqq 0$, $\sum_{g \in G} \varphi(g) = 1$, it follows that for an invariant mean m, we also have $m(\varphi * f) = m(f)$. We shall need this property for arbitrary G, for the mean in Lemma 7.2.5.

7.2.7 Lemma: *If m is a G-invariant mean on $UCB(G)$, then for $\varphi \in P(G)$, $f \in UCB(G)$, we have $m(\varphi * f) = m(f)$.*

Proof: [*Greenleaf* 1] The result will follow from our discussion of the barycenter construction following the statement of Proposition 4.1.4 and a formal interpretation of the averaging effect of convolution as a barycenter construction. Namely, we first remark that it suffices to prove the result if $\mathrm{supp}(\varphi)$ is compact. Fix such a φ and fix $f \in UCB(G)$. Define a map $F: G \to UCB(G)$ by $F(g) = \pi(g)f$. Since f is uniformly continuous, F is continuous. Let μ be the Haar measure on G and $d\nu = \varphi d\mu$, so that ν is a probability measure on G with compact support. Let $K = \mathrm{supp}(\nu)(= \mathrm{supp}(\varphi))$. Then $F(K) \subset UCB(G)$ is norm compact, and hence so is its convex hull. The measure $F_*(\nu)$ is thus a probability measure supported on a (separable) compact convex set in $UCB(G)$. Thus, we can let $h = b(F_*(\nu))$ be the barycenter, $h \in UCB(G)$. We recall that h is characterized by the condition that for all $\lambda \in UCB(G)^*$, $\lambda(h) = \int \lambda(y) d(F_*\nu)(y)$. We claim that in fact $h = \varphi * f$. To see this, for $x \in G$, let $\lambda_x \in UCB(G)^*$ be evaluation at x. Then

$$\begin{aligned} h(x) = \lambda_x(h) &= \int \lambda_x(y) d(F_*\nu)(y) \\ &= \int \lambda_x(F(g)) d\nu(g) \\ &= \int f(g^{-1}x)\varphi(g) d\mu(g) \\ &= (\varphi * f)(x). \end{aligned}$$

To see that $m(\varphi * f) = m(f)$, we can now simply observe that since $m \in UCB(G)^*$,

$$\begin{aligned} m(h) = \int m(y) d(F_* \nu)(y) &= \int m(F(g)) d\nu(g) \\ &= \int m(f) d\nu(g) \qquad \text{by } G\text{-invariance of } m \\ &= m(f). \end{aligned}$$

We can now obtain an invariant mean on $L^\infty(G)$.

7.2.8 Lemma: *If G is amenable, then there exists a G-invariant mean m on $L^\infty(G)$ which satisfies the further condition that $m(\varphi * f) = m(f)$ for all $\varphi \in P(G), f \in L^\infty(G)$.*

Proof: Let m be an invariant mean on $UCB(G)$ (Lemma 7.2.5). Fix any $\psi \in P(G)$ and for $f \in L^\infty(G)$, define $\tilde{m}(f) = m(\psi * f)$. (This is defined by 7.2.6a.) Since m is a mean, so is $\tilde{m}$. We claim that if $\varphi \in P(G)$, $\tilde{m}(\varphi * f) = \tilde{m}(f)$. Namely, if $\{e_n\}$ is an approximate identity (7.2.6c),

$$\begin{aligned} \tilde{m}(\varphi * f) &= \lim \tilde{m}(\varphi * e_n * f) \\ &= \lim m(\psi * \varphi * e_n * f) \\ &= \lim m(e_n * f) \quad \text{by Lemma 7.2.7 and the fact that } e_n * f \in UCB(G) \text{ by 7.2.6a.} \\ &= \lim m(\psi * e_n * f) \quad \text{for the same reason} \\ &= m(\psi * f) \\ &= \tilde{m}(f). \end{aligned}$$

To see G-invariance of $\tilde{m}$, let $g \in G, f \in L^\infty(G)$. Fix $\varphi \in P(G)$. Then $\tilde{m}(\pi(g)f) = \tilde{m}(\varphi * \pi(g)f)$ by the preceding paragraph. By 7.2.6d, $\varphi * \pi(g)f = \alpha * f$ where $\alpha(x) = \Delta(x)^{-1}\varphi(xg^{-1})$. However, $\alpha \in P(G)$, and hence, $\tilde{m}(\pi(g)f) = \tilde{m}(\alpha * f) = \tilde{m}(f)$ by the preceding paragraph, completing the proof.

With these preparations, we now proceed to the proof of Theorem 7.1.8 in general.

Proof of Theorem 7.1.8: In light of Theorem 7.2.4, it suffices to show that if G is amenable, then the regular representation almost has invariant vectors. Choose a G-invariant mean m as in Lemma 7.2.8. By Proposition 7.2.3(b), there is a net $\varphi_j \in P(G)$ such that $m_{\varphi_j} \to m$ in the weak-*-topology. In the discrete case, we next showed that G-invariance of m implies that for each $g \in G$, $(\pi(g)\varphi_j - \varphi_j) \to 0$ weakly in $L^1(G)$. Exactly the same argument applies in general, and moreover, the same argument shows that the fact that $m(\varphi * f) = m(f)$ for all $\varphi \in P(G)$,

$f \in L^\infty(G)$ implies that for each $\varphi \in P(G)$, $(\varphi * \varphi_j - \varphi_j) \to 0$ weakly in $L^1(G)$. The reason we use this latter condition rather than only use G-invariance is to be able to deal with the second problem we described above concerning generalizing 7.1.8 from discrete groups to general groups, namely the problem of obtaining control uniformly on compact sets in G rather than the pointwise control which sufficed in the discrete case.

As in the discrete case, we now pass to strong convergence. Let $E = \prod_{\varphi \in P(G)} L^1(G)$ and define $T: P(G) \to E$, by $(T\varphi)_\psi = \psi * \varphi - \varphi$. The fact that $\psi * \varphi_j - \varphi_j \to 0$ weakly for all $\psi \in P(G)$ implies that 0 is in the weak closure of $T(P(G))$. Since $T(P(G))$ is convex, it follows (as in the discrete case) that 0 is in the strong closure of $T(P(G))$, i.e. there is a net $\psi_j \in P(G)$ such that $\| \psi * \psi_j - \psi_j \|_1 \to 0$ for all $\psi \in P(G)$.

Now let $\varepsilon > 0$, $K \subset G$ compact. We claim there is an (ε, K)-invariant vector in $P(G) \subset L^1(G)$, i.e., $\varphi \in P(G)$ such that $\| \pi(g)\varphi - \varphi \|_1 < \varepsilon$ for all $g \in K$. In fact, choose any $\beta \in P(G)$. We claim that for some j, $\varphi = \beta * \psi_j$ satisfies this condition. There exists an open neighborhood E of $e \in G$ such that $\| \pi(g)\beta - \beta \|_1 < \varepsilon$ for all $g \in E$. Let $\varphi_E \in P(G)$ be the normalized characteristic function of E. Then it follows that $\| \varphi_E * \beta - \beta \|_1 < \varepsilon$ as well. Choose $x_1, \ldots, x_n \in G$ such that $K \subset \cup x_i E$. Then

$$\begin{aligned} \| \pi(g)(\beta * \psi_j) - \beta * \psi_j \|_1 &\leqq \| \pi(g)(\beta * \psi_j) - \varphi_{x_i E} * \beta * \psi_j \|_1 \\ &+ \| \varphi_{x_i E} * \beta * \psi_j - \psi_j \|_1 + \| \psi_j - \beta * \psi_j \|_1 . \end{aligned}$$

Since $\| \psi * \psi_j - \psi_j \|_1 \to 0$ for all $\psi \in P(G)$, for j sufficiently large the last two terms will be at most ε, independent of i. Thus, (using 7.2.6e),

$$\| \pi(g)(\beta * \psi_j) - \beta * \psi_j \|_1 \leqq \| \pi(g)\beta - \varphi_{x_i E} * \beta \|_1 \| \psi_j \|_1 + 2\varepsilon.$$

Furthermore,

$$\begin{aligned} \| \pi(g)\beta - \varphi_{x_i E} * \beta \|_1 &\leqq \| \pi(g)\beta - \pi(x_i)\beta \|_1 + \| \pi(x_i)\beta - \varphi_{x_i E} * \beta \|_1 \\ &\leqq \| \pi(x_i^{-1} g)\beta - \beta \|_1 + \| \beta - \pi(x_i^{-1})\varphi_{x_i E} * \beta \|_1 \\ &\leqq \| \pi(x_i^{-1} g)\beta - \beta \|_1 + \| \beta - \varphi_E * \beta \|_1 . \end{aligned}$$

Choosing x_i such that $g \in x_i E$, each of these terms is at most ε, and since $\| \psi_j \|_1 = 1$, we deduce that $\| \pi(g)(\beta * \psi_j) - (\beta * \psi_j) \|_1 \leqq 4\varepsilon$.

Finally, we can pass from L^1 to L^2 exactly as in the discrete case and this completes the proof of Theorem 7.1.8.

7.3 Unitary representations and semi-direct products, II

In proving the vanishing of matrix coefficients for unitary representations of simple Lie groups in Chapter 2, a basic role was played by an analysis of some of the representation theory for certain subgroups which were actually semi-direct products. For the proof of Kazhdan's result that simple Lie groups of higher $\mathbb{R}$-rank have the Kazhdan property (7.1.4), we will need further information on representations of semi-direct products, and we develop these results in this section.

We recall the context of section 2.3. Let G be locally compact and $A \subset G$ a normal abelian subgroup with $A \cong \mathbb{R}^n$. We then have the following basic connection between representations of G and the action of G on $\hat{A} \cong \hat{\mathbb{R}}^n$. If $H \subset G$ is a closed subgroup, and σ is a unitary representation of H, we denote the representation of G induced from σ (section 4.2) by $\mathrm{ind}_H^G(\sigma)$.

7.3.1 Theorem [*Mackey* 3]: *Suppose the G-action on $\hat{\mathbb{R}}^n$ is smooth. Then for any irreducible unitary representation π of G, there is a point $\lambda_0 \in \hat{\mathbb{R}}^n$ with G_{λ_0} its stabilizer in G, and an irreducible unitary representation σ of G_{λ_0} such that*

(i) $\pi = \mathrm{ind}_{G_{\lambda_0}}^G(\sigma)$,

(ii) $\sigma | \mathbb{R}^n = (\dim \sigma)\lambda_0$

(iii) $\pi | \mathbb{R}^n \cong \pi_{(\mu, \mathscr{H}_\lambda)}$ (section 2.3) where μ is a measure on $\hat{\mathbb{R}}^n$ corresponding to a quasi-invariant measure on the orbit G/G_{λ_0}.

Proof: We write $\pi | \mathbb{R}^n \cong \pi_{(\mu, \mathscr{H}_\lambda)}$ as in Proposition 2.3.3, and apply Proposition 2.3.5. Since π is irreducible, 2.3.5 implies that μ is quasi-invariant and ergodic under the G-action and $\dim \mathscr{H}_\lambda$ is essentially constant. Therefore, we can assume $\mathscr{H}_\lambda = \mathscr{H}_0$ a.e. for some fixed Hilbert space $\mathscr{H}_0$. By smoothness of G on $\hat{\mathbb{R}}^n$, μ is supported on a G-orbit, and hence we can identify $L^2(\hat{\mathbb{R}}^n, \mu, \{\mathscr{H}_\lambda\})$ with $L^2(G/G_{\lambda_0}, \mathscr{H}_0)$ for any $\lambda_0 \in \hat{\mathbb{R}}^n$ in the G-orbit supporting μ.

Let ω be the representation of G on $L^2(G/G_{\lambda_0}, \mathscr{H}_0)$ defined by inducing the identity representation of G_{λ_0} on $\mathscr{H}_0$, i.e., $(\omega(g)f)(\lambda) = \rho(\lambda, g)^{1/2} f(\lambda g)$ where $\rho: G/G_{\lambda_0} \times G \to \mathbb{R}^+$ is the Radon-Nikodym cocycle. (See the discussion preceding Proposition 4.2.20.) If $E \subset G/G_{\lambda_0}$, and $\mathscr{H}_E \subset L^2(G/G_{\lambda_0}, \mathscr{H}_0)$ is the subspace of functions supported on E, it follows that $\omega(g)\mathscr{H}_{E \cdot g} = \mathscr{H}_E$. Since we also have $\pi(g)\mathscr{H}_{E \cdot g} = \mathscr{H}_E$ by Proposition 2.3.5 (after switching to a right action in 2.3.5), it follows that for all $E \subset G/G_{\lambda_0}$, $\pi(g)\omega(g)^{-1}\mathscr{H}_E = \mathscr{H}_E$. However, if (M, μ) is a (standard) measure space, and T is a unitary operator on $L^2(M, \mu, \mathscr{H}_0)$ such that $T\mathscr{H}_E = \mathscr{H}_E$ for all $E \subset M$, then there is a Borel function $M \to U(\mathscr{H}_0)$ $m \to U_m \in U(\mathscr{H}_0)$, such that $(Tf)(m) = U_m(f(m))$. (See [Dixmier 2], for example.) Thus, for each g, we have $[(\pi(g)\omega(g)^{-1})f](\lambda) = \alpha(\lambda, g)f(\lambda)$ for some $\alpha: G/G_{\lambda_0} \times G \to U(\mathscr{H}_0)$, and replacing f by $\omega(g)f$, we obtain $(\pi(g)f)(\lambda) = \alpha(\lambda, g)\rho(\lambda, g)^{1/2}f(\lambda g)$. Since π is a representation, it follows (as in section 4.2) that α is a cocycle.

By Proposition 4.2.15 there is a strict cocycle α' such that for all g, $\alpha(\lambda, g) = \alpha'(\lambda, g)$ a.e. It follows that α' corresponds to a representation $\sigma: G_{\lambda_0} \to U(\mathscr{H}_0)$ by Proposition 4.2.13, and by the definition of induced representation, $\pi = \mathrm{ind}_{G_{\lambda_0}}^{G}(\sigma)$. Since inducing preserves direct sums and π is irreducible, σ must be as well.

To see the second assertion of the theorem, we observe that if $g \in \mathbb{R}^n$, g acts trivially on $\hat{\mathbb{R}}^n$, and hence on G/G_{λ_0}. Thus, for $t \in \mathbb{R}^n$ we have $(\pi(t)f)(\lambda) = \alpha'(\lambda, t)f(\lambda)$ a.e. for any $f \in L^2(G/G_{\lambda_0}, \mathscr{H}_0)$. Identifying $L^2(G/G_{\lambda_0}, \mathscr{H}_0)$ with $L^2(\hat{\mathbb{R}}^n, \mu, \mathscr{H}_0)$, we also have $(\pi(t)f)(\lambda) = \lambda(t)f(\lambda)$ by definition of $\pi_{(\mu, \mathscr{H}_\lambda)}$. Thus, for each $t \in \mathbb{R}^n$, we have $\alpha'(\lambda, t) = \lambda(t)$ for almost all λ (identifying $\lambda(t) \in \mathbb{C}$ as a scalar operator on $\mathscr{H}_0$). By Fubini, for μ-almost all $\lambda \in \hat{\mathbb{R}}^n$ this equality is true of almost all $t \in \mathbb{R}^n$, and since for fixed λ, $\alpha'(\lambda, t)$ and $\lambda(t)$ are continuous in t, for almost all λ this equality holds for all t. (α' is continuous in t because α' is a strict cocycle, t acts trivially on $\hat{\mathbb{R}}^n$, and hence $t \to \alpha'(\lambda, t)$ is a measurable homomorphism and hence continuous (Appendix B)). Fix λ_0 in the orbit supporting μ such that $\alpha'(\lambda_0, t) = \lambda_0(t)$. By definition, $\sigma: G_{\lambda_0} \to U(\mathscr{H}_0)$ is given by $\sigma(g) = \alpha'(\lambda_0, g)$, and hence it follows that $\sigma | \mathbb{R}^n = (\dim \sigma) \cdot \lambda_0$.

7.3.2 Example: Let N be the Heisenberg group, i.e.

$$N = \left\{ g = \begin{bmatrix} 1 & a & c \\ 0 & 1 & b \\ 0 & 0 & 1 \end{bmatrix} \middle| \, a, b, c \in \mathbb{R} \right\}.$$

Let $A = \{g \in N \mid a = 0\}$, so that $A \cong \mathbb{R}^2$. Then $g \in N$ acts on A by

$$g^{-1} \begin{bmatrix} 1 & 0 & z \\ 0 & 1 & y \\ 0 & 0 & 1 \end{bmatrix} g = \begin{bmatrix} 1 & 0 & z\text{-}ay \\ 0 & 1 & y \\ 0 & 0 & 1 \end{bmatrix}.$$

If $(\alpha, \beta) \in \mathbb{R}^2$, so that $\lambda_{(\alpha,\beta)}(y, z) = e^{i(\alpha y + \beta z)}$ then identifying A with $\mathbb{R}^2$ we have

$$(g \cdot \lambda_{(\alpha,\beta)})(y, z) = \lambda_{(\alpha,\beta)}(y, z - ay) = e^{i(\alpha y + \beta z - \beta a y)} = \lambda_{(\alpha - \beta a, \beta)}(y, z).$$

Thus the N-action on $\hat{\mathbb{R}}^2 \cong \mathbb{R}^2$ is given by $g \cdot (\alpha, \beta) = (\alpha - \beta a, \beta)$. There are thus two types of orbits. Each point on the x-axis is an orbit, and each horizontal line except for the x-axis is an orbit. By 2.1.12 the N-action on $\hat{\mathbb{R}}^2$ is smooth.

If π is an irreducible representation of N with $\pi | A = \pi_{(\mu, \mathscr{H}_\lambda)}$ where μ is supported on a point on the x-axis, then $\pi | [N, N]$ is trivial, since $[N, N] = \{g \in N \mid a = b = 0\}$. Thus, π factors to an irreducible representation of $N/[N, N] \cong \mathbb{R}^2$, and therefore π is 1-dimensional. On the other hand if μ is supported on a horizontal line which is not the x-axis, then the stabilizer in N of any point in this orbit is just the group A. Thus $\pi = \mathrm{ind}_A^N(\sigma)$ where σ is an irreducible representation of A, and hence is one-dimensional. Therefore, in conclusion, we deduce that any irreducible representation of N is either 1-dimensional or is induced from a 1-dimensional representation of A.

7.3.3 Example: Let $H = SL(2, \mathbb{R}) \ltimes \mathbb{R}^2$ where the action of $SL(2, \mathbb{R})$ on $\mathbb{R}^2$ is given by usual matrix multiplication. The action on $\hat{\mathbb{R}}^2$ is just the adjoint action. Thus there are only two orbits of H acting on $\hat{\mathbb{R}}^2$, namely the origin and its complement. If π is an irreducible unitary representation of H such that $\pi | \mathbb{R}^2 \cong \pi_{(\mu, \mathscr{H}_\lambda)}$ with μ supported on the origin then $\pi | \mathbb{R}^2$ is trivial, and hence π factors to a representation of $SL(2, \mathbb{R})$. If μ is supported on the complement, the stabilizer of a point in the complement can be taken to be isomorphic to the Heisenberg group N. (This can easily be seen by realizing H as a subgroup of $SL(3, \mathbb{R})$. Namely, consider 3×3 matrices of the form

$$\left[\begin{array}{cc|c} \multicolumn{2}{c|}{\multirow{2}{*}{g}} & c \\ & & b \\ \hline 0 & 0 & 1 \end{array}\right]$$

where $g \in SL(2, \mathbb{R})$. This group is clearly isomorphic to H in such a way that $SL(2, \mathbb{R}) \subset H$ corresponds to matrices with $b = c = 0$, and $\mathbb{R}^2 \subset H$ to matrices with $g = I$. The stabilizer of $\chi \in \hat{\mathbb{R}}^2$, $\chi(c, b) = e^{ib}$, is readily seen to be N.) Thus in this case $\pi = \mathrm{ind}_N^H(\sigma)$ where σ is an irreducible representation of N. Therefore we deduce that if π is any irreducible unitary representation of H, either $\pi | \mathbb{R}^2$ is trivial, or π is induced from a representation of N.

7.3.4 Example: We can generalize 7.3.3 as follows. Let $G = SL(2, \mathbb{R})$ or $PSL(2, \mathbb{R})$ and suppose G acts on $\mathbb{R}^n$ by a rational representation such that the only fixed point under G is the origin. Let $H = G \ltimes \mathbb{R}^n$. If π is an irreducible unitary representation of G then either $\pi | \mathbb{R}^n$ is trivial or $\pi \cong \mathrm{ind}_{H_0}^H(\sigma)$ where $H_0 \subset H$ is an amenable subgroup. To see this, using Theorem 7.3.1 as in Example 7.3.3, and 4.1.6(b), it suffices to see that the stabilizer in G of any point in $\hat{\mathbb{R}}^n$ except the origin, is amenable. But since there are no fixed points other than the origin, any stabilizer must be (the real points of) an $\mathbb{R}$-algebraic group of dimension at most 2. Any such connected group is amenable, and thus the stabilizer is a finite extension of an amenable group, and hence is amenable (Corollary 4.1.7).

We remark that the same arguments yield analogous results in 7.3.1–7.3.4 if $\mathbb{R}$ is replaced by any local field of characteristic 0.

To apply results on semidirect products to showing that certain groups have Kazhdan's property, it will be useful to generalize somewhat the notion of a representation almost possessing an invariant vector. Namely, we think of almost possessing an invariant vector as almost containing the 1-dimensional identity representation, and we will extend this definition by replacing the 1-dimensional identity by an arbitrary unitary representation.

Let G be a locally compact group and π a unitary representation on a Hilbert space $\mathscr{H}_\pi$. Let $\{v_i | i = 1, \ldots, n\}$ be an orthonormal set in $\mathscr{H}_\pi$. Define a function $f_{\pi,\{v_i\}}: G \to M(n \times n, \mathbb{C})$(the $n \times n$ complex matrices) by

$$(f_{\pi,\{v_i\}}(g))_{ij} = \langle \pi(g)v_i | v_j \rangle.$$

We then call a function of the form $f_{\pi,\{v_i\}}$ an ($n \times n$) submatrix of π. If $\{v_i\}$ consists of a single element, then a 1×1 submatrix is of course just a matrix coefficient in the sense of chapter 2. For $(a_{ij}) \in M(n \times n, \mathbb{C})$ let $\|(a_{ij})\| = \max |a_{ij}|$.

7.3.5 Definition: (i) Let σ, π be unitary representations of G. Let $\varepsilon > 0$ and $K \subset G$ compact. We say that σ is (ε, K)-contained in π if for every $n \times n$ submatrix f of σ, there is an $n \times n$ submatrix h of π such that $\|f(g) - h(g)\| < \varepsilon$ for all $g \in K$. (ii) We say that σ is weakly contained in π and write $\sigma \prec \pi$, if σ is (ε, K)-contained in π for every ε, K.

Since for any unit vector $\|\pi(g)v - v\|^2 = 2 - (\langle \pi(g)v | v \rangle + \langle \pi(g^{-1})v | v \rangle)$, it follows that the 1-dimensional identity representation is weakly contained in π if and only if π almost has invariant vectors in the sense of 7.1.1. The notion of weak containment is due to [Fell 1, 2]. As an example, we can extend Theorem 7.1.8.

7.3.6 Proposition: *(a) Suppose G is amenable. Let π be the regular representation. Then for any representation σ of G, $\sigma \prec \infty \cdot \pi$.*
(b) Suppose G is a group with regular representation π. If $I \prec \infty \cdot \pi$, then G is amenable.

Proof: (a) By 7.1.8, for the 1-dimensional identity representation I, we have $I \prec \pi$. It follows directly from the definitions that this implies that $I \otimes \sigma \prec \pi \otimes \sigma$, i.e., $\sigma \prec \pi \otimes \sigma$. To prove the proposition, it then suffices to observe that $\pi \otimes \sigma \cong (\dim \sigma)\pi$. However, the map $U: L^2(G) \otimes \mathscr{H}_\sigma \to L^2(G, \mathscr{H}_\sigma)$ defined by $[U(f \otimes v)](g) = f(g)(\sigma(g)v)$ is easily seen to be a unitary equivalence of $\pi \otimes \sigma$ and the representation $\tilde{\pi}$ on $L^2(G, \mathscr{H}_\sigma)$ given by $(\tilde{\pi}(h)\varphi)(g) = \varphi(gh)$, and $\tilde{\pi} = (\dim \sigma)\pi$.
(b) For any $\varphi \in L^2(G)$, define $\lambda_\varphi(g) = \langle \pi(g)\varphi | \varphi \rangle$, so $\lambda_\varphi \in L^\infty(G)$. In the proof of Theorem 7.2.4, we began with the assumption that there was a sequence of unit vectors $\psi_j \in L^2(G)$ such that $\|\pi(g)\psi_j - \psi_j\|_2 \to 0$ as $j \to \infty$, uniformly on compact subsets of G, or equivalently $\lambda_{\psi_j}(g) \to 1$ uniformly on compact sets of G. However, if we examine the proof, we see that in fact it suffices to find unit vectors ψ_j such that $\lambda_{\psi_j}(g) \to 1$ only as elements of $(L^1(G))^*$, i.e., in the weak-*-topology in $L^\infty(G)$. For if we have weak-* convergence, we can define φ_j as in 7.2.4 and deduce

that $(g \to \| \pi(g^{-1})\varphi_j - \varphi_j \|_1)$ converges to 0 in the weak-*-topology, and then define m as in 7.2.4. We still have for $f \in L^\infty(G)$,

$$|m(\pi(g)f - f)| \leq \|f\|_\infty \overline{\lim_j} \| \pi(g^{-1})\varphi_j - \varphi_j \|_1,$$

and hence for any positive $\alpha \in L^1(G)$, $\int \alpha(g)|m(\pi(g)f - f)| = 0$. Thus $m(\pi(g)f - f) = 0$ and this shows the existence of an invariant mean. The remainder of the proof of Theorem 7.2.4 then carries through without change. It therefore suffices to show that $I \prec \infty \cdot \pi$ implies that we can find unit vectors ψ_j such that $\lambda_{\psi_j}(g) \to 1$ in the weak-*-topology on $L^\infty(G)$.

Let $A = \{\lambda_\varphi | \varphi \in L^2(G), \|\varphi\|_2 = 1\}$, so that A is a subset of $L^\infty(G)_1$, the unit ball in $L^\infty(G)$. Let $\bar{A}$ be the closure in the weak-*-topology, and $co(\bar{A})$ the closed convex hull of $\bar{A}$ in $L^\infty(G)_1$. Then a general result of functional analysis asserts that the extreme points of $co(\bar{A})$ all lie in $\bar{A}$. Now consider the statement $I \prec \infty \cdot \pi$. This asserts that there is a sequence $v_n \in \sum^\oplus L^2(G)$, $v_n = \sum_i^\oplus v_{ni}$, $v_{ni} \in L^2(G)$, $\|v_n\|^2 = \sum_i \|v_{ni}\|^2 = 1$, such that $\langle (\infty \cdot \pi)(g) v_n | v_n \rangle \to 1$ uniformly on compact sets in G. In other words, as $n \to \infty$, $\sum_i \langle \pi(g) v_{ni} | v_{ni} \rangle \to 1$ uniformly on compact sets, or $\sum_i \|v_{ni}\|^2 \lambda_{w_{ni}}(g) \to 1$ uniformly on compact sets where $w_{ni} = v_{ni}/\|v_{ni}\|$. Since for each n, $\sum \|v_{ni}\|^2 = 1$, it follows that for the function $1 \in L^\infty(G)$ we actually have $1 \in co(\bar{A})$. However, 1 is clearly an extreme point of $L^\infty(G)_1$, and hence is an extreme point of $co(\bar{A})$. By our remarks above, this implies $1 \in \bar{A}$, completing the proof.

We shall need the following two general results.

7.3.7 Proposition: *Suppose that $H \subset G$ is a closed subgroup and that σ, π are representations of H with $\sigma \prec \pi$. Then $ind_H^G(\sigma) \prec ind_H^G(\pi)$.*

Proof: Let $\gamma: G/H \to G$ be a Borel section with $\overline{\gamma(K)}$ compact whenever K is compact. (Appendix A). We have a cocycle $\alpha: G/H \times G \to H$ given by $\alpha(y, g) = \gamma(y) g \gamma(yg)^{-1}$. We recall that $\mathrm{ind}_H^G(\sigma)$ is defined by $((\mathrm{ind}_H^G(\sigma)(g))f)(y) = \rho(y, g)^{1/2} \sigma(\alpha(y, g)) f(yg)$ where $f \in L^2(G/H, \mathscr{H}_\sigma)$ and ρ is the Radon-Nikodym derivative. (See section 4.2.) $\mathrm{Ind}_H^G(\pi)$ is of course defined similarly. For notational convenience, set $\mathrm{ind}_H^G(\pi) = U^\pi$, and similarly define U^σ. To show that any submatrix of U^σ can be approximated by submatrices of U^π, it suffices by standard approximation arguments to show this is true for a submatrix of U^σ defined by n (n finite) orthonormal functions $f_i: G/H \to \mathscr{H}_\sigma$ such that f_i is compactly supported and f_i takes on only finitely many values. Fix such $f_1, \ldots, f_n$ and let $K \subset G$ be compact and $\varepsilon > 0$.

Choose $Y \subset G/H$ compact such that $f_i = 0$ on $G/H - Y$ for all i. Then there is a compact set $K' \subset H$ such that for $y \in YK^{-1}$ and $g \in K$, $\alpha(y, g) \in K'$. Let $M = \max_i \{ \| f_i(G/H) \|_\infty, 1\}$. Let $V \subset \mathscr{H}_\sigma$ be the finite-dimensional subspace spanned by $\bigcup_i f_i(G/H)$. Choose an orthonormal basis of V and let φ be the corresponding submatrix of σ. Let ψ be a submatrix of π such that $\| \psi(h) - \varphi(h) \| < \varepsilon/M^2 n$ for all $h \in K'$. Let $W \subset \mathscr{H}_\pi$ be the space spanned by the orthonormal vectors defining ψ. By mapping one orthonormal set to the other, we obtain a unitary mapping $T: V \to W$. An elementary calculation then shows that for any $x, v \in V$,

$$|\langle \sigma(h)x, v\rangle - \langle \pi(h)(Tx), (Tv)\rangle| \leqq \varepsilon \|x\| \, \|v\|/M^2 \qquad \text{for all } h \in K'.$$

Define $a_i: G/H \to \mathscr{H}_\pi$ by $a_i(y) = T(f_i(y))$. It then follows by taking $h = e$ in the last inequality (we can clearly assume $e \in K'$) that

$$\int_{G/H} |\langle f_i(y), f_j(y)\rangle - \langle a_i(y), a_j(y)\rangle| \leqq \varepsilon.$$

In particular, $a_i \in L^2(G/H, \mathscr{H}_\pi)$ and $|\langle a_i, a_j\rangle| \leqq \varepsilon$ for $i \neq j$ where the inner product is in $L^2(G/H, \mathscr{H}_\pi)$. Furthermore, since $\alpha(y, g) \in K'$ for $g \in K$ and $y \in YK^{-1}$, we also see from the last inequality of the preceding paragraph that

$$|\langle U^\sigma(g) f_i, f_j\rangle - \langle U^\pi(g) a_i, a_j\rangle| \leqq \varepsilon \qquad \text{for each } i, j,$$

for all $g \in K$. We would therefore be done except that $\{a_i\}$ are not necessarily orthonormal. However, since $|\langle a_i, a_j\rangle| \leqq \varepsilon$ for all $i \neq j$, and $\|a_i\| = 1$, by perturbing $\{a_i\}$ we can assume we have the required approximation with $\{a_i\}$ orthonormal. This completes the proof.

7.3.8 Proposition: *Suppose π_x are unitary representations of G and $\pi = \int_X^\oplus \pi_x$ (section 2.3). If λ is a representation with $\pi_x \prec \lambda$ for almost all x, then $\pi \prec \infty \cdot \lambda$.*

Proof: It is straightforward that $\pi \prec \int^\oplus \lambda_x$ where $\lambda_x = \lambda$. However, if $\lambda_x = \lambda$, then $\int^\oplus \lambda_x \cong n \cdot \lambda$ for $n = \dim L^2(X)$.

We can now present a result concerning unitary representations of certain semidirect products that will arise in the next section in the proof of Kazhdan's theorem.

7.3.9 Theorem: *Let* $G = SL(2, \mathbb{R})$ *or* $PSL(2, \mathbb{R})$ *and suppose* G *acts on* $\mathbb{R}^n$ *by a rational representation so that the only* G*-invariant vector is the origin. Let* $H = G \ltimes \mathbb{R}^n$ *be the associated semidirect product. Let* π *be a unitary representation of* H *so that* $\pi | \mathbb{R}^n$ *has no non-trivial invariant vectors. Then* $\pi \prec \infty \cdot \lambda$ *where* λ *is the regular representation of* H.

Proof: Write $\pi = \int^{\oplus} \pi_x$ where π_x are irreducible (2.3.1). Since $\pi | \mathbb{R}^n$ has no non-trivial invariant vectors, for almost all x neither does $\pi_x | \mathbb{R}^n$ by (2.3.2). By 7.3.8, it suffices to prove the result for all π_x, i.e., we may assume π is irreducible. By Example 7.3.4, $\pi \cong \mathrm{ind}_{H_0}^{H}(\sigma)$ where $H_0 \subset H$ is amenable. It then follows from Proposition 7.3.6 that $\sigma \prec \infty \cdot \lambda_{H_0}$ where λ_{H_0} is the regular representation of H_0. We then conclude from 7.3.7 that $\pi \prec \mathrm{ind}_{H_0}^{H}(\infty \cdot \lambda_{H_0}) = \infty \cdot \mathrm{ind}_{H_0}^{H}(\lambda_{H_0})$. However, in general the induced representation of a regular representation is the regular representation, so $\pi \prec \infty \cdot \lambda$.

7.4 Kazhdan's property for semisimple groups

The main purpose of this section is to prove Kazhdan's theorem (7.1.4). We begin with the following simple remark.

7.4.1 Lemma: *If* $0 \to K \to G \to H \to 0$ *is an exact sequence of locally compact groups, and* K, H *both have the Kazhdan property, so does* G.

Proof: Suppose π is a unitary representation of G which almost has invariant vectors. Then $\pi | K$ has the same property and since K has Kazhdan's property, $\pi | K$ has non-trivial invariant vectors. Since K is normal in G, the space of $\pi(K)$ invariant vectors is a $\pi(G)$-invariant subspace. Let $\pi = \pi_0 \oplus \pi_1$, where π_0 is the subrepresentation of G on the space of all $\pi(K)$-invariant vectors. Now π_1 cannot almost have invariant vectors, for otherwise $\pi_1 | K$ would have invariant vectors which contradicts the choice of π_0. Therefore, since π almost has invariant vectors, it follows that π_0 must have this property as well. But π_0 factors to a representation of H, and since H has the Kazhdan property, $\pi_0(H)$ must have non-trivial invariant vectors. Thus, so does $\pi_0(G)$ (and hence $\pi(G)$).

7.4.2 Theorem [*Kazhdan* 1]: *Let* G *be a connected semisimple Lie group with finite center such that every simple factor of* G *has* $\mathbb{R}$*-rank at least* 2. *Then* G *has the Kazhdan property.*

Proof: By Lemma 7.4.1, it suffices to prove this if G is simple, with trivial center, and $\mathbb{R}$-*rank*$(G) \geqq 2$. We can therefore assume G is one of the form $\underline{G}^0_{\mathbb{R}}$ where $\underline{G}$ is $\mathbb{R}$-algebraic. Any such group contains a subgroup H isomorphic to a semidirect product considered in Theorem 7.3.9. (This is clear for $SL(n, \mathbb{R})$, since $SL(3, \mathbb{R})$ contains $SL(2, \mathbb{R}) \ltimes \mathbb{R}^2$ as described in Example 7.3.3. In general, a suitable subgroup of a maximal proper parabolic subgroup will suffice.)

Let π be a unitary representation of G which almost has invariant vectors. Suppose, however, that π has no invariant vectors. By Moore's theorem (2.2.19), $\pi | \mathbb{R}^n$ has no invariant vectors (where we write $\mathbb{R}^n \subset H$ as in 7.3.9) and hence, by Theorem 7.3.9, $\pi | H \prec \infty \cdot \lambda_H$ where λ_H is the regular representation of H. Since π almost has invariant vectors, $I \prec \pi$ where I is the 1-dimensional identity representation, and hence $I \prec (\pi | H)$. Thus, $I \prec \infty \cdot \lambda_H$. By Proposition 7.3.6, this implies that H is amenable, which is impossible (using 4.1.6, 4.1.8) since $SL(2, \mathbb{R}) \subset H$ or $PSL(2, \mathbb{R}) \subset H$. This shows that π must have non-trivial invariant vectors, completing the proof of Theorem 7.4.2.

To deal with lattices, we have the following.

7.4.3 Proposition [*Kazhdan* 1]: *Suppose G has the Kazhdan property and that $\Gamma \subset G$ is a lattice subgroup. Then Γ has the Kazhdan property.*

Proof: Suppose π is a representation of Γ with $\pi \succ I$. Then by Proposition 7.3.7, $\operatorname{ind}_\Gamma^G(\pi) \succ \operatorname{ind}_\Gamma^G(I)$. Since G/Γ has finite invariant measure, the constant functions are in $L^2(G/\Gamma)$, and hence $I \leqq \operatorname{ind}_\Gamma^G(I)$. Thus, $I \prec \operatorname{ind}_\Gamma^G(\pi)$, and since G has the Kazhdan property, $\operatorname{ind}_\Gamma^G(\pi)$ has invariant vectors. It is well-known that this implies that π has invariant vectors. (To prove the last assertion, let π be a representation on $\mathscr{H}_\pi$ and let $\alpha: G/\Gamma \times G \to U(\mathscr{H}_\pi)$ be a cocycle corresponding to π (4.2.13). If $I \leqq \operatorname{ind}_\Gamma^G(\pi)$, then there is an α-invariant (L^2) function $\varphi: G/\Gamma \to \mathscr{H}_\pi$. Since G is transitive on G/Γ, we must have that on a conull set $\varphi \neq 0$, and since α is unitary, $\psi: G/\Gamma \to (\mathscr{H}_\pi)_1$, the unit ball in $\mathscr{H}_\pi$, defined by $\psi(x) = \varphi(x)/\|\varphi(x)\|$, will also be α-invariant. By 4.2.19 there is a $\pi(\Gamma)$-invariant vector in $(\mathscr{H}_\pi)_1$.)

Theorem 7.4.2 and Proposition 7.4.3 together yield the proof of Theorem 7.1.4. There are natural examples of groups which are not semisimple but which have Kazhdan's property. For example:

7.4.4 Theorem [*Wang* 1]: *For $n \geqq 3$, $G = SL(n, \mathbb{R}) \ltimes \mathbb{R}^n$ has Kazhdan's property, where $SL(n, \mathbb{R})$ acts on $\mathbb{R}^n$ by matrix multiplication.*

Proof: Let π be a representation of G on $\mathscr{H}_\pi$ with $\pi \succ I$. Suppose π has no invariant vectors. Let $\mathscr{H}_0 = \{v \in \mathscr{H}_\pi \mid \pi(g)v = v \text{ for } g \in \mathbb{R}^n\}$ and $\mathscr{H}_1 = \mathscr{H}_0^\perp$. Since $\mathbb{R}^n \subset G$ is a normal subgroup $\mathscr{H}_0$ and $\mathscr{H}_1$ are invariant under $\pi(G)$, and we let π_0 and π_1 be the corresponding subrepresentations. From the fact that $\pi \succ I$, it follows that $\pi_i \succ I$ for either $i = 0$ or 1. If $\pi_0 \succ I$, then $(\pi_0 | SL(n, \mathbb{R})) \succ I$, and by Theorem 7.4.2, there are $\pi_0 | SL(n, \mathbb{R})$ invariant vectors in $\mathscr{H}_0$, which implies $\pi \geqq I$.

Suppose on the other hand, that $\pi_1 \succ I$. Let $\pi_1 = \int^\oplus \pi_x$ where π_x is irreducible (2.3.1). Since $\pi_1 | \mathbb{R}^n$ has no invariant vectors, it follows that for almost all x, $\pi_x | \mathbb{R}^n$ has no invariant vectors (2.3.2). The action of $SL(n, \mathbb{R})$ on $\hat{\mathbb{R}}^n$ has two orbits, namely the origin and its complement. Applying Theorem 7.3.1 to the representation π_x, since $\pi_x | \mathbb{R}^n$ has no invariant vectors (almost all x), it follows that for almost all x, the orbit that arises must be the complement of the origin. Therefore, for almost all x, $\pi_x = \operatorname{ind}_{G_0}^G(\sigma_x)$ where G_0 is the stabilizer in G of a non-0 point in $\hat{\mathbb{R}}^n$. However, Lebesgue measure on $\hat{\mathbb{R}}^n$ is an invariant measure on G/G_0, and since this measure is not finite, it follows that $\pi_x | SL(n, \mathbb{R})$ does not have non-trivial invariant vectors. Thus $\pi_1 | SL(n, \mathbb{R})$ does not have non-trivial invariant vectors. However, $\pi_1 \succ I$, so $(\pi_1 | SL(n, \mathbb{R})) \succ I$, and by Theorem 7.4.2, $\pi_1(SL(n, \mathbb{R}))$ has non-trivial invariant vectors. This contradiction shows that $\pi_1 \succ I$ is impossible, and this completes the proof.

7.4.5 Corollary [*Wang* 1]: *For* $n \geqq 3$, $SL(n, Z) \ltimes Z^n$ *has Kazhdan's property.*

Proof: 7.4.3 and 7.4.4.

8 Normal Subgroups of Lattices

8.1 Margulis' finiteness theorem – statement and first steps of proof

The point of this chapter is to prove the following finiteness theorem of Margulis.

8.1.1 Theorem [*Margulis* 6]: *Let G be a connected semisimple Lie group with finite center and no compact factors, and let $\Gamma \subset G$ be an irreducible lattice. Assume $\mathbb{R}$-rank$(G) \geqq 2$. Let $N \subset \Gamma$ be a normal subgroup such that Γ/N is not amenable. Then $N \subset Z(G)$ (the center of G) and in particular N is finite.*

Combined with Corollary 7.1.12 and Theorem 7.1.13, this yields the following result.

8.1.2 Theorem (*Margulis-Kazhdan*): *Let G be a connected semisimple Lie group with finite center and no compact factors, $\Gamma \subset G$ an irreducible lattice. Assume $\mathbb{R}$-rank$(G) \geqq 2$. Let $N \subset \Gamma$ be a normal subgroup. Then either*

(*a*) *$N \subset Z(G)$, and so N is finite; or*
(*b*) *Γ/N is finite.*

Remark: We recall that while we proved Corollary 7.1.12 in Chapter 7, we did not present a proof of 7.1.13. Thus, after we prove 8.1.1, we shall have a complete proof of 8.1.2 when the $\mathbb{R}$-rank of every simple factor of G is at least 2. The remaining case is complete once 7.1.13 is proved, and we again refer the reader to [Margulis 7] for this proof.

Earlier results in this direction were obtained by [Raghunathan 4] by different methods.

Margulis' proof of Theorem 8.1.1 is based on his discovery of a fundamental measure theoretic property of the action of Γ on G/P where $P \subset G$ is a minimal parabolic subgroup. Namely, suppose $P' \supset P$ is another parabolic subgroup. Then clearly there is a measure preserving Γ-map $G/P \to G/P'$. Margulis discovered that under suitable hypotheses, any Γ-space X for which a measure class preserving Γ-map $G/P \to X$ exists must be of this form.

8.1.3 Theorem [*Margulis* 6]: *Let G be a connected semisimple Lie group with no compact factors, trivial center, and $\mathbb{R}$-rank$(G) \geqq 2$. Let $P \subset G$ be a minimal parabolic subgroup and $\Gamma \subset G$ an irreducible lattice. Suppose (X, μ) is a (standard) measurable Γ-space and that there is a measure class preserving Γ-map $\varphi: G/P \to X$ (possibly defined only a.e.). Then there is a parabolic subgroup $P' \supset P$ so that as Γ-spaces, G/P' and X are isomorphic in such a way that φ corresponds (a.e.) to the natural Γ-map $G/P \to G/P'$. (We assume G/P, G/P' to have the G-invariant measure class.)*

Let us see how this result implies the finiteness theorem.

Proof of Theorem 8.1.1 from Theorem 8.1.3: With G as in 8.1.1, dividing G by its center, we may assume that G is center free and wish to show that N is trivial. Since Γ/N is not amenable, there is a compact metric Γ/N-space X without an invariant probability measure. We may also consider X as a compact metric Γ-space (on which N acts trivially). The Γ-action on G/P is amenable (Example 4.3.8), and hence by Proposition 4.3.9 there is a measurable Γ-map (defined a.e.) $\varphi: G/P \to M(X)$ (where, as usual, $M(X)$ is the space of probability measures on X). Let μ be a probability measure on G/P quasi-invariant under G. Then $\varphi*(\mu)$ is a measure on $M(X)$ quasi-invariant under Γ, and $\varphi:(G/P, \mu) \to (M(X), \varphi*(\mu))$ is a measure class preserving Γ-map. By Theorem 8.1.3, there is a parabolic subgroup $P' \supset P$ such that $(M(X), \varphi*(\mu)) \cong G/P'$ as Γ-spaces. If $P' = G$, then Γ would have a fixed point in $M(X)$, which would contradict the choice of X. Therefore $P' \neq G$. Recall that N acts trivially on X, hence on $M(X)$, and therefore each element of N fixes almost all points in G/P'. Since N acts continuously on G/P', this implies that N acts trivially on G/P', so that $N \subset \bigcap_{g \in G} gP'g^{-1}$, and $H = \bigcap_{g \in G} gP'g^{-1}$ is a proper normal subgroup.

Let $G = \prod_{i \in I} G_i$ be the decomposition of G into a product of simple groups. Then $H = \prod_{i \in J} G_i$, where $J \subset I$ is a proper subset. Since $N \subset H$, N is normalized by $\prod_{I-J} G_i$. But since N is normalized by $\prod_{I-J} G_i$ and by Γ, it is normalized by the product of these groups which is dense in G by irreducibility. Thus, $N \subset G$ is normal, and since N is discrete and G has trivial center, N is trivial. This completes the proof.

Before turning to the proof of Theorem 8.1.3, it will be convenient to reformulate it slightly. We recall that if (X, μ) is a measure space, by its measure algebra (or its Boolean σ-algebra), $B(X)$, we mean the space of measurable subsets of X, two sets being identified if they differ by a null set. We can identify

$B(X)$ as a subset of the unit ball in $L^\infty(X)$, namely $B(X) = \{ f \in L^\infty(X) | f^2 = f(a.e.)\}$. Then $B(X)$ is a weak-*-closed subset of $L^\infty(X)$. The induced topology on $B(X)$ is independent of the particular measure on X within its measure class. If $\varphi:(X, \mu) \to (Y, \nu)$ is a measure class preserving map, we obtain a map $\varphi^*: B(Y) \to B(X)$ which is injective, and φ^* is a Boolean operation preserving homeomorphism onto its image. Conversely, it is a basic fact of measure theory that if (X, μ), (Y, ν) are standard Borel spaces and $\lambda: B(Y) \to B(X)$ is an injective continuous Boolean operation preserving map, then there is an a.e. defined measure class preserving map $\varphi:(X, \mu) \to (Y, \nu)$ such that $\varphi^* = \lambda$. Furthermore, λ is a bijection if and only if φ is a measure space isomorphism. In addition, if $B \subset B(X)$ is any closed subset which is also closed under the Boolean operations, then there is a standard measure space (Y, ν) and a measure class preserving map $\varphi: X \to Y$ such that $B = \varphi^*(B(Y))$.

If a locally compact group G acts on (X, μ), then G acts continuously on $B(X)$. If $\varphi: X \to Y$ is a measure class preserving G-map, then φ^* is also a G-map. Conversely, suppose G acts continuously on $B(X)$ so as to preserve the Boolean operations. Then there is a measure class preserving G-action on (a conull subset of) X which induces the given action on $B(X)$. This is straightforward if G is discrete but involves some delicate measure theory in general. (See Appendix B.) If X, Y are G-spaces and $\lambda: B(Y) \to B(X)$ is a continuous Boolean G-map, then there is a conull G-invariant subset $X_0 \subset X$ and a measure class preserving G-map $\varphi: X_0 \to Y$ such that $\varphi^* = \lambda$. (Once again, see Appendix B for a proof.)

Given these remarks, we can reformulate Theorem 8.1.3 as follows.

8.1.4 Theorem [*Margulis* 6]: *Let G, P, Γ as in* 8.1.3. *Suppose $B \subset B(G/P)$ is a closed Boolean subspace (i.e., closed in topology, and closed under the Boolean operations.) If B is Γ-invariant, then B is G-invariant.*

Let us see how to deduce 8.1.3 from 8.1.4. Let $B = \varphi^*(B(X))$, where φ, X as in 8.1.3. Then B is Γ-invariant, and hence by 8.1.4, B is G-invariant. Since φ^* is a Boolean homeomorphism, G acts on $B(X)$, and by our above remarks, G acts on (almost all of X) in such a way that the restriction to Γ is equal (a.e.) to the original Γ-action. Since $\varphi^*: B(X) \to B(G/P)$ is now a G-map, there is a measure class preserving G-map $\psi: G/P \to X$, and since $\psi^* = \varphi^*$, $\varphi = \psi$ a.e. Furthermore, G is clearly essentially transitive on X since it is transitive on G/P, and hence as G-spaces, $X \cong G/P'$ for some subgroup P'. Since we have a surjective G-map $G/P \to G/P'$, we can suppose (replacing P' by a conjugate subgroup if necessary) that $P \subset P'$. Finally, since the restriction of the G-action on X to Γ is the same a.e. as the original Γ-action, (and in particular is isomorphic to the original Γ-action), we have $X \cong G/P'$ as Γ-spaces, verifying 8.1.3.

Our task is therefore to prove Theorem 8.1.4, and we will do this in the following sections.

8.2 Contracting automorphisms of groups

In this section we examine some properties of contracting automorphisms that we will need in the proof of 8.1.4.

8.2.1 Definition: Let H be a locally compact, compactly generated (separable) group. An automorphism $A: H \to H$ is called contracting if for any compact neighborhood U of $e \in H$ and any compact subset $K \subset H$, there is an integer N such that for $n \geqq N$, $A^n(K) \subset U$.

8.2.2 Example: (a) Let $0 < \alpha < 1$ and define $A: \mathbb{R}^n \to \mathbb{R}^n$ by $Ax = \alpha x$.
(b) Let g be an $r \times r$ diagonal matrix with diagonal entries $g_{ii} = \lambda_i$, where $\lambda_i < \lambda_{i+1}$. Let U be the unipotent group of upper triangular matrices with all diagonal entries equal to 1, and let $\bar{U}$ be the transpose of U, i.e., the corresponding lower triangular matrices. Let $A: GL(n, \mathbb{R}) \to GL(n, \mathbb{R})$ be $A = \text{Int}(g)$, i.e., $A(h) = ghg^{-1}$. Then both U and $\bar{U}$ are normalized by g, so that A defines an automorphism of both U and $\bar{U}$. If $h = (h_{ij})$, then $(Ah)_{ij} = \lambda_i \lambda_j^{-1} h_{ij}$, and hence A is a contracting automorphism of U. Similarly, $(A^{-1}h)_{ij} = \lambda_i^{-1} \lambda_j h_{ij}$, and so A^{-1} is a contracting automorphism of $\bar{U}$.

8.2.3 Example: The preceding example can be generalized as follows. Fix an integer $r \geqq 2$, and positive integers $r_1 < r_2 < \cdots < r_n = r$. Choose real numbers $\lambda_1 < \lambda_2 < \cdots < \lambda_n$ and let s be the diagonal $r \times r$ matrix given by $s_{ii} = \lambda_i$ for $r_{i-1} < i \leqq r_i$ (where we take $r_0 = 0$). Let V be the subgroup of the upper triangular unipotent group U (of 8.2.2(b)) given by $V = \{u \in U \mid u_{ij} = 0 \text{ if } i \neq j \text{ and if there exists } m \text{ such that } r_{m-1} < i, j \leqq r_m\}$. Then Int$(s)$ (i.e., $h \to shs^{-1}$) is a contracting automorphism of V, and Int$(s)^{-1}$ is a contracting automorphism of $\bar{V}$, the transpose of V.

Using the structure theory of semisimple groups, we can find an analogue of Example 8.2.3 in any connected semisimple Lie group with finite center and no compact factors. To express this, let us rephrase some of the structure in the above example. Let P_0 be the stabilizer in $SL(r, \mathbb{R})$ of the flag $0 \subset W_1 \subset W_2 \subset \cdots \subset W_n = R^r$ where $W_k = [e_1, \ldots, e_{r_k}]$, the linear span of the first r_k standard basis vectors. Then P_0 is a parabolic subgroup of $SL(r, \mathbb{R})$, and V is the unipotent radical of P_0.

If we let $R = \{g \in P_0 \mid gY_k = Y_k$ for all k where $Y_k = [e_{r_{k-1}+1}, \ldots, e_{r_k}]\}$, then P_0 is the semidirect product $P_0 = R \ltimes V$ and R is reductive (so R is semisimple after dividing by its center). We also remark that for s the diagonal matrix above, $s \in Z(R)$, the center of R. The transposed subgroup $\bar{P}_0$ is an opposite parabolic in the sense that $P_0 \cap \bar{P}_0 = R$. The subgroup $\bar{V}$ is the unipotent radical of $\bar{P}_0$. We have thus seen the validity of the following assertion for $G = SL(r, \mathbb{R})$, and which is true in general from structure theory.

8.2.4 Proposition: *Let G be a connected semisimple Lie group with finite center and no compact factors. Let A be a maximal connected* $\mathbb{R}$*-split abelian subgroup of G and P a proper parabolic subgroup of G with* $P \supset A$. *Let* $V \subset P$ *be the unipotent radical of P,* $P = R \ltimes V$ *where R is the reductive component of P containing A, and* $\bar{V}$ *the unipotent radical of the opposite parabolic* $\bar{P}$. *Then there exists* $s \in A$, *with* $s \neq e$ *and* $s \in Z(R)$ *such that the inner automorphism Int(s) is contracting on V and* $Int(s)^{-1}$ *is contracting on* $\bar{V}$.

We now return to the general situation and make the following simple observation.

8.2.5 Proposition: *Suppose that A is an automorphism of H and that* A^{-1} *is contracting. Let* $E \subset H$ *be a measurable subset. Then*
(a) if E contains a neighborhood of $e \in H$, *then as* $n \to \infty$, $A^n(E) \to H$ *in measure;*
(b) if E misses a neighborhood of $e \in H$, *then as* $n \to \infty$, $A^n(E) \to \emptyset$ *in measure.*

Proof: We recall (see the beginning of section 3.3) that $E_n \to E$ in measure means that on any subset F of finite measure we have $\mu((E_n \cap F)\Delta(E \cap F)) \to 0$. Consider assertion (a). If F is any compact set, since A^{-1} is a contraction, for n sufficiently large $A^{-n}(F) \subset E$, so $F \subset A^n(E)$. Since any set of finite measure differs from a compact set by a set of small measure, it follows that $A^n(E) \to H$ in measure. Similarly, to see (*b*), we note that if F is any compact set, and E misses a neighborhood of e, then for n sufficiently large $A^n(E) \cap F = \emptyset$.

The main result we will need about contracting automorphisms is a generalization of 8.2.5 to arbitrary measurable sets, without the assumption that the set either contains or misses a neighborhood of the identity. This will follow by a suitable approximation of an arbitrary set by open sets. In $\mathbb{R}^n$, a useful approximation of this kind is the classical Lebesgue theorem on density points.

We recall that if $E \subset \mathbb{R}^n$ is a measurable set and $x \in E$, then x is called a point of density for E if $\lim_{r \to 0} \frac{\mu(B(x,r) \cap E)}{\mu(B(x,r))} = 1$ where μ is Lebesgue measure and $B(x,r)$ is the ball of radius r about x. Then Lebesgue's theorem asserts that for any measurable E, almost every point of E is a point of density.

An examination of the proof of this theorem (see [Stein 1], for example), shows that it is true in the following more general context. (The sets $B_{x,n}$ below play the role of $B(x, 2^{-n})$ in $\mathbb{R}^m$.)

8.2.6 Theorem (*Lebesgue*): *Suppose X is a locally compact separable metrizable space, and that μ is a σ-finite measure on X which is positive on open sets and finite on compact sets. Suppose that for each $x \in X$, we have a decreasing sequence, $B_{x,n}$, of relatively compact open neighborhoods of x, forming a basis for the open sets at x, such that*

(*i*) *$\mu(B_{x,n-1})/\mu(B_{x,n})$ is constant independent of x, n.*
(*ii*) *If $n \leqq p$ and $B_{x,n} \cap B_{y,p} \neq \emptyset$, then $B_{y,p} \subset B_{x,n-2}$.*

If $E \subset X$ is any measurable set, then for almost all $x \in E$,

$$\lim_{n \to \infty} \mu(B_{x,n} \cap E)/\mu(B_{x,n}) = 1.$$

Of course, if (X, μ) is $\mathbb{R}^m$ with Lebesgue measure, as we remarked above $B_{x,n} = B(x, 2^{-n})$ satisfy these conditions. More generally, the following result shows that we can obtain such sets from more general contracting automorphisms.

8.2.7 Proposition: *Let A be a contracting automorphism of H, and suppose that $U \subset H$ is a relatively compact open neighborhood of the identity with $U = U^{-1}$ and such that $A(U^2) \subset U$. (By U^2 we mean $U \cdot U$.) For $x \in H$, let $B_{x,n} = xA^n(U)$. Then these sets satisfy the hypotheses of Theorem 8.2.6, where μ is left Haar measure on H.*

Proof: To see that they satisfy condition (i) of 8.2.6, recall first that if A is an automorphism of H and μ is Haar measure on H, then $A_*(\mu)$ is also a Haar measure and hence for some constant c, $A_*(\mu) = c\mu$. Then (i) follows from the fact that

$$\frac{\mu(B_{x,n-1})}{\mu(B_{x,n})} = \frac{\mu(xA^{n-1}(U))}{\mu(xA^n(U))} = \frac{\mu(A^{n-1}(U))}{\mu(A(A^{n-1}(U))} = c.$$

To verify (ii), suppose that there exists $z \in yA^p(U) \cap xA^n(U)$, with $n \leqq p$. Let $w \in yA^p(U)$. Then

$$z^{-1}w \in A^p(U)y^{-1}yA^p(U) = A^p(U^2) = A^{p-1}(A(U^2)) \subset A^{p-1}(U) \subset A^{n-1}(U),$$

i.e., $w \in zA^{n-1}(U)$. Since $z \in xA^n(U)$, it follows that

$$\begin{aligned} w \in xA^n(U)A^{n-1}(U) &\subset xA^{n-1}(U)A^{n-1}(U) \\ &\subset xA^{n-1}(U^2) \\ &\subset xA^{n-2}(A(U^2)) \\ &\subset xA^{n-2}(U). \end{aligned}$$

Thus, $yA^p(U) \subset xA^{n-2}(U)$, completing the proof.

Remarks: (1) When $H = \mathbb{R}^n$, $Ax = x/2$, and U is the unit ball, we recover the classical example.

(2) Given any contracting automorphism, replacing A by A^N for some $N \geqq 0$, we can assume we have $A(U^2) \subset U$ for any fixed relatively compact symmetric neighborhood of the origin.

(3) If $A(U^2) \subset U$, then $A((U^n)^2) \subset U^n$ for any n.

Finally, we return to the promised generalization of Proposition 8.2.5.

8.2.8 Proposition [*Margulis* 6]: *Let A be an automorphism of the compactly generated locally compact (separable) group H. Suppose A^{-1} is contracting. Then, replacing A by some fixed positive power of A if necessary, we have the following property. If $E \subset H$ is any measurable set, then for almost all $h \in H$,*

(i) $A^n(hE)$ converges in measure to H if $e \in hE$.
(ii) $A^n(hE)$ converges in measure to $\varnothing$ if $e \notin hE$

Proof: By remark (2) above, we can assume $A^{-1}(U^2) \subset U$ where $U \subset H$ is a relatively compact symmetric open neighborhood of $e \in H$, such that $\bigcup_{n \geqq 1} U^n = H$. To see (i) it suffices to see that for any measurable set $E \subset H$ and any open set $V \subset H$, of finite measure, we have

$$(*)\qquad \lim_{n\to\infty} \mu(A^n(hE)\cap V) = \mu(V) \qquad \text{for almost all } h\in E^{-1}.$$

However, if we apply (*) with E replaced by $H - E$, we see that (*) implies (ii) as well. To verify (*), it suffices to verify it for arbitrary measurable E and $V = U^r$ for $r \geqq 1$.

Thus, we wish to show that for almost all $h\in E^{-1}$

$$\lim_{n\to\infty} \mu(A^n(hE)\cap U^r)/\mu(U^r) = 1.$$

As we observed in the proof of 8.2.7, $A_*\mu = c\mu$, so this is equivalent to

$$\lim_{n\to\infty} \mu(hE\cap A^{-n}(U^r))/\mu(A^{-n}(U^r)) = 1,$$

i.e., for almost all $h\in E^{-1}$, (i.e., a.e. $h^{-1}\in E$),

$$\lim_{n\to\infty} \mu(E\cap h^{-1}A^{-n}(U^r))/\mu(h^{-1}A^{-n}(U^r)) = 1.$$

However, since A^{-1} is contracting, the validity of this equation follows from 8.2.6, 8.2.7, and remark (3) above. This proves 8.2.8.

In our application of 8.2.8, the group H will actually be appearing as a factor in a semidirect product. Therefore, it will be useful to formulate a consequence of 8.2.8 in that context.

Thus, suppose $G = H \rtimes L$, and that A is an automorphism of G which restricts to automorphisms of both H and L. We shall assume that A^{-1} is contracting on H and that A is the identity on L. Let $E\subset G$ be a measurable subset. We want to identify $\lim A^n(E)$ in the spirit of 8.2.8. For each $y\in L$, let $E_y = \{x\in H \mid xy\in E\}$. By 8.2.8, for each y, we have that for a.e. $h\in H$, $A^n(hE_y)\to H$ (or $\varnothing$) in measure on H if $e\in hE_y$ (or $e\notin hE_y$). We observe that $e\in hE_y$ if and only if $y\in hE$. By Fubini, we have for almost all $h\in H$, that for almost all y, $A^n(hE_y)\to H$ (or $\varnothing$) in measure on H if $y\in hE$ (or $y\notin hE$).

Once again using Fubini's theorem (by integrating over y) we deduce that for almost all $h\in H$,

$$A^n(hE)\to H\cdot(hE\cap L) \text{ in measure on } G.$$

Suppose we have $h\in H$ for which this is true, and let $y\in L$ be arbitrary. Then $yA^n(hE)\to yH\cdot(hE\cap L)$, and since A is trivial on L and y normalizes H, we deduce the following.

8.2.9 Corollary: *Let $G = H \rtimes L$ and A be as above. Then (possibly after replacing A by some positive power of A), for any $E \subset G$, we have that for almost all $g \in G$, $A^n(gE) \to \psi(gE)$ in measure on G, where $\psi(F) = H \cdot (F \cap L)$ for $F \subset G$.*

8.3 Completion of the proof – equivariant measurable quotients of flag varieties

In this section we present Margulis' proof of Theorem 8.1.4. As with the proof of the superrigidity theorems in Chapter 5, we shall use the fact that up to null sets, we can identify G/P with a unipotent subgroup of G (Lemma 5.1.4). While this was useful in chapter 5 for showing that a suitable measurable map on G/P was rational, here it will be useful as we will be able to apply the results of section 8.2 on contracting automorphisms.

We shall begin by reviewing some structural facts about semisimple groups (with the hypotheses in 8.1.4), presenting $SL(n, \mathbb{R})$ as an example. In this section we shall be taking G/H to be $\{gH\}$, so that G acts on the left on G/H.

Let $A \subset G$ be a maximal connected $\mathbb{R}$-split abelian subgroup (so for $SL(n, \mathbb{R})$ we can take A to be the positive diagonals.) Let P be a minimal parabolic subgroup containing A, and V the unipotent radical of A. (Thus for $SL(n, \mathbb{R})$, we can take P to be the upper triangular matrices and V the subgroup of P consisting of all matrices in P with all diagonal entries equal to 1.) Let $\bar{P}$ be the opposite parabolic, with unipotent radical $\bar{V}$, (for $SL(n, \mathbb{R})$, the corresponding lower triangular matrices.) Then (Lemma 5.1.4), the restriction of the natural map $G \to G/P$ to $\bar{V}$ defines a rational isomorphism of $\bar{V}$ with an open conull subset of G/P.

Now let $P_0 \supset P$ be another parabolic. (For $SL(n, \mathbb{R})$, take P_0 to be the stabilizer of a flag as in the discussion preceding 8.2.4.) Let V_0 be the unipotent radical of P_0, and $\bar{P}_0$, $\bar{V}_0$ the corresponding opposite subgroups. One shows as in Lemma 5.1.4 that the natural map $\bar{V}_0 \to G/P_0$ is also a rational isomorphism onto an open conull subset of G/P_0. In 8.1.4, we are interested in the map $G/P \to G/P_0$, and hence we wish to express this in terms of $\bar{V}$, $\bar{V}_0$. In other words, we want to identify the map $\bar{V} \to \bar{V}_0$ making the following diagram commute.

(*)
$$\begin{array}{ccc} \bar{V} & \longrightarrow & G/P \\ \downarrow & & \downarrow \\ \bar{V}_0 & \longrightarrow & G/P_0 \end{array}$$

Let R_0 be the reductive component of P_0 containing A, so $P_0 = R_0 \ltimes V_0$,

$\bar{P}_0 = R_0 \ltimes \bar{V}_0$. (For $SL(n, \mathbb{R})$, this group is also described in the discussion preceding 8.2.4.) Let $\bar{L}_0 = R_0 \cap \bar{V} (= P_0 \cap \bar{V})$. (For $P_0 \subset SL(n, \mathbb{R})$, as above,

$$\bar{L}_0 = \{g \in R_0 \mid g_{ii} = 1 \text{ and } g_{ij} = 0 \text{ if } i < j\}.)$$

Then $\bar{V}_0, \bar{L}_0 \subset \bar{V}$, and in fact $\bar{V} = \bar{V}_0 \rtimes \bar{L}_0$. Thus, the map $\bar{V} \to \bar{V}_0$ we needed above is simply the projection of $\bar{V}$ onto $\bar{V}_0$ given by this semidirect product decomposition.

Let $i: \bar{V} \to G/P$ be the natural map, which is a measure space isomorphism. For $g \in G$, the action of g on G/P will then give us an a.e. defined measure space automorphism of $\bar{V}$. More precisely, if $g \in G$, then for almost all $x \in \bar{V}$, $gi(x) \in i(\bar{V})$. Therefore, we define $g \circ x \in \bar{V}$ via the equation $g \cdot i(x) = i(g \circ x)$, so that for each g this is defined a.e. For certain elements of G, we can be more explicit. We observe that

(i) If $g \in \bar{V}$, then $g \circ x = g \cdot x$ for all $x \in \bar{V}$.
(ii) If $g \in P \cap \bar{P}$ (i.e., $g \in R$ where $P = R \ltimes V$), then $g \circ x = gxg^{-1}$.

To see (ii), note that
$$\begin{aligned} i(gxg^{-1}) &= gxg^{-1}P \\ &= gxP \\ &= gi(x). \end{aligned}$$

By the equation defining $g \circ x$, we obtain (ii). We shall identify the G-action on $B(G/P)$ with this induced action on $B(\bar{V})$.

Now let $B \subset B(\bar{V}) = B(G/P)$ be a Γ-invariant closed Boolean subspace, and suppose $E \subset \bar{V}$ with $E \in B$. Let $g \in G$. Of course, if this implied $g \cdot E \in B$, we would be done. Our first reduction of the proof is to show that it suffices to show that $g \cdot \tilde{E} \in B$ where $\tilde{E}$ is a certain subset related to E. More precisely, suppose $E \subset \bar{V}$, and let $P_0 \supset P$ be as above, so that $\bar{V} = \bar{V}_0 \rtimes \bar{L}_0$. Define $\psi_0(E) = \bar{V}_0 \cdot (E \cap \bar{L}_0)$ (Cf. 8.2.9.).

8.3.1 Lemma: *To prove Theorem* 8.1.4, *it suffices to prove that for any* $E \in B \subset B(\bar{V})$, $g \in G$ *and parabolic* P_0, $G \supsetneq P_0 \supsetneq P$, *we have* $g \cdot \psi_0(vE) \in B$ *for almost all* $v \in \bar{V}$.

Proof: We continue to identify $B(\bar{V}) \cong B(G/P)$. If $P_0 \supset P$ we shall also identify $B(G/P_0)$ as a subspace of $B(G/P)$ via the natural map $G/P \to G/P_0$. In light of our discussion above, this identification is identified with the realization of $B(\bar{V}_0)$ as a subspace of $B(\bar{V})$ induced by the projection $\bar{V} \to \bar{V}_0$.

Let B' be the largest G-invariant closed Boolean subspace of B, so we have $B' \subset B \subset B(G/P)$. By our remarks in section 8.1 B' can be identified with $B(G/P')$

where $P' \supset P$. We may have $P' = G$ (so that B' is trivial), but we can assume $P' \neq P$ for otherwise we are clearly done. We claim that $B' = B$. Suppose not, and choose $E \subset G/P$ with $E \in B$ but $E \notin B(G/P')$.

Let $P_1, \ldots, P_m$ be the standard parabolics with $P' \supset P_i \underset{\neq}{\supset} P$, and P_i minimal with respect to this property. As above, write $P_i = R_i \ltimes V_i$, $\bar{L}_i = R_i \cap \bar{V}$. The structure theory of semisimple group implies that P' is generated by $\{P_i\}$. (This is a straightforward exercise for $G = SL(n, \mathbb{R})$.) We have the map $G \to G/P$, and we can thus identify $B(G/P)$ as a subspace of $B(G)$, and since $E \in B(G/P)$, we have, as an element of $B(G)$, that $E \cdot g = E$ for all $g \in P$. On the other hand, since $E \notin B(G/P')$, we cannot have $E \cdot g = E$ for all $g \in P'$. Thus, for some i, there is an element $g \in P_i$ such that $E \cdot g \neq E$ (modulo null sets). In other words, $E \in B(G/P)$ is not in the image of $B(G/P_i)$ under the embedding $B(G/P_i) \to B(G/P)$ defined by the natural projection. In terms of the diagram (*) above, with $P_0 = P_i$, this means that considering $E \in B(\bar{V})$, we have $E \cdot \bar{L}_i \neq E$.

We now consider $E \in B(\bar{V})$, and recall that $\bar{V} = \bar{V}_i \rtimes \bar{L}_i$. Since $E \cdot \bar{L}_i \neq E$, it follows from Fubini's theorem that for a set of $w \in \bar{V}_i$ of positive measure, $\{y \in \bar{L}_i \mid wy \in E\}$ is neither null nor conull in $\bar{L}_i$, i.e., $\bar{L}_i \cap w^{-1}E$ is neither null nor conull in $\bar{L}_i$. For such a $w \in \bar{V}_i$, we clearly have for any $v \in w\bar{L}_i$ that $\bar{L}_i \cap v^{-1}E$ is neither null nor conull in $\bar{L}_i$. Hence, we deduce that for a set of $v \in \bar{V}$ of positive measure, $\psi_i(vE) \in B(\bar{V})$ is not right $\bar{L}_i$-invariant, hence not P'-invariant. In other words, under the identification in diagram (*), $\psi_i(vE)$ is not in the image of $B(G/P_i) \subset B(G/P)$, and in particular is not in the image of $B(G/P')$ (which is B'). However, $\mathbb{R}$-rank$(G) \geqq 2$, and hence $P_i \neq G$. Therefore, by the hypotheses of the lemma, we can find $v \in \bar{V}$ with $\psi_i(vE) \notin B'$, but $g \cdot \psi_i(vE) \in B$ for all $g \in G$. Let $\tilde{B}$ be the closed Boolean subspace generated by $\{g\psi_i(vE) \mid g \in G\}$ and B'. Then $\tilde{B} \subset B$ and $\tilde{B}$ is G-invariant. However, $\psi_i(vE) \in \tilde{B} - B'$, so this contradicts the maximality assumption on B', proving Lemma 8.3.1.

The completion of the proof of Theorem 8.1.4, i.e., verification of the conditions in Lemma 8.3.1, will follow from the following two lemmas.

8.3.2 Lemma: *Let P_0 be a parabolic with $G \underset{\neq}{\supset} P_0 \supset P$. Write $P_0 = R_0 \rtimes V_0$ as above, and as in* 8.2.4, *choose non-trivial $s \in Z(R_0)$, the center of R_0, (with $s \in A$ as in* 8.2.4) *such that $Int(s)$ is contracting on V_0 and $Int(s)^{-1}$ is contracting on $\bar{V}_0$. Then for any measurable set $E \subset \bar{V}$, we have (possibly replacing s by s^N for some $N \geqq 1$), for almost all $v \in \bar{V}$, as $n \to \infty$, $Int(s)^n(vE) \to \psi_0(vE)$ in measure on $\bar{V}$.*

Proof: Apply Corollary 8.2.9.

The second lemma we will need is an ergodicity type of statement.

8.3.3 Lemma: *Let s be as in* 8.3.2. *Then for almost all* $v \in \bar{V}$, $\{\gamma v^{-1} s^{-n} | \gamma \in \Gamma, n \geqq 1\}$ *is dense in G.*

This lemma is close to being a consequence of Moore's theorem (2.2.6). Namely, the latter implies that the integer action defined by powers of s acts ergodically on G/Γ, and hence (by 2.1.7), for almost all $g \in G$, $\{\gamma g s^{-n} | \gamma \in \Gamma, n \in Z\}$ is dense in G. Thus 8.3.3 differs from this last assertion in two respects. First, in 8.3.3 we consider only positive powers of s^{-1}, and second we have an assertion about almost all $v \in \bar{V}$. The latter is of course a lower dimensional space than G, and therefore such an assertion is not an immediate consequence of Moore's theorem. We shall show how to obtain 8.3.3 from Moore's theorem, but first show why 8.3.2 and 8.3.3 suffice to prove Theorem 8.1.4.

Proof of Theorem 8.1.4: We verify the condition of Lemma 8.3.1. Let $E \subset \bar{V}$ with $E \in B$. Then for almost all v, the conclusions of 8.3.2 and 8.3.3 are valid. Fix $g \in G$. Then by 8.3.3, we can write $g = \lim_j \gamma_j v^{-1} s^{-n_j}$ where $n_j \geqq 1$. Furthermore, since the conclusion of 8.3.3 is valid if we replace s by some positive power of s, we can assume $n_j \to \infty$ as $j \to \infty$. Let $g_j = \gamma_j v^{-1} s^{-n_j}$, so that $\gamma_j = g_j s^{n_j} v$. Recalling that we are denoting the G action on $\bar{V}$ and $B(\bar{V})$ by $\circ$, we have $\gamma_j \circ E = (g_j s^{n_j} v) \circ E = g_j \circ (s^{n_j} v E s^{-n_j})$ by (i), (ii) in the discussion preceding 8.3.1. As $j \to \infty$, $g_j \to g$ and $s^{n_j} v E s^{-n_j} \to \psi_0(vE)$ in measure on $\bar{V}$. We thus have $\gamma_j \circ E \to g \circ \psi_0(vE)$ in measure, and hence in the weak-*-topology on $B(\bar{V})$. However, since $E \in B$ and B is closed, we deduce that $g \circ \psi_0(vE) \in B$ and by Lemma 8.3.1, this completes the proof of Theorem 8.1.4.

It therefore remains only to prove Lemma 8.3.3. We begin with a general observation to deal with the fact that we restrict ourselves to $n \geqq 1$ in 8.3.3.

8.3.4 Proposition: *Suppose the group of integers Z acts ergodically and with a finite invariant measure on a space* (S, μ). *Then for any* $Y \subset S$ *with* $\mu(Y) > 0$, $\bigcup_{n \geqq 1} (Y \cdot n)$ *is conull.*

Proof: Let $W = \bigcup_{n \geqq 1} (Y \cdot n)$. Then $W \cdot 1 \subset W$. Since the action is measure preserving and the measure is finite, $W \cdot 1 = W$. Thus W is essentially invariant under the Z-action, and by ergodicity W is conull.

8.3.5 Corollary: *With the notation of* 8.3.4, *if S is also a separable metric space and μ is positive on open sets, then for almost all $x \in S$, $\{x \cdot (-n) | n \geqq 1\}$ is dense in S.*

Proof: The proof of 2.1.7 applies in this case as well.

Proof of Lemma 8.3.3: Let $W = \{g \in G | \{\gamma g s^{-n} | \gamma \in \Gamma, n \geqq 1\}$ is dense in $G\}$. By Moore's theorem (2.2.6), $\{s^{-n}, n \in Z\}$ is ergodic on G/Γ, and hence by Corollary 8.3.5, W is conull. Recall that the multiplication map $\bar{V} \times P \to G$ is an injective map onto a conull subset of G. For each $v \in \bar{V}$, let $Y_v = \{p \in P | vp \in W\}$, and let $U = \{v \in \bar{V} | Y_v$ is conull in $P\}$. Then by Fubini, U is conull in $\bar{V}$. To prove the lemma, it suffices to show that $e \in Y_v$ for all $v \in U$.

Thus, we fix $v \in U$. Then we can choose $p_k \in Y_v$ such that $p_k \to e$. By the choice of W, we then have that for each $k \geqq 1$, $\{\gamma v p_k s^{-n} | \gamma \in \Gamma, n \geqq 1\}$ is dense in G, i.e., $\{\gamma v s^{-n} s^n p_k s^{-n} | \gamma \in \Gamma, n \geqq 1\}$ is dense in G. Recall that we can write $P = R \ltimes V$, and that R is the centralizer of A in G. (In $SL(n, \mathbb{R})$, R is just the subgroup of all diagonal matrices.) Let $p_k = u_k r_k$, $r_k \in R$, $u_k \in V$. Since $s \in A$, s commutes with r_k, and hence for each $k \geqq 1$ we have $\{\gamma v s^{-n} s^n u_k s^{-n} | \gamma \in \Gamma, n \geqq 1\}$ is dense in G. Since $p_k \to e$, we also have $u_k \to e$.

Suppose $e \notin Y_v$, so that $\{\gamma v s^{-n} | n \geqq 1\}$ misses an open set $A \subset \bar{V}$. Then we can find an open set $B \subset A$ and an open symmetric neighborhood C of the origin in $\bar{V}$ such that $BC \subset A$. Thus $B \cap (X - A)C = \varnothing$. However, $\{u_k\}$ is relatively compact since $u_k \to e$. Thus, for n sufficiently large, $\mathrm{Int}(s)^n(\{u_k\}) \subset C$ since $\mathrm{Int}(s)$ is contracting on V. For any n, there are only finitely many elements of $\mathrm{Int}(s)^n(\{u_k\})$ which do not lie in C, and it follows that for some k (in fact for all sufficiently large k), $\{\mathrm{Int}(s^n) u_k | n \geqq 1\} \subset C$. But since $\{\gamma v s^{-n} | n \geqq 1\} \subset X - A$, and $(X - A)C \cap B = \varnothing$, it follows that for some k, $\{\gamma v s^{-n} s^n u_k s^{-n} | n \geqq 1\}$ does not intersect B. This is impossible because this set is dense, contradicting the assumption that $e \notin Y_v$. This completes the proof of the lemma.

9 Further Results on Ergodic Actions

9.1 Cocycles and Kazhdan's property

If a group has the Kazhdan property, then there are significant restrictions on the type of cocycles certain actions of the group can have. We begin with the following.

9.1.1 Theorem: *Let G be a group with the Kazhdan property. Suppose S is an ergodic G-space with invariant probability measure. Let $\alpha: S \times G \to H$ be a cocycle where H is an amenable group. Then α is equivalent to a cocycle into a compact subgroup. (In particular, if $H = \mathbb{R}^n \times Z^m$, α is trivial.)*

For G discrete, this was first shown in [Schmidt 2], [Zimmer 7].
We preface the proof with the following general observation.

9.1.2 Lemma: *Let G, H be locally compact, S an ergodic G-space and $\alpha: S \times G \to H$ a cocycle. Let π be the regular representation of H. Then α is equivalent to a cocycle into a compact subgroup of H if and only if there is a $\pi \circ \alpha$-invariant function $\varphi: S \to L^2(H)_1$, the norm one vectors in $L^2(H)$. (Definition 4.2.17.)*

Proof: If $\alpha \sim \beta$ where $\beta(S \times G) \subset K$, $K \subset H$ a compact group, then there is a $\pi(K)$-invariant vector in $L^2(H)_1$ (simply because K is compact), and hence a $\pi \circ \beta$-invariant function $S \to L^2(H)_1$. Therefore, there is such a $\pi \circ \alpha$-invariant function (4.2.18c).

To see the converse, we first observe that for each $\varphi \in L^2(H)$, $\pi(h)\varphi \to 0$ in the weak-*-topology as $h \to \infty$ in H. (To see this, suppose $\psi \in L^2(H)$. Then we can find compact sets A, B such that $\| \chi_A \varphi - \varphi \|$ and $\| \chi_B \psi - \psi \|$ are small. But as $h \to \infty$, we can assume $Ah \cap B = \varnothing$, and hence $\langle \pi(h)\varphi | \psi \rangle$ is small.) Thus, each H-orbit in $L^2(H)$ is norm closed, so that the H action on $L^2(H)_1$ is smooth (2.1.12). Furthermore, since $\pi(h)\varphi \to 0$ in the weak-*-topology as $h \to \infty$, the stabilizer of any $\varphi \neq 0$ must be compact. (Cf. the comments following the statement of Theorem 2.2.20.) The required converse now follows immediately from the cocycle reduction lemma (5.2.11).

Proof of Theorem 9.1.1: Let π be the regular representation of H. Then $\pi \circ \alpha$ is a cocycle $S \times G \to U(L^2(H))$, and we can form the induced representation, say σ, of G on $L^2(S; L^2(H))$; namely $(\sigma(g)f)(s) = \pi(\alpha(s, g))f(sg)$. We know from Theorem 7.1.8 that π almost has invariant vectors. We claim that σ does as well.

Let $K \subset G$ be a compact symmetric set and $\varepsilon > 0$. Let ν denote Haar measure on G. If $A \subset H$ is compact and $g \in G$, let

$$S(g, A) = \{s \in S \mid \alpha(s, g) \in A, \ \alpha(s, g^{-1}) \in A\}.$$

Since α is measurable and H is σ-compact, there is a compact set $A \subset H$ such that

$$(\mu \times \nu)(\{s, g) \in S \times K \mid \alpha(s, g) \in A, \ \alpha(s, g^{-1}) \in A\}) \geqq (1 - \varepsilon/3)(\nu(K)).$$

By Fubini's theorem, $\nu(K_0) > 0$, where

$$K_0 = \{g \in K \mid \mu(S(g, A)) \geqq 1 - \varepsilon/3\}.$$

Clearly K_0 is symmetric. Thus $K_0^2 = K_0 K_0^{-1}$ contains a neighborhood W of the identity (Appendix B). By the cocycle identity $\alpha(s, gh) = \alpha(s, g)\alpha(sg, h)$ and the fact that μ is G-invariant, for any $y \in K_0^2$ we have $\mu(S(y, A^2)) \geqq 1 - 2\varepsilon/3$. Now choose $g_i \in G$, $i = 1, \ldots, n$, such that $\cup g_i W \supset K$. Then there is a compact set $B \subset H$ such that $\mu(S(g_i, B)) > 1 - \varepsilon/3$ for all i. It follows as above from the cocycle identity that $\mu(S(g, BA^2)) > 1 - \varepsilon$ for all $g \in K$.

Since π almost has invariant vectors, we can choose a unit vector $x \in L^2(H)$ such that $\| \pi(h)x - x \| < \varepsilon$ for all $h \in BA^2$. Define $\varphi \in L^2(S; L^2(H))$ by $\varphi(s) = x$ for all $s \in S$. Then for $g \in K$, we have

$$\mu(\{s \in S \mid \| \pi(\alpha(s, g))\varphi(sg) - \varphi(s) \| < \varepsilon\}) > 1 - \varepsilon.$$

Therefore an elementary computation shows that $\| \sigma(g)\varphi - \varphi \| < \lambda(\varepsilon)$ for all $g \in K$ where $\lambda(\varepsilon) \to 0$ as $\varepsilon \to 0$. This verifies our assertion that σ almost has invariant vectors.

Since G has the Kazhdan property, this implies that σ has a non-trivial invariant vector. In other words, letting H act in $L^2(H)$ via π, there is an α-invariant function $f: S \to L^2(H)$. Finally, we remark that we can assume that for all $s, f(s) \in L^2(H)_1$ by ergodicity. More precisely, since $f \neq 0, f(s) \neq 0$ on a set of positive measure and since $\alpha(s, g) \cdot f(sg) = f(s)$ a.e., ergodicity implies that $\| f(s) \|$ is essentially constant. The theorem now follows from Lemma 9.1.2.

We remark that this theorem is false if the assumption of finite invariant measure is eliminated.

9.1.3 Corollary: *Let $G = \prod G_i$, G_i a connected simple non-compact Lie group with finite center. Let S be an ergodic G-space with finite invariant measure. Assume that G has the Kazhdan property, e.g., $\mathbb{R}$-rank$(G_i) \geqq 2$ for all i. Suppose $\alpha: S \times G \to H$ is a cocycle with H amenable. Then α is equivalent to a cocycle into a compact subgroup of H. (In particular, if $H = \mathbb{R}^n \times Z^m$, α is trivial.)*

Proof: This follows from 9.1.1, and 7.1.4.

Kazhdan's property has significant implications for the kind of actions such a group can have on a compact manifold. (See [Zimmer 17].) The geometric background required for most of these results makes them beyond the scope of this book. However, we shall be applying Theorem 9.1.1 in section 9.4 to study entropy of actions of semisimple Lie groups or their lattices on compact manifolds. Here, we only present a very elementary illustration of the geometric implications of the Kazhdan property, showing that groups with this property do not have volume preserving hyperbolic actions. We recall the latter notion.

Let M be a compact manifold and $f: M \to M$ a diffeomorphism. There are a variety of conditions of f that describe "hyperbolic" behavior. Here, as an illustration we consider the following notion. See for example [Katok 1] for a discussion of notions of hyperbolicity. Call f weakly hyperbolic if there is a df-invariant splitting of the tangent bundle of M into measurable subbundles, $TM = E_1 + E_2$ such that for all $v \in E_1$, $\| df^n v \| \to 0$ as $n \to \infty$ and for all $v \in E_2$, $\| df^{-n} v \| \to 0$ as $n \to \infty$. Since M is compact, this is independent of the Riemannian metric.

If Γ is a group acting on M by diffeomorphisms, call the action weakly hyperbolic if the tangent bundle splits into $d\Gamma$-invariant subbundles as above, and with respect to this splitting, there is some $\gamma \in \Gamma$ that is weakly hyperbolic.

9.1.4 Proposition: *Suppose Γ is a group with the Kazhdan property. Then Γ does not have any weakly hyperbolic volume preserving ergodic actions on a compact manifold.*

Proof: Suppose we had such an action. We can measurably choose orthogonal bases for E_i, and hence df^n acting on each E_i will give us a cocycle $\alpha_i: S \times \Gamma \to GL(n_i, \mathbb{R})$, and for some $\gamma_0 \in \Gamma$, and $i = 1$ or 2, $\| \alpha_i(s, \gamma_0^n) v \| \to 0$ for all $v \in \mathbb{R}^{n_i}$. (Cf. Example 4.2.3.) Let us simply denote α_i by α. Let $\beta: S \times \Gamma \to \mathbb{R}$ be $\beta(s, \gamma) = \log|\det \alpha(s, \gamma)|$. Then for all s, we have $\beta(s, \gamma_0^n) \to -\infty$ as $n \to \infty$. However, by Theorem 9.1.1, β is trivial, i.e., there is a function $\varphi: S \to \mathbb{R}$ such that $\beta(s, \gamma) = \varphi(s) - \varphi(s\gamma)$. Choose a subset $A \subset S$ of positive measure on which

φ is bounded. We will clearly have a contradiction to the fact that $\beta(s, \gamma_0^n) \to -\infty$ if we can show that for almost all $s \in A$, there is a sequence $n_i \to \infty$ such that $s \cdot \gamma_0^{n_i} \in A$. However, this follows from the following classical result.

9.1.5 Lemma: (*Poincaré recurrence*): *Let $T: S \to S$ be a measure space isomorphism preserving the finite measure μ. Let $A \subset S$ with $\mu(A) > 0$. Then for almost all $x \in A$, $T^n(x) \in A$ for infinitely many positive n.*

Proof: Let $A_N = \{x \in A \mid \text{for all } n \geqq N, T^n x \notin A\}$. It suffices to show $\mu(A_N) = 0$ for all N. Suppose not. Consider the sets $T^{rN}(A_N)$ where r is a positive integer. These are all of the same measure since T preserves measure, and hence there are r_1, r_2, (say $r_1 < r_2$) such that $T^{r_1 N}(A_N) \cap T^{r_2 N}(A_N) \neq \varnothing$. Therefore, there exist $x, y \in A_N$ such that $T^{r_1 N} x = T^{r_2 N} y$, i.e., $T^{(r_2 - r_1)N} y = x$. But since $x, y \in A_N \subset A$, this is impossible.

As with amenability, one can extend the notion of the Kazhdan property for a group to that of the Kazhdan property for an action. Here, we shall only indicate, without proof, some developments in this direction, for simplicity of exposition restricting our attention to discrete groups.

9.1.6 Definition [Zimmer 7]: (a) Let Γ be a discrete group, (S, μ) an ergodic Γ-space, μ a quasi-invariant probability measure. Let $\varepsilon > 0$ and $K \subset \Gamma$ a finite subset of Γ. Let $\mathscr{H}$ be a Hilbert space and $\alpha: S \times \Gamma \to U(\mathscr{H})$ be a cocycle. Let $F_1(S, \mathscr{H}) = \{\varphi \in F(S, \mathscr{H}) \mid \|\varphi\|_\infty = 1\}$. A function $\varphi \in F_1(S, \mathscr{H})$ is called (ε, K)-invariant if $\mu\{s \in S \mid |\langle \alpha(s, \gamma)\varphi(s\gamma) \mid \varphi(s) \rangle - 1| < \varepsilon\} \geqq 1 - \varepsilon$ for all $\gamma \in K$.
(b) We say that α almost has invariant functions if for all (ε, K), there exists an (ε, K)-invariant $f \in F_1(S, \mathscr{H})$.
(c) We say that the action of Γ on S has the Kazhdan property if any cocycle $\alpha: S \times \Gamma \to U(\mathscr{H})$ which almost has invariant functions actually has an invariant function $\varphi \in F_1(S, \mathscr{H})$.

We then have the following results. We refer the reader to [Zimmer 7] for proofs.

9.1.7 Theorem [*Zimmer* 7]:
(*a*) *If Γ has the Kazhdan property, and S has a finite Γ-invariant measure, then the action of Γ on S has the Kazhdan property.* (*This is false in general without the assumption of finite invariant measure.*)

(b) If the action of Γ *on* S *has the Kazhdan property, finite invariant measure and is weakly mixing* (*i.e.,* $L^2(S) \ominus \mathbb{C}$ *has no finite dimensional invariant subspaces*), *then* Γ *has the Kazhdan property.* (*This is also false in general without the assumption of finite invariant measure.*)
(c) If the action of Γ *on* S *has the Kazhdan property, then any cocycle into an amenable group is equivalent to one into a compact subgroup.* (*Cf. Theorem* 9.1.1.)
(d) For essentially free actions, the Kazhdan property is an invariant of orbit equivalence.

As a consequence of 9.1.7 and 7.4.5, we have:

9.1.8 Corollary: *For* $n \geqq 3$, *let* $\Gamma_1 = SL(m, Z) \times Z^m$ *and* $\Gamma_2 = SL(n, Z) \ltimes Z^n$ *where the action is matrix multiplication. Then* Γ_1 *and* Γ_2 *do not have essentially free, weakly mixing actions with finite invariant measure which are orbit equivalent.*

9.2 The algebraic hull of a cocycle

In this section we introduce a useful invariant of a cocycle into an algebraic group. We shall be applying this in Section 9.3 to prove a basic property about orbit equivalence for actions of lattices in higher rank semisimple Lie groups and in Section 9.4 to the study of entropy of the actions of these groups on compact manifolds. We will be concerned with the following question.

Suppose G, H are locally compact, S an ergodic G-space and $\alpha: S \times G \to H$ is a cocycle. Is there a unique (up to conjugacy) smallest subgroup L of H such that α is equivalent to a cocycle taking values in L? In general the answer is no. However, if H is algebraic and we restrict attention to algebraic subgroups, then the answer is yes.

9.2.1 Proposition: *Let k be a local field of characteristic* 0, *H an algebraic k-group and* $\alpha: S \times G \to H_k$. *Then there is an algebraic k-group* $L \subset H$ *such that* α *is equivalent to a cocycle taking values in* L_k, *but is not equivalent to a cocycle taking values in any proper subgroup of* L_k *of the form* M_k, *where M is also a k-group. The group* L_k *is unique up to conjugacy in* H_k.

Proof: Such a minimal L_k exists by the descending chain condition on algebraic groups. To see it is unique up to conjugacy, suppose we have two such minimal groups, L, J. Then by 4.2.18(b), we have α-invariant functions $\varphi: S \to H_k/L_k$,

$\psi: S \to H_k/J_k$. Then $\lambda = (\varphi, \psi): S \to H_k/L_k \times H_k/J_k$ is also α-invariant. However, the action of H_k on $H_k/L_k \times H_k/J_k$ is smooth (3.1.3). Therefore, by the cocycle reduction lemma (5.2.11), α is equivalent to a cocycle into the stabilizer of a point in $H_k/L_k \times H_k/J_k$. In any orbit, we can choose the stabilizer to be of the form $L_k \cap hJ_kh^{-1}$ where $h \in H_k$. By the minimality property of L_k, $L_k \subset hJ_kh^{-1}$. Similarly, J_k is contained in a conjugate of L_k, and this implies that J_k and L_k are conjugate.

9.2.2 Definition: The group L_k (or more precisely its conjugacy class) in 9.2.1 is called the algebraic hull of α.

A useful example is the following

9.2.3 Theorem [*Zimmer* 4]: *Suppose $\alpha: S \times G \to H_{\mathbb{R}}$ where H is a connected $\mathbb{R}$-group. If the G action on S is amenable, then the algebraic hull of α is amenable.*

Proof: Let N be the radical of H, so that $Q = H/N$ semisimple. Let P be a minimal $\mathbb{R}$-parabolic subgroup of Q. Then by amenability there is an α-invariant function $S \to M(Q_{\mathbb{R}}/P_{\mathbb{R}})$, the latter being the space of probability measures on $Q_{\mathbb{R}}/P_{\mathbb{R}}$. The action of $Q_{\mathbb{R}}$ on $M(Q_{\mathbb{R}}/P_{\mathbb{R}})$ is smooth by Corollary 3.2.17 and the stabilizers are amenable and algebraic by Corollary 3.2.22. Since $N_{\mathbb{R}}$ acts trivially on $Q_{\mathbb{R}}/P_{\mathbb{R}}$, the $H_{\mathbb{R}}$-action on $M(Q_{\mathbb{R}}/P_{\mathbb{R}})$ is obviously also smooth, and since $N_{\mathbb{R}}$ is amenable, the stabilizer in $H_{\mathbb{R}}$ will also be amenable (4.1.6). The result now follows from the cocycle reduction lemma (5.2.11).

We remark that virtually the same proof shows (using the fact that any connected group has a normal compact subgroup with a Lie quotient):

9.2.4 Corollary: *Let $\alpha: S \times G \to H$ where H is connected and the G action on S is amenable. Then α is equivalent to a cocycle into an amenable subgroup of H.*

We also remark that if H is discrete, this result is no longer true. For a counterexample, see [Zimmer 4]. When 9.2.4 is applied to the cocycle $\alpha: S \times G \to G$, $\alpha(s, g) = g$, where G is connected, it provides the basic step in the proof of the following result.

9.2.5 Theorem: *Suppose G is connected and that S is an amenable ergodic G-space. Then S is induced from an action of an amenable subgroup of G.*

Proof: Applying 9.2.4 to $\alpha(s, g) = g$, we can deduce from Example 4.2.18 that there is a G-map $S \to G/H$ for some amenable subgroup H. One can show that, quite generally, this implies that S is induced from an H-action. We refer the reader to [Zimmer 4] for details.

We recall that for free actions, amenability is an invariant of orbit equivalence (4.3.10). On the other hand, the condition that an action be induced from a certain subgroup, or from a certain type of subgroup, is a feature that depends on the action, not only on the action up to orbit equivalence. Thus, Theorem 9.2.5 has in common with the rigidity theorem for actions (5.2.1) that it deduces structural results about the action from features of orbit equivalence. We also recall that in 9.2.5 (in contrast to the rigidity theorem) the actions will not be actions with finite invariant measure (4.3.3).

We also remark that one of the hypotheses in the superrigidity theorem for cocycles (5.2.5) is actually a hypothesis about the algebraic hull of a cocycle. Namely the assumption that α is not equivalent to a cocycle into a subgroup of the form L_k where $L \subset H$ is a proper algebraic k-group is of course just the assumption that the algebraic hull of α is in fact H_k.

We will need the following result in Section 9.4.

9.2.6 Proposition: *Let $\alpha: S \times G \to H_k$ where H is a k-group and suppose the algebraic hull of α is H_k. Let H^0 be the connected component of H. Then the skew product action (see the discussion preceding* 4.2.21) *of G on $X = S \times_\alpha H_k/(H^0)_k$ is ergodic. Further, if we let $\tilde{\alpha}: X \times G \to H_k$ be $\tilde{\alpha}(s, [h], g) = \alpha(s, g)$, then the algebraic hull of $\tilde{\alpha}$ is $(H^0)_k$.*

Proof: Let $A \subset X$ be an invariant set of positive measure. For each $s \in S$, let $A_s = \{y \in H_k/(H^0)_k \mid (s, y) \in A\}$. Then A_s is not empty on a set of positive measure, and since A is invariant under the skew product action, for each $g \in G$, $A_s \cdot \alpha(s, g) = A_{sg}$. By ergodicity of G on S, A_s is not empty on a conull set of s, and so $s \to A_s$ is an α-invariant map into the set of non-empty finite subsets of $H_k/(H^0)_k$. By the cocycle reduction lemma (5.2.11), it follows that α is equivalent to a cocycle into a subgroup of H_k leaving a finite subset $F \subset H_k/(H^0)_k$ invariant, and such that A_s is in the H_k-orbit of F for almost all s. However, since the algebraic hull of α is H_k, it follows that $F = H_k/(H^0)_k$. Thus, $A_s = H_k/(H^0)_k$ a.e., so that A is conull. This establishes ergodicity. Essentially by definition, the projection map $X \to H_k/(H^0)_k$ onto the second coordinate is an $\tilde{\alpha}$-invariant map. Thus by 4.2.18(b), $\tilde{\alpha}$ is equivalent to a cocycle into $(H^0)_k$. Therefore, it only remains to prove the assertion that $\tilde{\alpha}$ is not equivalent to a cocycle into L_k,

$L \underset{\neq}{\subset} H^0$ a k-subgroup.

Suppose $\tilde{\alpha}$ is equivalent to a cocycle into L_k. Then (4.2.18(b)), there is an $\tilde{\alpha}$-invariant map $\varphi: S \times H_k/(H^0)_k \to H_k/L_k$, i.e., for each $g \in G$, we have

$$\varphi(s, y)\alpha(s, g) = \varphi((s, y)\cdot g) = \varphi(sg, y\alpha(s, g)) \quad \text{for almost all } (s, y).$$

Let $E = F(H_k/(H^0)_k, H_k/L_k)$ be the space of functions, so that H_k acts on E by $(f\cdot h)(y) = f(yh^{-1})\cdot h$. Furthermore, this action is smooth. (To see this, first observe that we can identify E as a subset of the space of all subsets of $V = H_k/(H^0)_k \times H_k/L_k$ of cardinality d, where d is the order of $H_k/(H^0)_k$. Thus to see that H_k acts smoothly on E, it suffices to see that H_k acts smoothly on $V^d/S(d)$ where $S(d)$ is the symmetric group on d letters. However, H_k is smooth on V^d by 3.1.3, i.e., $H_k \backslash V^d$ is countably separated. Hence, so is $H_k \backslash (V^d/S(d)) = (H_k \backslash V^d)/(S(d)$ by 2.1.21.) Define $\Phi: S \to E$ by $(\Phi(s))(y) = \varphi(s, y)$. Then the above consequence of $\tilde{\alpha}$-invariance implies that $\Phi(sg) = \Phi(s)\cdot\alpha(s, g)$. In other words, Φ is an α-invariant function. As we have seen that H_k is smooth on E, the cocycle reduction lemma (5.2.11) applies and $\alpha \sim \beta$ where β takes values in the stabilizer of some element of E. However, any such stabilizer clearly leaves a finite subset of H_k/L_k invariant. On the other hand, the stabilizer is the set of k-points of a k-group, and since the algebraic hull of α is H_k, the stabilizer must actually be H_k. Therefore, we deduce that H_k actually leaves a finite subset of H_k/L_k invariant, i.e., H_k/L_k is finite. Thus, $L \supset H^0$, completing the proof.

9.3 Actions of lattices and product actions

The rigidity theorem for ergodic actions (5.2.1) is one major respect in which ergodic actions of semisimple groups exhibit behavior markedly different from actions of amenable groups. The cohomology result (9.1.1) is another example of a marked difference between the semisimple and amenable cases. In this section we wish to describe another such result.

9.3.1 Theorem [*Zimmer* 13]: *Let G be the real points of a connected simple algebraic $\mathbb{R}$-group, and $\Gamma \subset G$ a lattice. Let S be an essentially free ergodic Γ-space with finite invariant measure. Suppose, for $i = 1, 2$, that Γ_i are discrete groups acting ergodically, with finite invariant measure, on (S_i, μ_i), and that the Γ-action on S is orbit equivalent to the $\Gamma_1 \times \Gamma_2$-action on $S_1 \times S_2$. Then either S_1 or S_2 is finite (modulo null sets).*

We have stated this for real groups but the same result is true if $\mathbb{R}$ is replaced by any local field of characteristic 0. A similar type of result is also true for actions of G itself, as well as for certain actions of irreducible lattices in semisimple groups and for actions of certain discrete groups arising as fundamental groups of manifolds of negative curvature. We refer the reader to [Zimmer 13] for these and other generalizations. We also observe that this result stands in sharp contrast to the case when Γ is amenable. In fact, by 4.3.12 and 4.3.16, if Γ is amenable and acts properly ergodically and with finite invariant measure on S, then Γ is orbit equivalent to the $\Gamma \times \Gamma$-action on $S \times S$. As with many other results we have presented, our proof of 9.3.1 will depend upon the behavior of measures on the variety G/P, where P is a minimal parabolic subgroup.

In the proof of Theorem 9.3.1, we will need the following general result.

9.3.2 Proposition: *Suppose Γ is a discrete group acting properly ergodically with invariant probability measure on (S, μ). Then there is a discrete amenable group H which acts ergodically on S, preserves μ, and such that for (almost) all $s \in S$, $sH \subset s\Gamma$.*

Proof: Since all our groups are countable, there is no problem in ignoring null sets and we shall do so throughout the proof. We first claim the following. Suppose $A, B \subset S$ where A, B are measurable disjoint sets and $\mu(A) = \mu(B)$. Then there exists a measure space isomorphism $T: A \to B$ such that for all $x \in A$, x and Tx lie in the same Γ-orbit. To prove this we define a sequence of subsets $A_N \subset A$, $B_N \subset B$ inductively as follows. Let $A_1, B_1 = \varnothing$. For $N \geqq 2$, assume $A_1, \ldots, A_{N-1}, B_1, \ldots, B_{N-1}$ have been defined. Consider subsets D of $A - A_{N-1}$ of measure at least $1/N$ for which there is a measure space isomorphism $T: D \to E$ where $E \subset B - B_{N-1}$, with x and Tx in the same Γ-orbit. For a fixed N, there is clearly a maximal finite number of such triples (D_j, T_j, E_j) for which all D_j, E_j are mutually disjoint. Choose such a maximal collection of triples and let $\tilde{A}_{N-1} = \cup D_j$, $\tilde{B}_{N-1} = \cup E_j$. Define $A_N = A_{N-1} \cup \tilde{A}_{N-1}$, $B_N = B_{N-1} \cup \tilde{B}_{N-1}$. Let $A_\infty = \cup A_N$, $B_\infty = \cup B_N$. By our construction of A_N, B_N, we clearly have the existence of a measure space isomorphism $T: A_\infty \to B_\infty$ with x, Tx in the same Γ-orbit. If $A_\infty = A$, then we are done. If not, let $D \subset A - A_\infty$ with $\mu(D) > 0$. Since T is measure preserving $B_\infty \neq B$, and we can choose $E \subset B - B_\infty$ of positive measure. Since Γ acts ergodically $D \cdot \Gamma \supset E$, and hence there is a subset $D_0 \subset D$ of positive measure and $\gamma \in \Gamma$ such that $D_0 \cdot \gamma \subset E$. However, if N is such that $1/N < \mu(D_0)$, this contradicts the maximality of A_N. This shows $A = A_\infty$, verifying our assertion.

Before proceeding, we introduce some terminology. If $A \subset S$ is measurable, $P = \{B_1, \ldots, B_r\}$ is a partition of A into disjoint measurable sets B_i of equal measure, and F is a finite group of measure space isomorphisms of A onto itself, let us say that F is P-admissible if: for all $T \in F$ and all i, $TB_i = B_j$ for some j; F acts simply transitively on the elements of P; and for all $x \in A$ and $T \in F$, x and Tx are the same Γ-orbit. The assertion of the preceding paragraph implies that for any partition P of $A \subset S$ into r sets of equal measure there is a Z/rZ-action on A which is P-admissible. If $D \subset A$ is a measurable set and P is a partition of A, we let $d(D, P) = \inf\{\mu(D \Delta B) \mid B$ is a union (possibly empty) of subsets belonging to $P\}$. We remark that if D, $E \subset A$, then

$$d(D, P) \leqq d(E, P) + \mu(D \Delta E).$$

Now let $\{A_i\}$ be a sequence of measurable subsets of S which are dense in the measure algebra of S and such that for each i, $A_i = A_n$ for infinitely many $n > i$. We claim that we can inductively construct for each i the following:
(1) a partition P_i of S into 2^{n_i} mutually disjoint measurable subsets of equal measure, for some $n_i \geqq i$;
(2) a P_i-admissible action on S of a finite abelian group F_i; such that
(3) $P_{i+1} \supset P_i$ (i.e. P_{i+1} is a "finer" partition);
(4) $F_{i+1} \supset F_i$;
(5) $d(A_i, P_i) < 1/2^i$.
For $i = 1$, any partition into two sets of measure 1/2 suffices. We now construct P_{i+1} from P_i.

Let $P_i = \{D_1, \ldots, D_{2^{n_i}}\}$. We can find a number $m > n_i$ and for each j a partition $Q_j = \{C_{jk}\}$ of D_j such that:
(6) $\mu(C_{jk}) = 1/2^m$ for all j, k; and
(7) for each j, $d(A_{i+1} \cap D_j, Q_j) < 2^{-(i+2+n_i)}$.
We can then choose a partition R of D_1 such that:
(8) for some $n_{i+1} > m$ every set in R has measure $2^{-(n_{i+1})}$; and
(9) if $T_j \in F_i$ is the unique element with $T_j(D_1) = D_j$, then
$d(T_j^{-1}(C_{jk}), R) < 2^{-(i+2+n_i+m)}$ for any j, k.
Define P_{i+1} to be the partition of S into sets of the form $T(E)$, where $E \in R$ and $T \in F_i$. Since F_i is measure preserving, P_{i+1} is a partition of S into $2^{n_{i+1}}$ sets of equal measure. It follows from (9) that for all j, k, $d(C_{jk}, P_{i+1}) < 2^{-(i+2+n_i+m)}$, and hence for any set $E \subset D_j$ which is a disjoint union of elements of Q_j that

$$d(E, P_{i+1}) < 2^m 2^{-(i+2+n_i+m)} = 2^{-(i+2+n_i)}.$$

From this inequality and (7) we deduce that for all j,

$$d(A_{i+1} \cap D_j, P_{i+1}) < 2 \cdot 2^{-(i+2+n_i)} = 2^{-(i+1+n_i)}.$$

Thus $d(A_{i+1}, P_{i+1}) < 2^{n_i} 2^{-(i+1+n_i)} = 2^{-(i+1)}$. This establishes condition (5). To complete the induction, we construct a P_{i+1}-admissible finite abelian group F_{i+1} acting on S such that (4) is also satisfied.

By our earlier remarks there is an R-admissible action of Z/rZ on D_1 where $r = 2^{n_{i+1}-n_i}$. Let $\lambda \in Z/rZ$, $j > 1$, and $T_j \in F_i$ be as in (9). Let λ act on D_j by $T_j \lambda T_j^{-1}$. This extends the Z/rZ action to all of S and clearly Z/rZ commutes with F_i. Thus, letting F_{i+1} be the group generated by Z/rZ and F_i, the induction argument is complete.

Let $H = \cup F_i$. Then H is abelian and hence amenable. Furthermore $sH \subset s\Gamma$ and so to prove the proposition it suffices to see that H acts ergodically on S.

Suppose $A \subset S$ is H-invariant. Since $F_i \subset H$ transitively permutes the elements of P_i, it follows for B, $C \in P_i$, we have $\mu(A \cap B) = \mu(A \cap C)$. Therefore $\mu(A \cap B) = 2^{-n_i}\mu(A) = \mu(B)\mu(A)$. It follows that $\mu(A \cap B) = \mu(B)\mu(A)$ for any set B which is a union of elements of $\cup P_i$. From (5) and the fact that $A_i = A_n$ for infinitely many $n > i$, we deduce that for each i, $\mu(A \cap A_i) = \mu(A_i)\mu(A)$. Since $\{A_i\}$ is dense in the measure algebra of S, we obtain $\mu(A \cap B) = \mu(B)\mu(A)$ for all measurable B. Hence, letting $B = A$, we conclude that A is either null or conull, establishing ergodicity.

We now turn to the proof of 9.3.1.

Proof of Theorem 9.3.1: Throughout this proof, "algebraic group" will mean the set of real points of an algebraic $\mathbb{R}$-group. If S_i is not finite, then Γ_i acts properly ergodically on S_i. Choose an amenable subgroup A_i acting on S_i as in 9.3.2, i.e., with $xA_i \subset x\Gamma_i$ for $x \in S_i$. Let $\theta: S \to S_1 \times S_2$ be the orbit equivalence of the Γ-action with the $\Gamma_1 \times \Gamma_2$ action. Let $\alpha: (S_1 \times S_2) \times \Gamma_1 \times \Gamma_2 \to \Gamma \subset G$ be the cocycle corresponding to the orbit equivalence θ^{-1} (Example 4.2.8), i.e., for $x \in S_1 \times S_2, h \in \Gamma_1 \times \Gamma_2, \theta^{-1}(x)\alpha(x,h) = \theta^{-1}(xh)$. By Theorem 9.2.3 (and Example 4.2.18), there is an amenable algebraic subgroup $H \subset G$ and an $\alpha | S_1 \times S_2 \times A_1 \times A_2$-invariant function $\varphi: S_1 \times S_2 \to G/H$. As the following lemma shows, to provide the desired contradiction, it actually suffices to show that there is such an α-invariant function.

9.3.3 Lemma: *To prove the theorem, it suffices to show that there is a proper algebraic subgroup $L \subset G$ and an α-invariant function $\varphi: S_1 \times S_2 \to G/L$.*

Proof: Let $\psi: S \to G/L$ be $\psi = \varphi \circ \theta$. For $s \in S$, $\gamma \in \Gamma$, $h \in \Gamma_1 \times \Gamma_2$ such that $\theta(s) \cdot h = \theta(s\gamma)$, we have by the definition of α that $\alpha(\theta(s), h) = \gamma$, and hence that

$$\psi(s\gamma) = \varphi(\theta(s\gamma)) = \varphi(\theta(s)\cdot h) = \varphi(\theta(s))\cdot\alpha(\theta(s), h) \quad \text{[by } \alpha\text{-invariance of } \varphi].$$
$$= \psi(s)\cdot\gamma$$

In other words, $\psi: S \to G/L$ is a Γ-map. If μ is the Γ-invariant measure on S, then $\psi_*(\mu)$ will be a Γ-invariant probability measure on G/L. We can then define a G-invariant probability measure on G/L as follows. Let m be the G-invariant measure on G/Γ. For $A \subset G/L$ measurable, $(\psi_*\mu)(Ag^{-1})$ depends only on the image of g in G/Γ, and thus we can define $\nu(A) = \int (\psi_*(\mu))(Ag^{-1})dm$. This is then clearly a G-invariant probability measure on G/L. However, since L is a proper algebraic subgroup, this is impossible by the Borel density theorem (3.2.5). This proves the lemma.

Therefore, we need to pass from the existence of an $\alpha | S_1 \times S_2 \times A_1 \times A_2$-invariant function to the existence of a suitable α-invariant function. We shall do this by developing a uniqueness property for $\alpha | S_1 \times S_2 \times A_1 \times A_2$-invariant maps.

Suppose that for each $s \in S_1$ we have an algebraic subgroup $H_s \subset G$ such that $\{(s, g) | g \in H_s\}$ is a measurable subset of $S_1 \times G$. Let $p_s: G \to G/H_s$ be the natural projection. Let $\varphi: S_1 \times S_2 \to G$ be a Borel map. Let us say that $(\varphi, \{H_s\})$ is α-admissible if for almost all $s \in S_1$, $p_s \circ \varphi$ is an $\alpha | \{s\} \times S_2 \times A_2$-invariant function. For $s \in S_1$, $h \in \Gamma_1$, $x \in S_2$, and $a \in A_2$, we have from the cocycle relation that

$$\alpha(s, x, h, e)\alpha(sh, x, e, a) = \alpha(s, x, e, a)\alpha(s, xa, h, e).$$

For a fixed $h \in \Gamma_1$, and α-admissible $(\varphi, \{H_s\})$, define $h^*(\varphi, \{H_s\})$ to be the pair $(\psi, \tilde{H}_s)$ where $\psi(s, x) = \varphi(s\cdot h, x)\cdot\alpha(sh, x, h^{-1}, e)$ and $\tilde{H}_s = H_{sh}$. The above consequence of the cocycle relation is readily seen to imply that $h^*(\varphi, H_s)$ is also α-admissible for any $h \in \Gamma_1$. Furthermore, it is clear that if $(\varphi_i, \{H_s^i\})$, $i = 1, 2$, are α-admissible, and $S_1 = B \cup C$ is a disjoint decomposition into measurable sets, then defining φ to be φ_1 on $B \times S_2$ and φ_2 on $C \times S_2$, and defining H_s similarly from H_s^i, we still obtain an α-admissible pair. It follows from these remarks and ergodicity of the Γ_1-action on S_1 that there exists an α-admissible $(\varphi, \{H_s\})$ such that $\dim H_s$ is essentially constant over $s \in S_1$ and that for any admissible $(\psi, \{J_s\})$, we have $\dim H_s \leqq \dim J_s$ a.e. The following observation is basic to the proof of Theorem 9.3.1.

9.3.4 Lemma: *Suppose $(\varphi, \{H_s\})$, $(\psi, \{J_s\})$ are α-admissible and both satisfy the above minimality of dimension property. Then*
(i) H_s^0 and J_s^0 are conjugate for almost all s.
(ii) The conjugacy class of H_s^0 is essentially constant in s. In particular, we can

choose H_s satisfying the minimality of dimension property with $H_s^0 = H^0$, independent of s.
(iii) With such a choice, let $p: G \to G/N(H^0)$, where $N(H^0)$ is the normalizer of H^0 (and is therefore algebraic). Suppose we also have $J_s^0 = H^0$ a.e., so that H_s, $J_s \subset N(H^0)$. Then $p \circ \varphi = p \circ \psi$.

Proof: Consider the map $(\varphi, \psi): S_1 \times S_2 \to G$. For each $s \in S_1$, let $p_s: G \to G/H_s$, $q_s: G \to G/J_s$ be the projections. Then $(p_s \circ \varphi, q_s \circ \psi)$ yields an $\alpha | \{s\} \times S_2 \times A_2$-invariant map $\{s\} \times S_2 \to G/H_s \times G/J_s$. Since A_2 is ergodic on S_2, and G-orbits are locally closed in $G/H_s \times G/J_s$ (3.1.3), it follows from the cocycle reduction lemma (5.2.11) and Example 4.2.18(b), as in the proof of 9.2.1, that $(p_s \circ \varphi, q_s \circ \psi)(\{s\} \times S_2)$ is (a.e.) contained in a single G-orbit in $G/H_s \times G/J_s$. The stabilizer of such an orbit can be chosen to be of the form $L_s = H_s \cap bJ_sb^{-1}$ for some $b \in G$. It is a technical exercise to see that $s \to L_s$ can be chosen measurably, and hence $((\varphi, \psi), \{L_s\})$ is also α-admissible. By the dimension property, $\dim L_s = \dim H_s = \dim(J_s)$ a.e., and hence H_s^0 and J_s^0 are conjugate.

To see that the conjugacy class of H_s^0 is independent of $s \in S_1$, observe first that if $h \in \Gamma_1$, $h^*(\varphi, \{H_s\})$ will also satisfy the minimality of dimension condition, and so by the above H_s^0 and H_{sh}^0 are conjugate. We now employ an argument of [Auslander-Moore 1]. Let $L(H_s)$ denote the Lie algebra of H_s, so that $L(H_s) \subset L(G)$ is a linear subspace, and the dimension of this subspace is essentially constant for $s \in S_1$, say of dimension d. Since H_s^0 and H_{sh}^0 are conjugate, it follows that $L(H_s)$ and $L(H_{sh})$ are in the same $Ad(G)$-orbit for the action of $Ad(G)$ on the Grassmann variety V of d-planes in $L(G)$. However, the action of $Ad(G)$ on V is smooth (3.1.3), and thus $V/Ad(G)$ is countably separated. Since $L(H_s)$ and $L(H_{sh})$ are equal when projected to $V/(Ad(G))$, the map $s \to$ (projection of $L(H_s)$ in $V/\mathrm{Ad}(G)$) is an essentially Γ_1-invariant function on an ergodic Γ_1-space taking values in a countably separated Borel space. It follows that this map is essentially constant, or equivalently, on a conull set in S_1, all $L(H_s)$ are in the same $Ad(G)$-orbit. This implies that on a conull set in S_1, all H_s^0 are conjugate in G. The usual correspondence of cosets of a subgroup and that of a conjugate subgroup then enable us to choose an admissible $(\varphi, \{H_s\})$ satisfying the minimality condition with $H_s^0 = H^0$ a.e.

Finally, suppose we also have $J_s^0 = H^0$ a.e. As above, the image of the map (φ, ψ) will lie in a single G-orbit, with the stabilizer of a point in the orbit of the form $L_s = H_s \cap bJ_sb^{-1}$, for some $b \in G$. Since $\dim H_s = \dim J_s = \dim L_s$, $bH^0b^{-1} = H^0$, i.e., $b \in N(H^0)$. Thus the image of the relevant orbit of $G/H_s \times G/J_s$ in $G/N(H^0) \times G/N(H^0)$ contains a point on the diagonal, and hence is contained in the diagonal. Therefore, the image of $(p \circ \varphi, p \circ \psi)$ is essentially contained in the diagonal, and this proves the final assertion of the lemma.

9.3.5 Lemma: *Let H_s as in the conclusion of Lemma 9.3.4. Then $H^0 \neq \{e\}$.*

Proof: If it were, each H_s would be finite. Fix $s \in S_1$ so that all required conditions which are known to hold a.e. hold on almost all $\{s\} \times S_2$. In particular, $\varphi_s: S_2 \to G/H_s$ given by $\varphi_s(x) = (p_s \circ \varphi)(s, x)$ is an $\alpha | \{s\} \times S_2 \times A_2$-invariant function. We can choose a compact set $B \subset G/H_s$ such that $\varphi_s^{-1}(B)$ has positive measure. For $x \in \varphi_s^{-1}(B)$, since θ is an orbit equivalence, we can choose $\gamma_n \in \Gamma$, $\gamma_n \to \infty$ in Γ, such that $\theta^{-1}(x), \theta^{-1}(x)\gamma_n \in \theta^{-1}(\varphi_s^{-1}(B))$. Let $\psi = \varphi \circ \theta$. Then the same calculation as in the beginning of the proof of Lemma 9.3.3 shows that $\psi(\theta^{-1}(x)\gamma_n) = \psi(\theta^{-1}(x))\gamma_n$. However, $\theta^{-1}(x)\gamma_n \in \theta^{-1}(\varphi_s^{-1}(B))$, so $\psi(\theta^{-1}(x)\gamma_n) \in B$ for all n. On the other hand, since $H_s \subset G$ is finite, if $\gamma_n \in \Gamma$, $\gamma_n \to \infty$, then for any $y \in G/H_s$, $y\gamma_n \to \infty$, i.e., leaves every compact subset of G/H_s. In particular, this is true for $y = \psi(\theta^{-1}(x))$, which is a contradiction. This proves Lemma 9.3.5.

We now complete the proof of Theorem 9.3.1. As in Lemma 9.3.4 choose $(\varphi, \{H_s\})$ α-admissible, satisfying the minimality of dimension condition, and $H_s^0 = H^0$. As above, let $h \in \Gamma_1$. As observed preceding Lemma 9.3.4, $h^*(\varphi, \{H_s\})$ will also be α-admissible. It follows from the final conclusion of Lemma 9.3.4 and the definition of $h^*(\varphi, \{H_s\})$ that $p(\varphi(sh, x)) \cdot \alpha(sh, x, h, e)) = p(\varphi(s, x))$. In other words, $p \circ \varphi: S_1 \times S_2 \to G/N(H^0)$ is an $\alpha | S_1 \times \{x\} \times \Gamma_1$-invariant map for a.e. $x \in S_2$. By our construction at the beginning of the proof, $H^0 \neq G$, and by Lemma 9.3.5, $H^0 \neq \{e\}$, and so $N(H^0) \neq G$. We can now repeat our entire argument working with $\alpha | S_1 \times \{x\} \times \Gamma_1$-invariant functions. Lemmas 9.3.4 and 9.3.5 hold in this situation as well, and repeating the arguments immediately above using $\gamma \in \Gamma_2$ rather than $h \in \Gamma_1$, we deduce the existence of an α-invariant map $\varphi: S_1 \times S_2 \to G/L$ for some non-trivial proper algebraic subgroup $L \subset G$. By Lemma 9.3.3, this completes the proof.

9.4 Rigidity and entropy

In Section 5.2 we applied the superrigidity theorem for cocycles (5.2.5) to the cocycles arising from an orbit equivalence to deduce the rigidity theorem for ergodic actions. Theorem 5.2.5 can also be applied to the derivative cocycle for an action on a manifold (Example 4.2.3) or to other associated geometrically defined cocycles. This is a basic step in proving results about actions of semisimple groups or their lattices on compact manifolds. The full proofs, however, require some significant geometric background and so we shall not present these results here. As an introduction to this subject, we shall present an application of superrigidity to the computation of entropy, a basic measure theoretic invariant of the diffeomorphisms defined by single elements of the acting group. This particular consequence of superrigidity was first observed in [Furstenberg 5]. For further developments of the applications of superrigidity to actions on manifolds, we refer the reader to [Zimmer 15, 18, 19].

We shall first review the concept of entropy for an integer action. We will not provide proofs, but refer the reader to [Billingsley 1] and [Ornstein 2]. We assume that (S, μ) is a Z-space, where μ is a Z-invariant probability measure. Let $T: S \to S$ be given by $Ts = s \cdot 1$. To this transformation, we will associate a non-negative real number h(T), called the (Kolmogorov-Sinai) entropy of T.

We begin by seeking a quantitative measure of randomness. Namely, suppose $p = (p_1, \ldots, p_k)$ is a probability measure on $\{1, \ldots, k\}$, i.e., $0 \leqq p_i \leqq 1$, $\Sigma p_i = 1$. We seek a number $H(p)$, which we want to measure the randomness of p, satisfying the following axioms:

(i) H is continuous in p, $H(p) \geqq 0$, H is symmetric, and $H(p, 0) = H(p)$.
(ii) $H(1, 0, \ldots, 0) = 0$. (No randomness.)
(iii) For each k, $H(p_1, \ldots, p_k)$ achieves a maximum at $p_i = 1/k$ for all i.
(iv) If $(p_1, \ldots, p_k)$, $(q_1, \ldots, q_k)$ are probability vectors, then

$$H(p_1, \ldots, p_{k-1}, p_k q_1, \ldots, p_k q_r) = H(p_1, \ldots, p_k) + p_k H(q_1, \ldots, q_r).$$

We then have:

9.4.1 Theorem [*Khinchin* 1]: *Up to a scalar multiple, the only function satisfying (i)–(iv) is*

$$H(p_1, \ldots, p_k) = -\Sigma p_i \log p_i.$$

$H(p)$ is called the entropy of p.

We now let (S, μ) be a standard measure space, $\mu(S) = 1$. By a partition $\mathscr{A}$ of S we mean a collection $\mathscr{A} = \{A_1, \ldots, A_k\}$, where $A_i \subset S$ are mutually disjoint measurable sets and $\cup A_i = S$. Define $H(\mathscr{A}) = H(\mu(A_1), \ldots, \mu(A_k))$. If $\mathscr{A}$, $\mathscr{B}$ are two partitions, let $\mathscr{A} \vee \mathscr{B}$ be the partition $\mathscr{A} \vee \mathscr{B} = \{A_i \cap B_j\}$. It is then easy to check from property (iv) above that $H(\mathscr{A}) \leqq H(\mathscr{A} \vee \mathscr{B}) \leqq H(\mathscr{A}) + H(\mathscr{B})$. If $\mathscr{A} = \mathscr{B}$, then clearly $H(\mathscr{A} \vee \mathscr{B}) = H(\mathscr{A})$. Hence if $\mathscr{B}$ is close to $\mathscr{A}$, $H(\mathscr{A} \vee \mathscr{B})$ will be close to $H(\mathscr{A})$. On the other hand, suppose $\mathscr{A}$ and $\mathscr{B}$ are independent partitions, i.e., $\mu(A_i \cap B_j) = \mu(A_i)\mu(B_j)$ (for example, partitions from partitions in factors of a product space.) Then we have $H(\mathscr{A} \vee \mathscr{B}) = H(\mathscr{A}) + H(\mathscr{B})$.

Recall the map $T: S \to S$. The entropy of T will measure the rate of growth of $H(\mathscr{A} \vee T(\mathscr{A}) \vee \ldots \vee T^n(\mathscr{A}))$.

9.4.2 Definition: (i) Set $h(T, \mathscr{A}) = \overline{\lim}\, H(\mathscr{A} \vee T(\mathscr{A}) \vee \ldots \vee T^{n-1}(\mathscr{A}))/n$.
(ii) Set $h(T) = \sup\{h(T, \mathscr{A}) \mid A \text{ is a partition of } S\}$.

$h(T)$ is called the entropy of T (or of the Z-action), and is clearly an invariant of the Z-action up to conjugacy. One can compute $h(T)$ because of the following:

9.4.3 Theorem (*Sinai*): *If $\mathscr{A}$ is a generating partition for the Z-action, i.e., the smallest closed Boolean subalgebra of the measure algebra $B(S)$ that contains $\mathscr{A}$ and is Z-invariant is $B(S)$ itself, then $h(T) = h(T, \mathscr{A})$.*

9.4.4 Example: (a) Let (X, p) be a finite probability space. Let $\Omega = \sum_{-\infty}^{\infty} (X, p)$, and define $T:\Omega \to \Omega$ by $(Tw)_n = w_{n+1}$. This is called the Bernoulli shift on the state space (X, p). Then $h(T) = H(p)$. (In particular, $h(T)$ can take any positive value.)
(b) If T is a rotation of the circle, then $h(T) = 0$.
(c) Let $A \subset SL(2, \mathbb{R})$ be the positive diagonal matrices, and $\Gamma \subset SL(2, \mathbb{R})$ a cocompact lattice. For $a \in A$, let $h(a)$ be the entropy of a acting on $SL(2, \mathbb{R})/\Gamma$. Then up to scale, $h(a) = \log\|a\|$. Thus, $h(T)$ can take any non-negative value for T a diffeomorphism of a compact manifold.

We now state two fundamental results that describe the relationship of entropy to Bernoulli shifts.

9.4.5 Theorem [*Ornstein* 1]: *Any two Bernoulli shifts with the same entropy define conjugate Z-actions.*

9.4.6 Theorem [*Sinai* 1]: *Suppose S is a Z-space with entropy h. Then there is a Bernoulli shift with entropy h, say on Ω, and a measure preserving Z-map $S \to \Omega$.*

For a C^2-diffeomorphism, the entropy may be computed from properties of the derivative cocycle on the tangent bundle. To describe this result, we first describe some general results concerning the asymptotic behavior of cocycles of Z-actions.

If A is any $n \times n$ matrix, then $\lim_{n\to\infty} \frac{1}{n}\log^+\|A^n\|$ exists and is equal to $\max\{\log^+|\lambda| \mid \lambda$ is an eigenvalue of $A\}$. If (S, μ) is a Z-space, and $\alpha: S \times Z \to GL(n, \mathbb{C})$ then $\alpha(s, n) = \alpha(s, 1)\alpha(s\cdot 1, 1), \ldots, \alpha(s\cdot n - 1, 1)$, and so is naturally a product of n (unequal in general) matrices. For cocycles which are sufficiently regular, one can obtain a similar result on the behavior of $\log^+ \|\alpha(s, n)\|$. Namely, let us call a cocycle α "tempered" if $\|\alpha(\cdot, n)\| \in L^\infty(S, \mu)$ for each n. Define $(e(\alpha))(s) = \lim_{n\to\infty} \frac{1}{n}\log^+ \|\alpha(s, n)\|$ if it exists.

9.4.7 Theorem [*Furstenberg-Kesten* 1]: *Let (S, μ) be a Z-space where μ is an invariant probability measure. Suppose $\alpha: S \times Z \to GL(n, \mathbb{C})$ is a tempered cocycle. Then: (i) $e(\alpha)(s)$ exists for almost all $s \in A$, and $e(\alpha)$ is essentially Z-invariant. Hence, if Z acts ergodically, $e(\alpha)$ is constant.*
(ii) If α and β are tempered and $\alpha \sim \beta$, then $e(\alpha)(s) = e(\beta)(s)$ (a.e.).

For a proof of 9.4.7, see [Kingman 1] and [Derriennic 1].

A basic result on the entropy of diffeomorphisms is the following theorem of Pesin. If the integers Z act on a compact manifold M by diffeomorphisms, compactness implies that $\|df^n_x\|$ is bounded over $x \in M$ for each fixed n (for any choice of Riemannian metric) and hence that the derivative cocycle (Example 4.2.3) is tempered (since we can measurably trivialize the tangent bundle with orthonormal sections).

9.4.8 Theorem [*Pesin* 1]: *Suppose the integers act on a compact manifold by C^2- diffeomorphisms so as to be volume preserving. Set $Ts = s\cdot 1$. Let α be the derivative cocycle, and let $\Lambda^p(\alpha)$ be the p-th exterior power of α. Then the entropy is given by $h(T) = \int [\max_p e(\Lambda^p(\alpha))(s)]ds$. In particular, in the ergodic case*

$$h(T) = \max_p e(\Lambda^p(\alpha)).$$

Pesin's result depends in part upon a sharpening of 9.4.7 due to [Oseledec 1] called the "multiplicative ergodic theorem". Before describing this, we make some remarks about a single matrix $A \in GL(n, \mathbb{C})$. Let λ_i, $i = 1, \ldots, k$ be the distinct real numbers of the form $\log^+|\lambda|$ where λ is an eigenvalue of A, and assume $\lambda_1 < \lambda_2 \ldots < \lambda_k$. Then there are subspaces $0 \underset{\neq}{\subset} V_1 \subset V_2 \subset \ldots \subset V_k = \mathbb{C}^n$ such that for $v \in V_i$ but $v \notin V_{i-1}$, we have $\lim_{n\to\infty} \frac{1}{n}\log^+ \|A^n v\| = \lambda_i$. Oseledec's theorem generalizes this spectral type decomposition to tempered cocycles.

9.4.9 Theorem [*Oseledec* 1]: *Let (S, μ) and α be as in 9.4.7 and assume the Z-action is ergodic. Then there exist $\lambda_1, \ldots, \lambda_k \in \mathbb{R}$, $\lambda_1 < \lambda_2 < \ldots < \lambda_k$, positive integers $1 \leq n_1 < n_2 < \ldots < n_k = n$, and α-invariant measurable maps $V_i: S \to Gr_{n_i}(\mathbb{C}^n)$, the* *latter being the Grassmann variety of n_i-planes in $\mathbb{C}^n$, with $V_i(s) \subset V_{i+1}(s)$ such that for almost all $s \in S$, we have for all $v \in V_i(s)$, $v \notin V_{i-1}(s)$*

$$\lim_{n \to \infty} \frac{1}{n} \log^+ \| \alpha(s, n)v \| = \lambda_i.$$

Remark: Clearly $e(\alpha) = \lambda_k$. We also have $\max_p e(\Lambda^p(\alpha)) = \Sigma(n_i - n_{i-1})\lambda_i$. The λ_k are called characteristic (or Lyapunov) exponents of α.

For a more recent proof of 9.4.9, see [Raghunathan 2].

9.4.10 Example: If $\alpha(s, n) = A^n$ where $A \in GL(n, \mathbb{C})$, then Theorem 9.4.9 reduces to the observation we made following 9.4.8.

If we know that a tempered cocycle α takes values in an algebraic subgroup of $GL(n, \mathbb{C})$, we can say more about $e(\alpha)$. For example, if α takes values in a unipotent subgroup, then $e(\alpha) = 0$ (as do all the characteristic exponents). More generally, if $G \subset GL(n, \mathbb{C})$ is a connected algebraic subgroup, we can write $G = R \ltimes U$ where U is the unipotent radical and R is reductive. For a cocycle α taking values in G, let α_R be the reductive component, i.e., $\alpha_R = p \circ \alpha$ where $p: G \to R$ is the projection. Then we have $e(\alpha) = e(\alpha_R)$. We shall need the following generalization of this remark.

9.4.11 Proposition: *Suppose α is a tempered cocycle taking values in $GL(n, \mathbb{R})$. Let $H \subset GL(n, \mathbb{C})$ be an $\mathbb{R}$-group. Suppose H is connected, and write $H_{\mathbb{R}} = R \ltimes U$ where R is reductive and U is unipotent. Let β be a cocycle equivalent to α with $\beta(S \times Z) \subset H_{\mathbb{R}}$ and, as above, let β_R be the composition of β with projection onto R. Finally, assume β_R is also tempered. Then $e(\beta_R) = e(\alpha)$.*

In concluding these preliminaries, we mention that the entropy and the characteristic exponents of a diffeomorphism have important consequences for the study of classical dynamical properties of the diffeomorphism, for instance periodic points. See [Bowen 1] and [Katok 1] for example.

Now let G be a connected simple $\mathbb{R}$-group. Let $\tilde{G}$ be the algebraic universal covering group, and $\pi: \tilde{G}_{\mathbb{R}} \to SL(n, \mathbb{R})$ an irreducible rational representation. If $g_1, g_2 \in \tilde{G}_{\mathbb{R}}$ and $g_1 \equiv g_2$ in $G_{\mathbb{R}}$, then $g_1 = g_2 z$ where $z \in Z(\tilde{G}_{\mathbb{R}})$ (the center of $\tilde{G}_{\mathbb{R}}$). Since $Z(\tilde{G}_{\mathbb{R}})$ is finite, it follows that

$$\max\{|\lambda| \mid \lambda \text{ an eigenvalue of } \pi(g_1)\}$$
$$= \max\{|\lambda| \mid \lambda \text{ an eigenvalue of } \pi(g_2)\}.$$

Thus, for $g \in G_{\mathbb{R}}$, $\max\{|\lambda| \mid \lambda$ an eigenvalue of $\tilde{g}\}$ is well defined where $\tilde{g} \in \tilde{G}_{\mathbb{R}}$ is any element projecting to g.

We now present an application of the superrigidity theorem for cocycles (5.2.5). Combining 5.2.5 and 9.4.8, we obtain the following.

9.4.12 Theorem: *Let G be a connected almost simple $\mathbb{R}$-group with $\mathbb{R}$-rank$(G) \geqq 2$. Suppose $G^0_{\mathbb{R}}$ acts by C^2 volume preserving diffeomorphisms on a compact manifold M, and suppose the action is ergodic. For $g \in G^0_{\mathbb{R}}$, let $h(g)$ denote the entropy of the diffeomorphism defined by the action of g on M. Then either $h(g) = 0$ for all g, or for each $g \in G^0_{\mathbb{R}}$, there is an $\mathbb{R}$-rational representation $\pi:\tilde{G} \to SL(r, \mathbb{C})$ for some r, such that*

$$h(g) = \max\{\log(|\lambda|) \mid \lambda \textit{ is an eigenvalue of } \pi(\tilde{g})\}.$$

Further, if $n = \dim M$, π is an exterior power of a representation σ of $\tilde{G}$, where dim $\sigma \leqq n$.

9.4.13 Corollary: *For a fixed n, and fixed $g \in G^0_{\mathbb{R}}$, the set of possible values of $h(g)$ over all volume preserving, ergodic, C^2-actions of $G^0_{\mathbb{R}}$ on compact manifolds of dimension at most n is finite.*

Proof of 9.4.12: Let $\alpha: M \times G^0_{\mathbb{R}} \to GL(n, \mathbb{R}) \subset GL(n, \mathbb{C})$ be the derivative cocycle. For $g \in G^0_{\mathbb{R}}$ let $\alpha_g = \alpha | M \times \{g^n\}$. Thus $h(g) = \int \max_p e(\Lambda^p(\alpha_g))$ by Pesin's theorem. Now let $H_{\mathbb{R}} \subset GL(n, \mathbb{R})$ be the algebraic hull of α. As in 9.2.6, let $S = M \times_{\alpha} H_{\mathbb{R}}/(H^0)_{\mathbb{R}}$, and define $\beta = \tilde{\alpha}$ as in 9.2.6 as well. Since $H_{\mathbb{R}}/(H^0)_{\mathbb{R}}$ is finite, S has a finite invariant measure. Clearly for any p, $e(\Lambda^p(\alpha_g)) = e(\Lambda^p(\beta_g))$. By 9.2.6, the algebraic hull of β is $(H^0)_{\mathbb{R}}$.

Let $H^0 = L \ltimes U$ where L is reductive, U is unipotent, and L and U are $\mathbb{R}$-groups. Let $\gamma = \beta_L$ so that $\gamma: S \times G^0_{\mathbb{R}} \to L_{\mathbb{R}}$ is a cocycle with algebraic hull $L_{\mathbb{R}}$. Let $q: L_{\mathbb{R}} \to L_{\mathbb{R}}/[L, L]_{\mathbb{R}}$. Then the algebraic hull of $q \circ \gamma$ is $L_{\mathbb{R}}/[L, L]_{\mathbb{R}}$, and hence by Corollary 9.1.3 $L_{\mathbb{R}}/[L, L]_{\mathbb{R}}$ is compact. Since $L = [L, L] \cdot Z(L)$ where $Z(L)$ is the center of L, and $[L, L] \cap Z(L)$ is finite, it follows that $Z(L)_{\mathbb{R}}$ is also compact.

We can write the $\mathbb{R}$-group $L/Z(L)$ as a product of semisimple $\mathbb{R}$-groups $L/Z(L) = L_1 \times L_2$ such that $(L_2)_{\mathbb{R}}$ is compact, and $(L_1)_{\mathbb{R}}$ is center free and with no compact factors. Let $q_1: L_{\mathbb{R}} \to (L_1)_{\mathbb{R}}$ be the projection. Then $q_1 \circ \gamma: S \times G^0_{\mathbb{R}} \to (L_1)_{\mathbb{R}}$ is a cocycle with algebraic hull $(L_1)_{\mathbb{R}}$, and by the superrigidity theorem for cocycles (5.2.5), there is a rational homomorphism $\pi: G \to L_1$ defined over $\mathbb{R}$

such that $q_1 \circ \gamma \sim \alpha_{\pi|G^0_{\mathbb{R}}}$ where α_π is the cocycle $\alpha_\pi(s, g) = \pi(g)$. Thus, we can write, for $g \in G^0_{\mathbb{R}}$ and a.e. $s \in S$, $\varphi(s)q_1(\gamma(s, g))\varphi(sg)^{-1} = \pi(g)$ where $\varphi: S \to (L_1)_{\mathbb{R}}$ is Borel. We can choose a Borel section $f: (L_1)_{\mathbb{R}} \to L_{\mathbb{R}}$ of q_1. Then $q_1(f(\varphi(s))\gamma(s, g)f(\varphi(sg))^{-1}) = \pi(g)$. In other words, replacing γ by an equivalent cocycle $\delta: S \times G^0_{\mathbb{R}} \to L_{\mathbb{R}}$, we have $q_1(\delta(s, g)) = \pi(g)$.

We can consider π as a homomorphism $\pi: \tilde{G} \to L_1$, where $\tilde{G}$ is the (algebraic) universal covering group of G. Then we can lift this to a homomorphism $\tilde{\pi}: \tilde{G} \to [L, L] \subset GL(n, \mathbb{C})$ defined over $\mathbb{R}$. Thus, for each $g \in G^0_{\mathbb{R}}$, we have $q_1(\delta(s, g)) = q_1(\tilde{\pi}(\tilde{g}))$ for almost all s, where $\tilde{g} \in \tilde{G}_{\mathbb{R}}$ projects to g under the covering map. However, since $(L_2)_{\mathbb{R}}$ and $Z(L_2)_{\mathbb{R}}$ are compact, $(\ker q_1)_{\mathbb{R}}$ is compact. Thus, $\tilde{\pi}(\tilde{g}) = \delta(s, g) \cdot b$ where b is some element of the compact normal subgroup $(\ker q_1)_{\mathbb{R}}$ of $L_{\mathbb{R}}$. It follows that for any $g \in G^0_{\mathbb{R}}$, and any p, that $\Lambda^p \delta_g$ is tempered and that $e(\Lambda^p \delta_g) = \max\{\log|\lambda| \mid \lambda \text{ is an eigenvalue of } \Lambda^p(\tilde{\pi}(g))\}$.

Then $$h(g) = \int \max_p e(\Lambda^p \alpha_g) = \int \max_p e(\Lambda^p \beta_g)$$ by earlier observations,
$$= \int \max_p e(\Lambda^p \delta_p),$$ by Proposition 9.4.11.

This completes the proof.

Theorem 9.4.12 can in fact be extended to actions of lattices on manifolds. To see this, we first observe that the superrigidity theorem can be extended to actions of lattices by inducing. More precisely, we have the following.

9.4.14 Theorem (*A corollary of* 5.2.5): *Let $\Gamma \subset G^0_{\mathbb{R}}$ be a lattice where G is a connected almost $\mathbb{R}$-simple $\mathbb{R}$-group with $\mathbb{R}$-rank$(G) \geqq 2$. Let S be an ergodic Γ-space with finite invariant measure. Let H be a connected $\mathbb{R}$-simple Lie group (trivial center) with $H_{\mathbb{R}}$ not compact. Suppose $\alpha: S \times \Gamma \to H_{\mathbb{R}}$ is a cocycle with algebraic hull $H_{\mathbb{R}}$. Then there is an $\mathbb{R}$-rational homomorphism $\pi: G \to H$ such that $\alpha \sim \alpha_{\pi|\Gamma}$.*

Proof: Let $f: G^0_{\mathbb{R}}/\Gamma \times G^0_{\mathbb{R}} \to \Gamma$ be a strict cocycle corresponding to the identity map $\Gamma \to \Gamma$ (4.2.13). Define the $G^0_{\mathbb{R}}$-action induced from the Γ-action on S, $X = G^0_{\mathbb{R}}/\Gamma \times_f S$ (4.2.21). Define a cocycle $\beta: X \times G^0_{\mathbb{R}} \to H_{\mathbb{R}}$ by $\beta((y, s), g) = \alpha(s, f(y, g))$. Then X has a finite $G^0_{\mathbb{R}}$-invariant measure, and the algebraic hull of β is $H_{\mathbb{R}}$. To see this last assertion, assume that β is equivalent to a cocycle with algebraic hull $L_{\mathbb{R}} \subset H_{\mathbb{R}}$, i.e., there is a β-invariant function $\varphi: X \to H_{\mathbb{R}}/L_{\mathbb{R}}$ (4.2.18(b)). Let Γ act on $F(S, H_{\mathbb{R}}/L_{\mathbb{R}})$ by the α-twisted action. Under the identification $F(X, H_{\mathbb{R}}/L_{\mathbb{R}})$ with $F(G^0_{\mathbb{R}}/\Gamma, F(S, H_{\mathbb{R}}/L_{\mathbb{R}}))$, φ corresponds to an f-invariant element of $F(G^0_{\mathbb{R}}/\Gamma, F(S, H_{\mathbb{R}}/L_{\mathbb{R}}))$. By (4.2.19), there is a Γ-invariant element in $F(S, H_{\mathbb{R}}/L_{\mathbb{R}})$, i.e., an α-invariant function $S \to H_{\mathbb{R}}/L_{\mathbb{R}}$. Hence from 4.2.18(b), we deduce that $L_{\mathbb{R}} = H_{\mathbb{R}}$ since the algebraic hull of α is $H_{\mathbb{R}}$. This verifies that the algebraic hull of β is also $H_{\mathbb{R}}$. Thus, we can apply 5.2.5 and deduce that $\beta \sim \alpha_\pi$ where π is an $\mathbb{R}$-rational homomorphism $\pi: G \to H$. We claim that $\alpha \sim \alpha_{\pi|\Gamma}$.

If we knew that β was a strict cocycle and that β and α_π were strictly equivalent, then simply by writing out the equation for the equivalence we could deduce $\alpha \sim \alpha_{\pi|\Gamma}$. To deal with the almost everywhere conditions, we will reformulate matters so as to be able to apply (4.2.19).

Let $H \times H$ act on H by $(a, b)\cdot h = ahb^{-1}$. Then if α,β are cocycles on some space taking values in $H_{\mathbb{R}}$, then they are equivalent if and only if there is an (α, β)-invariant function into $H_{\mathbb{R}}$, where (α, β) is an $H_{\mathbb{R}} \times H_{\mathbb{R}}$-valued cocycle. Let $\tilde{\pi}: X \times G^0_{\mathbb{R}} \to H_{\mathbb{R}}$ be the cocycle $\tilde{\pi}((y, s), g) = \pi(f(s, g))$. Since $f(s, g)$ corresponds to the identity $\Gamma \to \Gamma$, $f(s, g) = \gamma(s)g\gamma(sg)^{-1}$, and hence it follows that $\tilde{\pi} \sim \alpha_\pi$. Thus, since we also have $\alpha_\pi \sim \beta$, it follows that there exists a $(\beta, \tilde{\pi})$-invariant function $\psi: X \to H_{\mathbb{R}}$. Once again, under the identification of $F(X, H_{\mathbb{R}})$ with $F(G^0_{\mathbb{R}}/\Gamma, F(S, H_{\mathbb{R}}))$, ψ corresponds to an f-invariant function $G^0_{\mathbb{R}}/\Gamma \to F(S, H_{\mathbb{R}})$ where Γ acts on $F(S, H_{\mathbb{R}})$ by $(\gamma\cdot h)(s) = \alpha(s, \gamma)h(s\gamma)\pi(\gamma)^{-1}$. By (4.2.19), there is a Γ-invariant function under this Γ-action, I.e., there is $h \in F(S, H_{\mathbb{R}})$ such that $h(s) = \alpha(s, \gamma)h(s\gamma)\pi(\gamma)^{-1}$. But this clearly implies that $\alpha \sim \alpha_{\pi|\Gamma}$, completing the proof.

Exactly the same proof as the proof of theorem 9.4.12 now yields the following.

9.4.15 Theorem: *Let G be a connected almost $\mathbb{R}$-simple $\mathbb{R}$-group with $\mathbb{R}$-rank$(G) \geq 2$. Let $\Gamma \subset G^0_{\mathbb{R}}$ be a lattice. Suppose Γ acts by C^2, volume-preserving diffeomorphisms on a compact manifold M, and suppose the action is ergodic. For $\gamma \in \Gamma$, let $h(\gamma)$ denote the entropy of the diffeomorphism defined by the action of γ on M. Then either $h(\gamma) = 0$ for all $\gamma \in \Gamma$, or for each γ there is an $\mathbb{R}$-rational representation $\pi: \tilde{G} \to SL(r, \mathbb{C})$ for some r, such that $h(\gamma) = max\{log(|\lambda|) | \lambda$ is an eigenvalue of $\pi(\tilde{\gamma})\}$. Further, if $n = dim\ M$, then π is an exterior power of a representation σ of $\tilde{G}$ where $dim\ \sigma \leq n$.*

9.4.16 Corollary: *The conclusion of* 9.4.13 *is true for actions of* Γ.

Theorem 9.4.15 puts strong restrictions on the type of actions Γ can have on low dimensional manifolds. For example:

9.4.17 Corollary: *For $n \geqq 3$, let $SL(n, Z)$ act ergodically by volume preserving C^2-diffeomorphisms of a compact manifold M. Suppose dim $M \leqq n-1$. Then $h(\gamma) = 0$ for all $\gamma \in SL(n, Z)$.*

Proof: $SL(n, \mathbb{R})$ has no non-trivial rational representation of dimension smaller than n. 9.4.15 then implies the result.

For another result of this type, see 10.1.10.

9.5 Complements

In this section we indicate without proof some further results concerning lattices and ergodic actions. We make no attempt at a survey but restrict ourselves to some results which directly complement those of the previous chapters.

A Ergodicity of restriction

Let $G = \Pi G_i$ where G_i is a simple, connected, non-compact Lie group with finite center, and let $\Gamma \subset G$ be an irreducible lattice. Let (S, μ) be an ergodic G-space, where μ is a quasi-invariant measure. The question arises as to when the restriction of the G-action to Γ will still be ergodic. If the G-action is essentially transitive, Moore's theorems (chapter 2) provide us with a complete answer. Namely, if $S = G/G_0$, then Γ is ergodic on S if and only if G_0 is not compact. Therefore, it remains only to consider the properly ergodic case.

9.5.1. Theorem [*Zimmer* 2]: *Suppose G, Γ as above, and that S is a properly ergodic G-space. Then the restriction of the G-action to Γ is still ergodic.*

B Dense orbits

In Chapter 2 we established a number of results concerning ergodicity of the action of H_1 on G/H_2, where $H_i \subset G$ are closed subgroups. In this case, ergodicity implies that almost every orbit is dense (2.1.7). The question then arises as to when one can deduce that every orbit is dense, or more generally, to identify the dense orbits. A group action on a topological space is called minimal if every orbit is dense. In analogy with 2.2.3, we have the following.

9.5.2 Proposition: *Let $H_1, H_2 \subset G$ be closed subgroups of a locally compact group. Then H_1 acts minimally on G/H_2 if and only if H_2 acts minimally on G/H_1.*

A general minimality assertion is the following.

9.5.3 Theorem [*Mostow* 2]: *Let G be a connected semisimple Lie group, $P \subset G$ a parabolic subgroup and $\Gamma \subset G$ a lattice. Then Γ acts minimally on G/P.*

If Γ is a uniform lattice, we have the following stronger assertion.

9.5.4 Theorem [*Veech* 1]: *Let G be (the real points of) a semisimple $\mathbb{R}$-algebraic group without compact factors and let $N \subset G$ be the unipotent radical of a minimal parabolic subgroup. If $\Gamma \subset G^0$ is a uniform lattice, then Γ acts minimally on G^0/N.*

For $G = SL(2, \mathbb{R})$, this result was first proved by [Hedlund 1]. Veech in fact proves a stronger theorem which in particular, when applied to $SL(n, \mathbb{R})$ yields the following. By a p-frame in $\mathbb{R}^n$ we mean an ordered linearly independent set in $\mathbb{R}^n$ with p elements.

9.5.5 Theorem [*Veech* 1]: *If $\Gamma \subset SL(n, \mathbb{R})$ is a uniform lattice, then for each p, $1 \leqq p \leqq n-1$, Γ acts minimally on the space of p-frames.*

For non-uniform lattices, 9.5.5 is of course no longer true. We call a p-frame irrational if its $\mathbb{R}$-linear span contains no elements of $\mathbb{Q}^n$ except the origin. We then have:

9.5.6 Theorem [*Dani-Raghavan*1]: *For $1 \leqq p \leqq n-1$, the orbit of any irrational p-frame under $SL(n, Z)$ is dense in the space of all p-frames.*

Questions involving minimality of actions and identifying minimal invariant (closed) subsets is closely related to the question of finding all locally finite invariant measures which are ergodic under the action. The reader is referred to [Furstenberg 3], [Veech 2], and the extensive work of [Dani 1–3] for results in this direction.

C Ratner's Rigidity Theorem

The rigidity results we have considered in Chapter 5 do not apply to a pair of subgroups in $pSL(2, \mathbb{R})$. In this case, we have the following theorem.

9.5.7 Theorem [*Ratner* 1]: *Let $N \subset PSL(2, \mathbb{R})$ be the 1-parameter upper triangular unipotent group. Suppose $\Gamma_1, \Gamma_2 \subset PSL(2, \mathbb{R})$ are lattices. If the N-actions on $PSL(2, \mathbb{R})/\Gamma_1$ and $PSL(2, \mathbb{R})/\Gamma_2$ are conjugate, then Γ_1 and Γ_2 are conjugate subgroups.*

D Orbit equivalence

Let M be a complete manifold of finite volume and constant negative sectional curvature. Then $\Gamma = \pi_1(M)$ can be realized as a lattice subgroup of $0(1, n)$. Hence, if $\Gamma' \subset G$ is a lattice where G is simple and $\mathbb{R}\text{-rank}(G) \geqq 2$, the results of section 5.2 imply that Γ and Γ' do not have orbit equivalent free ergodic actions with finite invariant measure. If we only assume that the curvature of M is negative and bounded away from 0, then we do not know that Γ is a lattice in $0(1, n)$ and hence section 5.2 does not apply. However, the non-orbit equivalence assertion is still valid.

9.5.8 Theorem [*Zimmer* 10]: *Let M be a complete manifold of finite volume and negative sectional curvature bounded away from 0. Let $\Gamma \subset \pi_1(M)$ be any subgroup, and let Γ' be as above. Then Γ and Γ' do not have orbit equivalent free ergodic actions with finite invariant measure.*

The proof of this result depends on the following generalization of Theorem 8.1.3. The essential ideas of the proof are similar to those of the proof of 8.1.3.

9.5.9 Theorem [*Zimmer* 10]: *Let S be an irreducible ergodic G-space with finite invariant measure, where G is a connected semisimple Lie group with finite center, no compact factors, and $\mathbb{R}$-rank at least 2. Let $P \subset G$ be a parabolic subgroup. Suppose X is a G-space with finite invariant measure for which we have measurable G-maps $S \times G/P \to X \to S$ whose composition is the projection. Then X is of the form $X = S \times G/P'$ where $P' \supset P$, and the maps are the natural ones.*

E Topological quotients of Γ-actions

The topological version of Theorem 8.1.3 is also true. Namely:

9.5.10 Theorem [*Dani* 4]: *Let G be a connected semisimple Lie group with finite center and* $\mathbb{R}$*-rank*$(G) \geqq 2$. *Let* $P \subset G$ *be a parabolic subgroup and* $\Gamma \subset G$ *an irreducible lattice. Suppose X is a compact Hausdorff* Γ*-space and there is a continuous surjective* Γ*-map* $\varphi: G/P \to X$. *Then there is a parabolic subgroup* $P' \supset P$ *such that the* Γ*-action on X is topologically conjugate to the* Γ*-action on* G/P' *in such a way that* φ *corresponds to the natural projection.*

This was first established in the special case $G = SL(n, \mathbb{R})$, $G/P = \mathbb{P}^{n-1}$, and Γ commensurable with $SL(n, Z)(n \geqq 3)$ in [Zimmer 11], using 9.5.6. A counterexample to 9.5.10 for lattices in groups of $\mathbb{R}$-rank one has been given by [Spatzier 1].

10 Generalizations to p-adic groups and S-arithmetic groups

In this chapter we indicate how to extend some of the results of the preceding chapters to lattices and actions of groups which are (finite) products of semisimple groups over possibly varying local fields. More precisely, for each prime $p \in \mathbb{Z}$ let $\mathbb{Q}_p$ be the field of p-adic numbers, and let $\mathbb{Q}_\infty = \mathbb{R}$. Let $V = \{\text{primes in } \mathbb{Z}\} \cup \{\infty\}$, and $S \subset V$ a finite subset. Suppose that for each $p \in S$, G_p is a connected, semisimple $\mathbb{Q}_p$-group such that $(G_p)_{\mathbb{Q}_p}$ has no compact factors. (I.e., if $H \subset G_p$ is a proper normal $\mathbb{Q}_p$-subgroup, then $(G_p)_{\mathbb{Q}_p}/H_{\mathbb{Q}_p}$ is not compact. We recall the standard fact from Lie theory that for $k = \mathbb{R}$, noncompactness of $G_\mathbb{R}$ for G semisimple is equivalent to $\mathbb{R}\text{-rank}(G) > 0$. That the same result is true for any local field of characteristic 0 is a result of Bruhat and Tits. See [Prasad 3] for a simple proof.) Let $G = \prod_{p \in S} (G_p)_{\mathbb{Q}_p}$, so that G is a locally compact group with the product topology.

Lattices and actions of such groups arise naturally via "arithmetic" constructions. For example, let us assume $\infty \in S$. Let $\mathbb{Z}_S$ be the subring of $\mathbb{Q}$ consisting of all rational numbers whose denominators have all prime factors lying in S. Clearly, if $S \neq \{\infty\}$, $\mathbb{Z}_S$ is dense in $\mathbb{R}$. However, we have a natural map $\mathbb{Z}_S \to \mathbb{Q}_p$ for each $p \in S$, and hence a natural map $\mathbb{Z}_S \to \prod_{p \in S} \mathbb{Q}_p$. The image of $\mathbb{Z}_S$ will then be discrete and in fact one can show that the image of $\mathbb{Z}_S$ is a (cocompact) lattice in $\prod_{p \in S} \mathbb{Q}_p$. More generally, suppose G is a connected algebraic $\mathbb{Q}$-group. We then have a natural embedding of $G_{\mathbb{Z}_S} \to \prod_{p \in S} G_{\mathbb{Q}_p}$, whose image is discrete. In fact, we have:

10.1.1 Theorem [*Borel* 6]: *If G is a connected semisimple $\mathbb{Q}$-group, then $G_{\mathbb{Z}_S}$ is a lattice in $\prod_{p \in S} G_{\mathbb{Q}_p}$.*

This is the prototype of an "S-arithmetic" group. (See Definition 10.1.11.)

Now let G be as in the first paragraph of this section. Any lattice in G is Zariski dense when projected to each $(G_p)_{\mathbb{Q}_p}$. (This follows from Theorem 3.2.5 and the remarks following its proof.) A lattice $\Gamma \subset G$ is called irreducible if ΓH is dense in G for all H of the form $(L)_{\mathbb{Q}_p}$ where $L \subset G_p$ is an infinite normal proper algebraic $\mathbb{Q}_p$-subgroup. Similarly, an ergodic measure class preserving action of G is called irreducible if the restriction to each such subgroup H is ergodic. Many of the results of the earlier chapters extend to this framework. In fact, the proofs we have given generalize in large measure as well once one has developed some further algebraic background concerning the structure of algebraic groups over local fields. We shall therefore first indicate (without proof) one such basic algebraic result. With this in hand, the extension of our earlier proofs to our current more general framework will then follow.

Suppose G is a connected semisimple k-group where k is a local field of characteristic 0. We define G^+ to be the subgroup of G_k generated by $\{U_k | U$ is the unipotent radical of some parabolic k-subgroup of $G\}$. Then $G^+ \subset G_k$ is a normal subgroup. (See [Tits 1] and [Borel–Tits 2] for a full discussion of G^+.) If G_k is compact (i.e., k-rank$(G) = 0$), then $G^+ = \{e\}$. On the other hand, for $k = \mathbb{R}$, and G almost $\mathbb{R}$-simple, if $G_{\mathbb{R}}$ is not compact then $G^+ = G^0_{\mathbb{R}}$. An analogous result in the case of an arbitrary k is given by the following theorem of Platonov verifying a conjecture of Kneser and Tits.

10.1.2 Theorem [*Platonov* 1] (*Kneser–Tits conjecture*): *Suppose k is a local field of characteristic 0 and G is a connected (algebraically) simply connected, almost k-simple k-group with k-rank$(G) > 0$. Then $G^+ = G_k$.*

If $f: G \to H$ is a k-isogeny, then $f(G^+) = H^+$. Hence, Platonov's theorem has the following consequence.

10.1.3 Corollary: *Suppose k is a local field of characteristic 0 and G is a connected almost k-simple k-group with k-rank$(G) > 0$. Then $[G_k : G^+] < \infty$.*

The proof of the vanishing of matrix coefficients for unitary representations of connected simple Lie groups (Theorem 2.2.20), then yields the following result.

10.1.4 Theorem [*Howe–Moore* 1] (*Cf.* 2.2.20): *With our assumptions as in* 10.1.3, *suppose π is a unitary representation of G_k so that $\pi | G^+$ has no invariant vectors. Then the matrix coefficients of π vanish at ∞ in G_k.*

This result enables us to establish analogues of all the versions of Moore's ergodicity theorem that appear in Chapter 2.

The following generalized versions of the superrigidity theorems (5.1.2, 5.2.5) then follow with essentially the same proof. With $G = \prod_{p \in S} (G_p)_{\mathbb{Q}_p}$ as above, we let $\operatorname{rank}(G) = \sum_S \mathbb{Q}_p\text{-rank}(G_p)$.

10.1.5 Theorem [*Margulis* 1]: *Suppose each G_p is (algebraically) simply connected, $(G_p)_{\mathbb{Q}_p}$ has no compact factors, and $\operatorname{rank}(G) \geqq 2$. Let $\Gamma \subset G$ be an irreducible lattice. Let k be $\mathbb{R}$, $\mathbb{C}$, or $\mathbb{Q}_r$ for r prime and H a connected k-simple k-group. Suppose $\pi: \Gamma \to H_k$ is a homomorphism with $\pi(\Gamma)$ Zariski dense in H. Then either:*
(a) $\overline{\pi(\Gamma)}$ is compact (where the closure is in the Hausdorff topology);
(b) For some $p \in S - \{\infty\}$, $k = \mathbb{Q}_p$, and there exists a k-rational surjection $\varphi: G_p \to H$ such that the following diagram commutes:

$$\begin{array}{ccc} \Gamma & \xrightarrow{\text{proj}} & (G_p)_k \\ & \underset{\pi}{\searrow} & \downarrow \varphi \\ & & H_k \end{array}$$

or c) $k = \mathbb{R}$ or $\mathbb{C}$ and there exists a k-rational surjection $G_\infty \to H$ such that the corresponding diagram as in (b) commutes.

10.1.6 Theorem (*Superrigidity for cocycles*): *Let G be as in Theorem* 10.1.5 *and suppose X is an irreducible G-space with finite invariant measure. Let k, H be as in* 10.1.5. *Suppose $\alpha: X \times G \to H_k$ is a cocycle that is not equivalent to a cocycle into a proper subgroup of H_k of the form L_k where $L \subset H$ is a k-subgroup. Then either:*
(a) α is equivalent to a cocycle taking values in a compact subgroup of H_k;
(b) For some $p \in S - \{\infty\}$, $k = \mathbb{Q}_p$, and there exist a k-rational surjection $\varphi: G_p \to H$ and a cocycle $\beta \sim \alpha$ such that the following diagram commutes:

$$\begin{array}{ccc} X \times G & \xrightarrow{\text{proj}} & (G_p)_k \\ & \underset{\beta}{\searrow} & \downarrow \varphi \\ & & H_k \end{array}$$

or (c) $k = \mathbb{R}$ or $\mathbb{C}$ and there exist a k-rational surjection and a cocycle $\beta \sim \alpha$ such that the corresponding diagram as in (b) commutes.

As in Chapter 5, from these results we deduce the following rigidity theorems.

10.1.7 Theorem [*Margulis* 1] [*Prasad* 2] (*Cf.* 5.1.1.): *Let S, S' be finite subsets of V. For $p \in S$, let G_p be a connected, adjoint, semisimple $\mathbb{Q}_p$ group so that $(G_p)_{\mathbb{Q}_p}$ has no compact factors. For each $p \in S'$, let G'_p be such a group. Let $G = \prod_{p \in S} (G_p)_{\mathbb{Q}_p}$, $G' = \prod_{p \in S'} (G'_p)_{\mathbb{Q}_p}$, and assume rank$(G) \geqq 2$. Let $\Gamma \subset G$, $\Gamma' \subset G'$ be lattices, with Γ irreducible. If $\pi:\Gamma \to \Gamma'$ is an isomorphism, then $S = S'$ and for each $p \in S$ there is a $\mathbb{Q}_p$-regular isomorphism $G_p \to G'_p$ such that the product map $G \to G'$ is an extension of π.*

Proof: Let $\tilde{G}_p$ be the algebraic universal covering of G_p and $\tilde{G} = \Pi(\tilde{G}_p)_{\mathbb{Q}_p}$. We have a natural map $q:\tilde{G} \to G$ whose image has finite index in G. (The map is not a priori surjective since we have passed to the k-points for the various fields k in question.) We can choose an irreducible lattice $\Lambda \subset \tilde{G}$ such that $q(\Lambda) \subset \Gamma$ is a normal subgroup of finite index. Then $\pi(q(\Lambda))$ is normal and of finite index in Γ' and hence is also a lattice. In particular, the projection of $\pi(q(\Lambda))$ onto each factor L of G' is Zariski dense. We can therefore apply super-rigidity (10.1.5) to the maps $\Lambda \to L_{\mathbb{Q}_p}$ given by composing $\pi \circ q$ with the projection onto $L_{\mathbb{Q}_p}$. For each p, we thus obtain a $\mathbb{Q}_p$-rational map $\tilde{\phi}_p:\tilde{G}_p \to G'_p$ such that the associated product map $\tilde{\phi} = \Pi\tilde{\phi}_p:\tilde{G} \to G'$ satisfies $\tilde{\phi}|\Lambda = \pi \circ q$. Since G'_p is an adjoint group, $\tilde{\phi}_p$ factors to a map $\phi_p:G_p \to G'_p$ and hence $\tilde{\phi}$ factors to a map $\phi:G \to G'$ such that $\phi|q(\Lambda) = \pi$. By Zariski density of the image of $\pi(q(\Lambda))$ in $(G'_p)_{\mathbb{Q}_p}$, $\phi_p(G_p)$ is Zariski dense in G'_p and hence ϕ_p is surjective. Since $q(\Lambda)$ is an irreducible lattice in G and $\phi(q(\Lambda))$ is discrete, $\dim(\ker \phi_p) = 0$, and since G_p is an adjoint group, ϕ_p is injective. Thus, ϕ_p is an isomorphism defined over $\mathbb{Q}_p$, and hence ϕ is an isomorphism. To prove the theorem it suffices to show that $\phi|\Gamma = \pi$. If $\gamma \in \Gamma$ and $\lambda \in q(\Lambda)$, we have $\phi(\gamma\lambda\gamma^{-1}) = \pi(\gamma\lambda\gamma^{-1})$ since $q(\Lambda) \subset \Gamma$ is normal and $\phi|q(\Lambda) = \pi$. Therefore $[\pi(\gamma)^{-1}\phi(\gamma)]\phi(\lambda)[\phi(\gamma)^{-1}\pi(\gamma)] = \pi(\lambda) = \phi(\lambda)$, i.e., $\pi(\gamma)^{-1}\phi(\gamma)$ commutes with $\phi(q(\Lambda))$. Since the projection of $\phi(q(\Lambda))$ onto each G_p is Zariski dense, the projection of $\pi(\gamma)^{-1}\phi(\gamma)$ commutes with all G_p. However, G_p is an adjoint group for all p, so we deduce that $\pi(\gamma)^{-1}\phi(\gamma) = e$, completing the proof.

As with Theorem 5.1.1, Theorem 10.1.7 is true under weaker hypotheses than rank$(G) \geqq 2$. See [Prasad 2] for full details.

Using 10.1.6 in a similar manner one can prove:

10.1.8 Theorem (*Cf.* 5.2.1): *Let S, S', G, G' be as in* 10.1.7. *Let (X, μ) (resp. (X', μ')) be an irreducible G-space (G'-space) which is essentially free and with finite invariant measure. If the G-action on X and the G' action on X' are orbit equivalent, then $G \cong G'$ and modulo this isomorphism, the actions are conjugate.*

10.1.9 Corollary (*Cf.* 5.2.2): *Suppose $p, q \in V$. Let $G(G')$ be an almost $\mathbb{Q}_p$-simple (almost $\mathbb{Q}_q$-simple) group, and $\Gamma \subset G_{\mathbb{Q}_p}(\Gamma' \subset G'_{\mathbb{Q}_q})$ a lattice. Assume $\mathbb{Q}_p$-rank $(G) \geqq 2$. If Γ and Γ' have orbit equivalent free ergodic actions with finite invariant measure, then $p = q$ and G and G' have the same adjoint group.*

Arguing as in section 9.4, we can also deduce the following property of lattices in p-adic groups acting on manifolds.

10.1.10 Theorem (*Cf.* 9.4.15): *Let G be an almost simple $\mathbb{Q}_p$-group where $p \in Z$ is a prime. Suppose $\Gamma \subset G_{\mathbb{Q}_p}$ is a lattice, and $\mathbb{Q}_p$-rank$(G) \geqq 2$. Let M be a compact manifold and suppose Γ acts by diffeomorphisms of M, preserving a smooth measure on M. Then for every $\gamma \in \Gamma$, we have $h(\gamma) = 0$, where h is the entropy with respect to the smooth measure.*

We remark that by a result of [Borel–Harder 1] every group of the form G_k where G is a semisimple k-group (k a local field of characteristic 0) admits a lattice subgroup.

Margulis's arithmeticity theorem (6.1.2) can be generalized to the present framework as well. We make the following definition.

10.1.11 Definition: Let S be a finite subset of V, and assume $\infty \in S$. For each $p \in S$ let G_p be a connected, adjoint, semisimple $\mathbb{Q}_p$-group for which every simple factor has $\mathbb{Q}_p$-rank > 0. Let $G = \prod_{p \in S} (G_p)_{\mathbb{Q}_p}$, and suppose $\Gamma \subset G$ is a lattice. We call Γ an S-arithmetic lattice if there exist:

(1) a semisimple algebraic $\mathbb{Q}$-group H;

(2) for each $p \in S$ a decomposition $H = H_{1,p} \times H_{2,p}$ where $H_{i,p}$ are semisimple $\mathbb{Q}_p$-groups;

(3) a compact open subgroup $K \subset \prod_{p \in S} (H_{2,p})_{\mathbb{Q}_p}$;

and (4) an isomorphism of topological groups $\prod_{p \in S} (H_{1,p})_{\mathbb{Q}_p} \to G$;

such that for $H^* = \prod_p (H_{1,p})_{\mathbb{Q}_p} \times K \subset \prod_p H_{\mathbb{Q}_p}$ and the associated projection $\phi: H^* \to G$, we have $\phi(H^* \cap H_{Z_S})$ and Γ are commensurable.

10.1.12 Theorem [*Margulis* 1] (*S-arithmeticity theorem*): *Let G be as in* 10.1.11 *and let $\Gamma \subset G$ be an irreducible lattice. Assume rank$(G) \geqq 2$. Then Γ is S-arithmetic.*

Proof: The proof is similar to that of the arithmeticity theorem (6.1.2) using 10.1.5 instead of 5.1.2. When convenient, we shall therefore refer the reader to the proof of 6.1.2. By passing to a subgroup of Γ of finite index we can assume $\Gamma = q(\Lambda)$ for some irreducible lattice $\Lambda \subset \tilde{G} = \Pi(\tilde{G}_p)_{\mathbb{Q}_p}$, where $\tilde{G}_p$ is the algebraic universal covering and $q:\tilde{G} \to G$ is the product of the covering maps. (Cf. the proof of 10.1.7.) Any homomorphism of Γ lifts to a homomorphism of Λ and we are therefore in a position to apply superrigidity (10.1.5).

Let Γ_∞ be the projection of Γ onto $(G_\infty)_{\mathbb{R}}$. By the irreducibility assumption Γ_∞ is dense in $(G_\infty)_{\mathbb{R}}$ and therefore Zariski dense in G_∞. Using 10.1.5 instead of 5.1.2, the argument of Lemma 6.1.6 shows that we may assume $Tr(Ad_{G_\infty}(\gamma))$ is algebraic for all $\gamma \in \Gamma_\infty$. The arguments of 6.1.7 and 6.1.8 apply verbatim, and we deduce that we may assume that G_∞ is defined over a real algebraic number field k such that $\Gamma_\infty \subset (G_\infty)_k$. Via the group $R_{k/\mathbb{Q}}(G_\infty)$, the paragraph following lemma 6.1.8 shows that we are reduced to the following situation. There is a connected semisimple adjoint $\mathbb{Q}$-group H, a surjective $\mathbb{R}$-map $p:H \to G_\infty$, and a homomorphism $\alpha:\Gamma_\infty \to H_{\mathbb{Q}}$ such that $\alpha(\Gamma_\infty)$ is Zariski dense in H and $p \circ \alpha$ is the identity.

Suppose $a \in V$, and that $L \subset H$ is a $\mathbb{Q}_a$-simple factor of $H_{\mathbb{Q}_a}$. We then have a homomorphism $\phi_L:\Gamma \to L_{\mathbb{Q}_a}$ defined by the composition

$$\Gamma \xrightarrow{p_\infty} \Gamma_\infty \xrightarrow{\alpha} H_{\mathbb{Q}} \longrightarrow H_{\mathbb{Q}_a} \xrightarrow{\text{proj}} L_{\mathbb{Q}_a}.$$

Since $\alpha(\Gamma_\infty)$ is Zariski dense in H, we clearly have $\phi_L(\Gamma)$ is Zariski dense in L. By Theorem 10.1.5 we deduce that:

(i) if $a \notin S$, $\phi_L(\Gamma)$ has compact closure;

(ii) if $a \in S$, either $\phi_L(\Gamma)$ has compact closure or ϕ_L extends to a homomorphism $\phi_L:G \to L$ of the form $\phi_L = \psi_L \circ p_a$ where $p_a:G \to (G_a)_{\mathbb{Q}_a}$ is projection and $\psi_L:G_a \to L$ is a $\mathbb{Q}_a$-rational surjection. By passing to a subgroup of finite index in Γ, we can assume that in case (ii), if $\phi_L(\Gamma)$ has compact closure, then it is contained in a compact open subgroup. (For $a = \infty$, this follows from the fact that $\phi_L(\Gamma)$ is Zariski dense and hence $L_{\mathbb{R}}$ itself would be compact. For $a \neq \infty$, this follows from the fact that $L_{\mathbb{Q}_a}$ admits compact open subgroups.)

Combining the maps ϕ_L in (ii), we deduce that for $a \in S$ we can write $H = H_{1,a} \times H_{2,a}$ as a product of $\mathbb{Q}_a$-groups, find a compact open subgroup $K_a \subset (H_{2,a})_{\mathbb{Q}_a}$, and a rational $\mathbb{Q}_a$-map $\phi_a:G_a \to (H_{1,a})$ such that, setting $K = \Pi K_a$, $H^* = \prod_a (H_{1,a})_{\mathbb{Q}_a} \times K \subset \prod_a H_{\mathbb{Q}_a}$, and $\phi = \Pi\phi_a:G \to \prod_a (H_{1,a})_{\mathbb{Q}_a}$, we have $(\alpha \circ p_\infty)(\Gamma) \subset H^*$ and $(\alpha \circ p_\infty) = (\phi, \lambda)$ where $\lambda:\Gamma \to K$. Each map $\phi_a:G_a \to (H_{1,a})$ is surjective by the argument in the next to last paragraph of the proof of Theorem 6.1.2, using now the Zariski density of $\alpha(\Gamma_\infty)$ in $H_{\mathbb{Q}_a}$. Furthermore, using case (i) above, the argument of the final paragraph of the proof of 6.1.2

shows that $\alpha(\Gamma_\infty) \cap H_{Z_S}$ is of finite index in $\alpha(\Gamma_\infty)$. In particular, $(\alpha \circ p_\infty)(\Gamma)$ is discrete. Since Γ is irreducible, and G is a product of adjoint groups, ϕ must be injective. Thus ϕ is an isomorphism. Applying the map

$$H^* \longrightarrow \prod_a (H_{1,a})_{\mathbb{Q}_a} \xrightarrow{\phi^{-1}} G$$

to the finite extension $((\alpha \circ p_\infty)(\Gamma) \cap H_{Z_S}) \subset (\alpha \circ p_\infty)(\Gamma)$, the argument of the final sentences of the proof of Theorem 6.1.2 completes the proof of Theorem 10.1.12.

In conclusion, we remark that the arguments of Chapters 7 and 8 apply in the present more general context as well. Thus, we have:

10.1.13 Theorem [*Margulis* 6] [*Kazhdan* 1] (*Cf.* 8.1.2): *Let S be a finite subset of V and for each $p \in S$, let G_p be a connected semisimple $\mathbb{Q}_p$-group. Let $G = \prod_{p \in S} (G_p)_{\mathbb{Q}_p}$ and let $\Gamma \subset G$ be an irreducible lattice. Assume rank$(G) \geqq 2$. If $N \subset \Gamma$ is a normal subgroup, then either $N \subset Z(G)$ (the center of G) and is finite, or Γ/N is finite.*

Appendices

A. Borel spaces

In this appendix we collect (largely without proof) some of the basic general facts about Borel spaces. An excellent presentation of much of this and related material is in [Arveson 1].

A Borel space is a set X together with a σ-algebra $\mathscr{B}$ of subsets of X, called the Borel sets of X. We also refer to $\mathscr{B}$ as a Borel structure on X. If $\mathscr{A}$ is any family of subsets of a set X, there is a unique smallest σ-algebra containing $\mathscr{A}$ which we call the Borel structure generated by $\mathscr{A}$. If X is a topological space, then X is a Borel space in a natural way, the Borel structure being that generated by the open (equivalently, closed) sets. If X is a Borel space and $A \subset X$ is any subset, then A is a Borel space with the Borel structure $\{A \cap B | B \subset X \text{ Borel}\}$. If X, Y are Borel spaces, then $f{:}X \to Y$ is called Borel if $A \subset Y$ Borel implies $f^{-1}(A) \subset X$ is Borel. X and Y are called isomorphic Borel spaces if there is a bijective Borel map $f{:}X \to Y$ with f^{-1} Borel.

A Borel space X is called countably separated if there is a countable family of Borel sets $\{A_i\}$ which separates points. X is called countably generated if there is a countable family $\{A_i\}$ which separates points and generates the Borel structure. Clearly any subset of a countably separated (generated) Borel space is countably separated (generated).

A.1 Proposition: (*i*) *X is countably separated if and only if there exists an injective Borel map $X \to [0, 1]$.*
(*ii*) *X is countably generated if and only if X is isomorphic to a subset of* $[0, 1]$.

Proof: let $\{A_i\}$ be a countable family of Borel sets which separates points. Let $\Omega = \prod_1^\infty \{0, 1\}$. Define $f{:}X \to \Omega$ by $f(a)_i = \chi_{A_i}(a)$ where χ_A is the characteristic function of A. Then f is an injective Borel map and $f(A_i)$ is Borel in $f(X)$. As is well-known, Ω is homeomorphic to the Cantor set in $[0, 1]$, and from these remarks the result follows.

A.2 Definition: A Borel space is called standard if it is isomorphic to a Borel subset of a complete separable metric space.

(We remark that in (ii) of A.1, X need not be isomorphic to a Borel subset of $[0, 1]$. To the author's knowledge, there is no purely "Borel theoretic" characterization of standard Borel spaces.) Standard Borel spaces of course include many spaces arising naturally in geometry and analysis, and they have a number of very strong regularity properties.

A.3 Theorem: *Any standard Borel space is either finite, isomorphic to Z, or isomorphic to $[0, 1]$.*

A.4 Theorem: *Suppose $f: X \to Y$ is an injective Borel map where X is standard and Y is countably separated. Then $f(X) \subset Y$ is a Borel set and $f: X \to f(X)$ is an isomorphism. (In particular, $f(X)$ is standard).*

If f is not injective in A.4, $f(X)$ may not be Borel. However, the following presents another general situation when it will be Borel.

A.5 Theorem [*Kallman* 1]: *Suppose X is a standard Borel space and Y is a separable topological space metrizable by a complete metric. Suppose $A \subset X \times Y$ is Borel and that for each $x \in X$, $A_x = \{y \in Y | (x, y) \in A\}$ is a countable union of compact sets. Let $f: X \times Y \to X$ be projection. Then $f(A)$ is Borel and there is a Borel section $f(A) \to A$ of the map $f|A$.*

A.6 Corollary: *Suppose X, Y are metrizable by complete separable metrics. Let $f: X \to Y$ be a Borel map such that for each $y \in Y$, $f^{-1}(y)$ is a countable union of compact sets. Then $f(X)$ is Borel and there is a Borel section $f(X) \to X$ of f.*

Proof: Apply A.5 to $A = \{(x, y) \in X \times Y | f(x) = y\}$ and the projection of $X \times Y$ onto Y.

From A.5 one can also deduce:

A.7 Theorem [*Kallman* 1]: *If X is a standard Borel G-space where G is a locally compact second countable group, and the action of G on X is smooth (i.e. X/G is countably separated), then X/G is standard and there is a Borel section $X/G \to X$ of the natural projection.*

A.8 Corollary: *If $H \subset G$ is a closed subgroup of a locally compact (second countable) group, then there is a Borel section $G/H \to G$.*

Remarks: A.5, A.7 are not in [Arveson 1]. See [Kallman 1] for the proof. A.8 was first observed by Mackey. If G is a Lie group then A.8 is a straightforward consequence of the implicit function theorem. In A.6, if X and Y are compact and f is continuous, the existence of a Borel section was first due to Federer and Morse. (See [Parthasarathy 1] for a proof in this case.)

Although the image of a standard space under a Borel map need not be standard in general, in the presence of a measure it is "almost standard." As usual, if X is a Borel space and μ is a measure defined on the Borel structure of X, a set $A \subset X$ is μ-null if there is a Borel set B with $\mu(B) = 0$ and $B \supset A$, and μ-conull if its complement is μ-null.

A.9 Theorem: *Let X be a standard Borel space, Y countably generated and $f: X \to Y$ a surjective Borel function. Suppose μ is a probability measure on Y. Then there is a conull Borel set $Z \subset Y$ such that Z is standard, and there is a Borel section $Z \to X$ of f.*

The existence of a section defined on a conull set as in A.9 is the von Neumann selection theorem.

If X is a compact metric space the presence of a measure also clarifies the relation of a Borel set to the topology.

A.10 Theorem: *Suppose X is a compact metric space with a probability measure. If $A \subset X$ is a Borel set, then there is a Borel set $B \subset A$ which is conull in A and is a countable union of compact sets.*

For a proof, see [Parthasarathy 1].

The analgoue of Theorem A.3 in the presence of a measure is also true.

A.11 Theorem: *Let X be a standard Borel space and μ a probability measure on X with no atoms. Then there is an isomorphism $f: X \to [0, 1]$ such that $f_*(\mu)$ is Lebesgue measure.*

We recall that if X is a standard Borel space with a probability measure, then $B(X)$ denotes the measure algebra of X. We can identify $B(X)$ as a closed subset of $L^\infty(X)$ with the weak-$*$-topology. If $f: X \to Y$ is a measure class preserving Borel map, there is an induced map $f^*: B(Y) \to B(X)$ which is continuous, Boolean operation preserving, and injective.

A.12 Theorem: *Suppose X, Y are standard Borel spaces with probability measures and $F:B(Y)\to B(X)$ is a continuous, Boolean operation preserving, injective map. Then there is a measure class preserving Borel map $f:X\to Y$ such that $f^* = F$.*

Given a continuous family of such maps F, the point maps can be chosen to vary in a Borel fashion.

A.13 Theorem: *Suppose Z is a space metrizable by a complete separable metric and that for each $z\in Z$, $F_z:B(Y)\to B(X)$ is a continuous, Boolean operation preserving, injective map. Assume also that the map $B(Y)\times Z\to B(X)$, $(A, z)\to F_z(A)$ is continuous. Then there is a Borel map $f:X\times Z\to Y$ such that for each $z\in Z$, $f_z(x)=f(x, z)$ is a measure class preserving Borel map with $f_z^* = F_z$.*

For a proof of A.12, A.13, see [Ramsay 1] (Theorems 2.1, 3.3).

B. Almost everywhere identities on groups

In this appendix we collect some facts concerning almost everywhere identities related to locally compact groups. Throughout, G will denote a locally compact second countable group with Haar measure m.

B.1 Proposition: *Let $H\subset G$ be a conull subset closed under multiplication. Then $H = G$.*

Proof: Since H is conull, so is H^{-1}. Let $x\in G$. Then $H^{-1}x$ is conull and hence so is $H^{-1}x\cap H$. Thus, in particular, there exists $h\in H$ such that $h^{-1}x\in H$. Hence $x\in hH\subset H$.

B.2 Theorem: *Suppose J is a topological group whose Borel structure is countably generated (e.g. J second countable). Let $\pi:G\to J$ be a Borel function such that for almost all $(x, y)\in G\times G$, $\pi(xy) = \pi(x)\pi(y)$. Then there is a Borel homomorphism $\pi_0:G\to J$ such that $\pi = \pi_0$ a.e.*

Proof: As a Borel space we can assume $J\subset[0, 1]$ by A.1. Let

$$H = \{y\in G|x\to\pi(x)^{-1}\pi(xy) \text{ is an essentially constant function on } G\}.$$

By Fubini, H is conull. Further, if $y, z\in H$, writing

$$\pi(x)^{-1}\pi(xyz) = \pi(x)^{-1}\pi(xy)\pi(xy)^{-1}\pi(xyz)$$

it is clear that $yz \in H$, and hence by B.1 that $H = G$. The same equation shows that if $\pi_0(y) \in J$ is such that $\pi(x)^{-1}\pi(xy) = \pi_0(y)$ for a.e. x, then $\pi_0: G \to J$ is a homomorphism. Finally π_0 is Borel by Fubini, since (remembering $J \subset [0, 1]$) $\pi_0(y) = \int \pi(x)^{-1}\pi(xy)\,dm'(x)$ where m' is a probability measure in the class of Haar measure.

B.3 Theorem (*Mackey*): *Suppose J is a second countable topological group and $\pi: G \to J$ is a measurable homomorphism. Then π is continuous.*

Proof: We can assume π is surjective. Let $U \subset J$ be an open neighborhood of the identity and $V \subset U$ a symmetric open neighborhood with $V^2 \subset U$. Let $\{y_n\}$ be a countable dense set in J and $\pi(g_n) = y_n$. We have $H = \cup V \cdot y_n$ and hence $G = \bigcup_n \pi^{-1}(V)g_n$. It follows that for some n, $m(\pi^{-1}(V)g_n) > 0$ so $m(\pi^{-1}(V)) > 0$. Since $\pi^{-1}(U) \supset \pi^{-1}(V)\pi^{-1}(V)$, it suffices to show:

B.4 Lemma: *If $A \subset G$ is a compact set in G with $m(A) > 0$, then $A^{-1}A$ contains a neighborhood of $e \in G$.*

Proof: If $Ax \cap A \neq \emptyset$, then $x \in A^{-1}A$. So it suffices to show that $\{x \in G \mid Ax \cap A \neq \emptyset\}$ contains a neighborhood of e. Since A is a compact set of positive measure, we can find an open set $W \supset A$ such that $m(W) < 2m(A)$. Again by compactness of A, we can find a symmetric neighborhood N of e such that $Ax \subset W$ for all $x \in N$. Since $m(Ax) = m(A)$ and $m(W) < 2m(A)$, we must clearly have $Ax \cap A \neq \emptyset$ for $x \in N$.

In 2.2.16, we showed that an essentially invariant function on a G-space agrees a.e. with an invariant function. We now consider the analogous fact for maps between G-spaces.

B.5 Proposition: *Suppose (X, μ) is a standard Borel G-space with quasi-invariant measure and that Y is a standard Borel G-space. Suppose $f: X \to Y$ is Borel and for each $g \in G$, $f(xg) = f(x)g$ for almost all x. Then there is a G-invariant conull Borel subset $X_0 \subset X$ and a Borel G-map $\tilde{f}: X_0 \to Y$ such that $f = \tilde{f}$ a.e.*

Proof: Let $X_0 = \{x \in X | g \mapsto f(xg)g^{-1}$ is essentially constant in $G\}$. For $x \in X_0$, let $\tilde{f}(x) \in Y$ be such that $f(xg)g^{-1} = \tilde{f}(x)$ for a.e. $g \in G$. As a Borel space we can assume $G \subset [0, 1]$ by A.1, and hence by Fubini X_0 is conull and $\tilde{f}$ is Borel (by an argument as in B.2). We have $f = \tilde{f}$ a.e., and from the expression $f(xhg)g^{-1} = [f(xhg)\cdot(hg)^{-1}]h$, we deduce that for $x \in X_0$, $h \in G$, we have $xh \in X_0$ and $\tilde{f}(xh) = \tilde{f}(x)h$.

Suppose now that (X, μ), (Y, ν) are standard G-spaces with quasi-invariant measure. Let $B(X, \mu)$, $B(Y, \nu)$ be the corresponding measure algebras. (Cf. 8.1).

B.6 Corollary: *Suppose $\varphi: B(Y, \nu) \to B(X, \mu)$ is a continuous injective Boolean homomorphism which is a G-map. Then there is a conull G-invariant Borel set $X_0 \subset X$ and a G-map $\tilde{f}: X_0 \to Y$ such that $\tilde{f}^* = \varphi$.*

Proof: By general measure theory there is a Borel map $f: X \to Y$ such that $f^* = \varphi$. Since φ is a G-map, f will satisfy the hypotheses of B.5.

B.7 Corollary: *Suppose (X, μ), (Y, ν) as in B.6 and that φ is a bijection. Then X and Y are conjugate G-spaces. I.e., there are conull, Borel, G-invariant sets $X_0 \subset X$, $Y_0 \subset Y$ and a measure class preserving Borel isomorphism $\tilde{f}: X_0 \to Y_0$ which is a G-map.*

Proof: Choose $\tilde{f}$ as in B.6. Since φ is surjective from general measure theory there is a conull Borel set $X_1 \subset X_0$ on which $\tilde{f}$ is injective. Let $X_2 = \{x \in X_1 | xg \in X_1$ for almost all $g \in G\}$. By Fubini's theorem, X_2 is conull and Borel. We claim that $\tilde{f}$ is injective on X_2G. Suppose $x, y \in X_2$ and $\tilde{f}(xa) = \tilde{f}(yb)$ for $a, b \in G$. Since $x, y \in X_2$, there is some $g \in G$ such that $xag, ybg \in X_1$. (In fact, this is true for almost all g.) Since $\tilde{f}$ is a G-map, we have $\tilde{f}(xag) = \tilde{f}(ybg)$, and by injectivity of $\tilde{f}$ on X_1, $xag = ybg$. Thus, $xa = yb$, establishing injectivity on X_2G. By A.4, it clearly suffices to see that X_2G contains a conull, G-invariant Borel set. This follows from the generally useful:

B.8 Lemma: *Suppose X is a standard Borel G-space with a quasi-invariant measure, and that $A \subset X$ is Borel. Then there is a Borel subset $B \subset A$ which is conull in A, and such that:*
(i) BG is Borel;

(ii) There is a Borel map $\varphi: BG \to G$ such that for all $x \in BG$, $x\varphi(x) \in B$, and for all $x \in B$, $\varphi(x) = e$.

Proof: By 2.1.19 and A.4, we can assume that X is compact metric and that G acts continuously on X. By A.10 there is a Borel set $B \subset A$ which is conull in A and is a countable union of compact sets B_i. Since G is also a countable union of compact sets, BG will be as well, and in particular BG is Borel. To see (ii), it suffices to construct for each compact $K \subset G$ and each i, a Borel map $\varphi: B_i K \to K$ such that $x \cdot \varphi(x) \in B_i$ for all $x \in B_i K$. However, this follows from A.5.

We now turn to the relationship between cocycles and strict cocycles, notions introduced in section 4.2. We recall that for transitive G-spaces, this relationship is clarified by 4.2.15. Here we deal with the general case.

B.9 Theorem: *Let S be a standard Borel G-space with quasi-invariant measure μ. Suppose H is a topological group whose Borel structure is countably generated (e.g. H second countable). Let $\alpha: S \times G \to H$ be a cocycle. Then there is a strict cocycle $\beta: S \times G \to H$ such that for all $g \in G$, $\beta(s, g) = \alpha(s, g)$ for almost all $s \in S$.*

Proof: Let $X = \{(s, g) \in S \times G | h \to \alpha(s, gh)\alpha(sg, h)^{-1}$ is an essentially constant H-valued function of $h \in G\}$. For $(s, g) \in X$, let $\tilde{\alpha}(s, g)$ be this constant. Then Fubini's theorem (recall $H \subset [0, 1]$ as a Borel space by A.1) shows that X is a conull Borel set, $\tilde{\alpha}: X \to H$ is Borel and $\tilde{\alpha} = \alpha$ a.e. Suppose $(s, g), (sg, a) \in X$. Then the relation

$$\alpha(s, gah)\alpha(sga, h)^{-1} = [\alpha(s, gah)\alpha(sg, ah)^{-1}][\alpha(sg, ah)\alpha(sga, h)^{-1}]$$

implies that $(s, ga) \in X$ and $\tilde{\alpha}(s, g)\tilde{\alpha}(sg, a) = \tilde{\alpha}(s, ga)$.

Now let $S_0 = \{s \in S|$ for almost all $g \varepsilon G$, $(s, g) \in X$ and $(sg, g^{-1}) \in X\}$. Since X is conull and Borel in $S \times G$ and $(s, g) \to (sg, g^{-1})$ is measure class preserving, it follows from Fubini's theorem that S_0 is a conull Borel subset of S. We now claim that if $s \in S_0$ and $a \in G$ with $sa \in S_0$, then $(s, a) \in X$. By the preceding paragraph it suffices to show that for some $g \in G$, we have $(s, ag), (sag, g^{-1}) \in X$. Since $s \in S_0$, $\{g|(s, ag) \in X\}$ is conull in G, and since $sa \in S_0$, $\{g|(sag, g^{-1}) \in X\}$ is conull in G. Therefore, our assertion follows. Summarizing we have produced a conull Borel set $S_0 \subset S$ such that if $s, sg, sgh \in S_0$ for $g, h \in G$, we have $\tilde{\alpha}(s, gh) = \tilde{\alpha}(s, g)\tilde{\alpha}(sg, h)$. By passing to a conull Borel subset, we can also assume S_0 satisfies the conditions of B.8.

Thus, $S_0 G$ is a conull Borel G-invariant set and we can find a Borel map

$\varphi: S_0 G \to G$ such that $\varphi(s) = e$ for $s \in S_0$ and $s \cdot \varphi(s) \in S_0$ for all $s \in S_0 G$. We now define $\beta: S_0 G \times G \to H$ by $\beta(s, g) = \tilde{\alpha}(s\varphi(s), \varphi(s)^{-1} g \varphi(sg))$. From the conclusion of the preceding paragraph, it follows that β is a strict cocycle. Finally, for $s \notin S_0 G$, define $\beta(s, g) = e$ for all g. Then $\beta: S \times G \to H$ is a strict cocycle and $\beta = \alpha$ a.e.

To complete the proof, let $G_0 = \{g \in G | \alpha(s, g) = \beta(s, g)$ for almost all $s\}$. By Fubini, $G_0 \subset G$ is conull and from the cocycle identity we deduce that G_0 is closed under multiplication. From B.1, it follows that $G_0 = G$.

We now show that a continuous G action on $B(X)$ actually defines a G action on points.

B.10 Theorem [*Mackey* 2]: *Let (X, μ) be a standard Borel space with a probability measure. Suppose G acts continuously on $B(X)$ so as to preserve the Boolean operations. Then there is a standard Borel G-space Y with quasi-invariant measure ν, conull Borel sets $X_0 \subset X$, $Y_0 \subset Y$, and a measure class preserving Borel isomorphism $\varphi: X_0 \to Y_0$ such that $\varphi^*: B(Y) \to B(X)$ is a homeomorphic G-map.*

Proof: By A.12, there is a Borel map $a: X \times G \to X$ such that for each $g \in G$ the map $a_g: X \to X$ given by $a_g(x) = a(x, g)$ is a measure class preserving map with a_g^* equal to the original action of g^{-1} on $B(X)$. It follows that a is almost an action. I.e.,

(i) for $g, h \in G$, $a(x, gh) = a(a(x, g), h)$ for almost all $x \in X$;

(ii) for each $g \in G$, $a(a(x, g), g^{-1}) = x$ a.e.

Choose an injective Borel function $f: X \to [0, 1]$. Define $h: X \to L^\infty(G)_1$ (the latter being the unit ball in $L^\infty(G)$, with the weak-$*$-topology) by $(h(x))(g) = f(a(x, g))$. Let $X_0 = \{x \in X | a(a(x, g), g^{-1}) = x$ for almost all $g \in G\}$. By (ii) and Fubini, X_0 is a conull Borel subset of X. If $x, y \in X$ and $h(x) = h(y)$, then for almost all $g \in G$, $a(x, g) = a(y, g)$. Thus if $x, y \in X_0$, we deduce $x = y$, i.e. h is injective on X_0. Then one checks that $(Y, \nu) = (L^\infty(G)_1, h_*(\mu))$ and $Y_0 = h(X_0)$ satisfy the required conditions. (For a proof verifying all details, see [Ramsay 1].)

References

ARVESON, W.

[1] Invitation to C^*-algebras. Springer, New York 1976.

AUSLANDER, L. – MOORE, C.C.

[1] Unitary representations of solvable Lie groups. Memoirs of A.M.S., No. 62 (1966).

BAILY, W.

[1] Introductory lectures on automorphic forms. Princeton Univ. Press, Princeton, N.J. 1973.

BERBERIAN, S.

[1] Measure and Integration. Macmillan, New York 1963.

BILLINGSLEY, P.

[1] Ergodic theory and information. Wiley, New York 1965.

BOREL, A.

[1] Linear algebraic groups. Benjamin, New York 1969.

[2] Introduction aux Groupes Arithmétiques. Hermann, Paris 1969.

[3] Density properties for certain subgroups of semisimple Lie groups without compact factors. Annals of Math. *72* (1960), 179–188.

[4] Compact Clifford-Klein forms of symmetric spaces. Topology *2* (1963), 111–122.

[5] Density and maximality of arithmetic subgroups. J. Reine Angew. Math. *224* (1966), 78–89.

[6] Some finiteness properties of adele groups over number fields. Publ. Math. IHES *16* (1963), 101–126.

BOREL, A. – HARDER, G.

[1] Existence of discrete cocompact subgroups of reductive groups over local fields. J. für Math. *298* (1978), 53–64.

BOREL, A. – HARISH-CHANDRA

[1] Arithmetic subgroups of algebraic groups. Annals of Math. *75* (1962), 485–535.

BOREL, A. – SERRE, J.-P.

[1] Théorèmes de finitude en cohomologie galoisienne. Comm. Math. Helv. *39* (1964), 111–164.

BOREL, A. – SPRINGER, T.

[1] Rationality properties of linear algebraic groups, in Algebraic groups and discrete subgroups. Proc. of Symp. Pure Math., IX, Amer. Math. Soc. 1966.

BOREL, A. – TITS, J.

[1] Groupes réductifs. Publ. Math. IHES *27* (1965), 55–150; Complements, ibid. *41* (1972), 253–276.

[2] Homomorphisms "abstraits" de groupes algébriques simples. Annals of Math. *97* (1973), 499–571.

BOURBAKI, N.

[1] Groupes de Lie et Algébres de Lie. Hermann, Paris 1968.

BOWEN, R.

[1] Periodic orbits for hyperbolic flows. Amer. J. Math. *94* (1972), 1–30.

BROOKS, R.

[1] The fundamental group and the spectrum of the Laplacian, Comm. Math. Helv. *56* (1981), 581–598.

BRUHAT, F. – TITS, J.

[1] Groupes réductifs sur un corps local. Publ. Math. IHES *41* (1972), 5–252.

CHEVALLEY, C.

[1] Théorie des Groupes de Lie, II: groupes algébriques. Hermann, Paris 1951.

CHOQUET, G.

[1] Lectures on Analysis, Vol. II. Benjamin, New York 1969.

CONNES, A. – FELDMAN, J. – WEISS, B.

[1] An amenable equivalence relation is generated by a single transformation. Ergodic Theory and Dynamic Systems *1* (1981), 431–450.

DANI, S.G.

[1] Invariant measures and minimal sets of horospherical flows. Invent. Math. *64* (1981), 357–385.

[2] On invariant measures, minimal sets, and a lemma of Margulis. Invent. Math. *51* (1979), 239–260.

[3] Invariant measures of horospherical flows on noncompact homogeneous spaces. Invent. Math. *47* (1978), 101–138.

[4] Continuous equivariant images of lattice actions on boundaries, Annals of Math. *119* (1984), 111–119.

DANI, S.G. – RAGHAVAN, S.

[1] Orbits of Euclidean frames under discrete linear groups. Israel J. Math. *36* (1980), 300–320.

DELAROCHE, C. – KIRILLOV, A.

[1] Sur les relations entre l'espace dual d'un groupe et la structure de ses sous-groupes fermés. Seminaire Bourbaki, No. 343, 1967/68.

DELIGNE, P. – MOSTOW, G.D.

[1] Monodromy of hypergeometric functions and non-lattice integral monodromy, preprint IHES.

DERRIENNIC, Y.

[1] Sur le théorème ergodique sous-additif. C.R. Acad. Sci. Paris *281* (1975), Série A, 985–988.

DIXMIER, J.

[1] Les C^*-algèbres et leur représentations. Gauthier-Villars, Paris 1964.

[2] Les algebres d'operateurs dans l'espaces Hilbertian. Gauthiers-Villars, Paris, 1969.

DYE, H.

[1] On groups of measure preserving transformations, I. Amer. J. Math. *81* (1959), 119–159.

[2] On groups of measure preserving transformations, II. Amer. J. Math. *85* (1963), 551–576.

EFFROS, E.G.

[1] Transformation groups and C^*-algebras. Annals of Math. *81* (1965), 38–55.

[2] Polish transformation groups and classification problems, to appear.

FELDMAN, J. – MOORE, C.C.

[1] Ergodic equivalence relations, cohomology, and von Neumann algebras, I. Trans. Amer. Math. Soc. *234* (1977), 289–324.

FELDMAN, J. – HAHN, P. – MOORE, C.C.

[1] Orbit structure and countable sections for actions of continuous groups. Advances in Math *28* (1978), 186–230.

FELL, J.M.G.

[1] Weak containment and induced representations of groups. Canadian J. Math. *14* (1962), 237–268.

[2] The dual spaces of C^*-algebras. Trans. AMS *94* (1960), 365–403.

FURSTENBERG, H.

[1] A Poisson formula for semisimple Lie groups. Annals of Math *77* (1963), 335–383.

[2] Boundary theory and stochastic processes in homogeneous spaces, in Harmonic analysis on homogeneous spaces. (Symposia on Pure and Applied Math.), Williamstown, Mass. 1972.

[3] The unique ergodicity of the horocycle flow, in Recent advances in topological dynamics. Springer Lecture Notes, No. 318.

[4] A note on Borel's density theorem. Proc. AMS *55* (1976), 209–212.

[5] Rigidity and cocycles for ergodic actions of semisimple groups [after G.A. MARGULIS and R. ZIMMER]. Seminaire Bourbaki, No. 559, 1979/80.

FURSTENBERG, H. – KESTEN, H.

[1] Products of random matrices. Ann. Math. Statistics *31* (1960), 457–469.

GLIMM, J.

[1] Locally compact transformation groups. Trans AMS *101* (1961), 124–138.

GREENLEAF, F.
[1] Invariant means on topological groups. van Nostrand, New York 1969.

GROMOV, M.
[1] Hyperbolic manifolds, groups, and actions, in Riemann surfaces and related topics, ed. I. Kra and B. Maskit. Annals of Math. Studies, No. 97. Princeton University Press 1981.

HAMACHI, T. – OSIKAWA, M.
[1] Ergodic groups of automorphisms and Krieger's theorems. Seminar on Math. Sci., No. 3 (1981), Keio Univ.

HEDLAND, G.
[1] Fuchsian groups and transitive horocycles. Duke Math. J. *2* (1936), 530–542.

HELGASON, S.
[1] Differential Geometry and Symmetric Spaces. Academic Press, New York 1962.

HOWE, R. – MOORE, C.C.
[1] Asymptotic properties of unitary representations. J. Functional Anal. *32* (1979), 72–96.

HULANICKI, A.
[1] Means and Følner conditions on locally compact groups. Studia Math. *27* (1966), 87–104.

HUMPHREYS, J.E.
[1] Linear algebraic groups. Springer, New York 1975.
[2] Arithmetic groups. Springer Lecture Notes, No. 789, 1980.

JACOBSON, N.
[1] Lie Algebras. Interscience, New York 1962.

KALLMAN, R.
[1] Certain quotient spaces are countably separated, III. J. Funct. Anal. *22* (1976), 225–241.

KATOK, A.
[1] Lyapunov exponents, entropy, and periodic orbits for diffeomorphisms. Publ. Math. IHES, Vol. 51.

KAZHDAN, D.
[1] Connection of the dual space of a group with the structure of its closed subgroups. Funct. Anal. Appl. *1* (1967), 63–65.

KHINCHIN, A.
[1] Mathematical Foundations of Information Theory. Dover, New York 1957.

KINGMAN, J.F.C.
[1] The ergodic theory of subadditive stochastic processes. J. Royal Statist. Soc. *B30* (1968), 499–510.

KRIEGER, W.
[1] On non-singular transformations of a measure space, I. Z. Wahr. *11* (1969), 83–97.
[2] On ergodic flows and the isomorphism of factors. Math. Ann. *223* (1976), 19–70.

KUO, H.-H.
[1] Gaussian measures in Banach spaces. Springer Lecture Notes, No. 463, 1975.

LOOMIS, L.H.
[1] Abstract harmonic analysis. van Nostrand, Princeton 1953.

MACKEY, G.W.
[1] Induced representations of locally compact groups, I. Annals of Math. *55* (1952), 101–139.
[2] Point realizations of transformation groups. Illinois J. Math. *6* (1962), 327–335.
[3] The theory of unitary group representations. University of Chicago Press, Chicago 1976.
[4] Unitary group representations in physics, probability, and number theory. Benjamin/Cummings, Reading, Mass. 1978.
[5] Ergodic theory and virtual groups. Math. Ann. *166* (1966), 187–207.

MAKAROV, V.
[1] On a certain class of discrete subgroups of Lobachevski space having infinite fundamental region of finite measure. Soviet Math. Dokl. *7* (1966), 328–331.

MARGULIS, G.A.
[1] Discrete groups of motions of manifolds of non-positive curvature. Amer. Math. Soc. Translations *109* (1977), 33–45.

[2] Arithmeticity of irreducible lattices in semisimple groups of rank greater than 1, appendix to Russian translation of M. RAGHUNATHAN, Discrete subgroups of Lie groups. Mir, Moscow 1977 (in Russian).
[3] Non-uniform lattices in semisimple algebraic groups, in Lie groups and their representations, ed. I.M. Gelfand. Wiley, New York 1975.
[4] Arithmeticity and finite dimensional representations of uniform lattices. Funct. Anal. Appl. *8* (1974), 258–259.
[5] Factor groups of discrete subgroups. Soviet Math. Dokl. *19* (1978), 1145–1149.
[6] Quotient groups of discrete subgroups and measure theory. Funct. Anal. Appl. *12* (1978), 295–305.
[7] Finiteness of quotient groups of discrete subgroups. Funct. Anal. Appl. *13* (1979), 178–187.

MOORE, C.C.
[1] Ergodicity of flows on homogeneous spaces. Amer. J. Math. *88* (1966), 154–178.
[2] Amenable subgroups of semisimple groups and proximal flows. Israel J. Math. *34* (1979), 121–138.

MOSTOW, G.D.
[1] Quasi-conformal mappings in *n*-space and the rigidity of hyperbolic space forms. Publ. Math. IHES *34* (1967), 53–104.
[2] Strong rigidity of locally symmetric spaces. Annals of Math. Studies, No. 78. Princeton Univ. Press, Princeton N.J. 1973.
[3] On a remarkable class of polyhedra in complex hyperbolic space. Pac. J. Math. *86* (1980), 171–276.

MOSTOW, G.D. – TAMAGAWA, T.
[1] On the compactness of arithmetically defined homogeneous spaces. Annals of Math. *76* (1961), 446–463.

MUNKHOLM, H.J.
[1] Simplices of maximal volume in hyperbolic space, Gromov's norm and Gromov's proof of Mostow's rigidity theorem (following Thurston). Proc. Symp. Alg. Top. Siegen, Germany 1979.

NAMIOKA, I.
[1] Følner's condition for amenable semigroups. Math. Scand. *15* (1964), 18–28.

OLSHANSKII, A.
[1] On the problem of the existence of an invariant mean on a group. Russian Math. Surveys *35* (1980), 180–181.

ORNSTEIN, D.
[1] Bernoulli shifts with the same entropy are isomorphic. Advances in Math. *4* (1970), 337–352.
[2] Ergodic theory, randomness, and dynamical systems. Yale University Press, New Haven 1974.
[3] On invariant measures. Bull. AMS *66* (1960).

OSELEDEC, V.I.
[1] A multiplicative ergodic theorem. Trans. Moscow Math. Soc. *19* (1968), 197–231.

PARTHASARATHY, K.R.
[1] Probability measures on metric spaces. Academic Press, New York 1967.

PESIN, YA.B.
[1] Lyapunov characteristic exponents and smooth ergodic theory. Russian Math. Surveys *32* (1977), 55–114.

PHELPS, R.
[1] Lectures on Choquet's theorem. van Nostrand, Princeton 1966.

PLATONOV, V.P.
[1] The problem of strong approximation and the Kneser-Tits hypothesis for algebraic groups. Math. USSR Izv. *3* (1969), 1139–1147; addendum, ibid. *4* (1970), 784–786.

PRASAD, G.
[1] Strong rigidity of Q-rank 1 lattices. Invent. Math. *21* (1973), 255–286.
[2] Lattices in semisimple groups over local fields. Advances in Math. Supplementary Studies, Vol. 6 (1979), 285–356.

[3] Elementary proof of a theorem of Tits and a theorem of Bruhat-Tits. Bull. Soc. Math. Fr. *110* (1982), 197–202.

RAGHUNATHAN, M.
[1] Discrete subgroups of Lie groups. Springer, New York 1972.
[2] A proof of Oseledec's multiplicative ergodic theorem. Israel J. Math. *32* (1979), 356–362.
[3] Discrete groups and Q-structures on semisimple Lie groups, in Discrete subgroups of Lie groups and applications to moduli. Tata Institute, Oxford Univ. Press, Bombay 1975.
[4] On the congruence subgroup problem. Publ. Math. IHES *46* (1976), 107–161.

RAMSAY, A.
[1] Virtual groups and group actions. Advances in Math. *6* (1971), 253–322.

RATNER, M.
[1] Rigidity of horocycle flows. Annals of Math. *115* (1982), 598–614.

ROSENLICHT, M.
[1] Some rationality questions on algebraic groups. Annali di Mat. *43* (1957), 25–50.

SAMUEL, P.
[1] Méthodes d'algèbre abstraite en géométrie algébrique. Springer, New York 1967.

SCHMIDT, K.
[1] Cocycles of ergodic transformation groups. Macmillan Co. of India 1977.
[2] Amenability, Kazhdan's property T, strong ergodicity, and invariant means for ergodic group actions. Erg. Theory and Dyn. Syst. *1* (1981), 223–236.

SEGAL, I.E.
[1] Ergodic subgroups of the orthogonal group of a real Hilbert space. Annals of Math. *66* (1957), 297–303.

SELBERG, A.
[1] On discontinuous groups in higher dimensional symmetric spaces, in International colloquium on function theory. Tata Institute, Bombay 1960.

SERIES, C.
[1] The Rohlin tower theorem and hyperfiniteness for actions of continuous groups. Israel J. Math. *30* (1978), 99–122.

SERRE, J.-P.
[1] Lie algebras and Lie groups. Benjamin, New York 1965.

SHAFAREVICH, I.R.
[1] Basic Algebraic Geometry. Springer, New York 1977.

SHERMAN, T.
[1] A weight theory for unitary representations. Canadian J. Math. *18* (1966), 159–168.

SINAI, YA.
[1] A weak isomorphism of transformations with invariant measure. Amer. Math. Soc. Translations, Ser. 2 *57* (1966), 123–143.

SPATZIER, R.
[1] On lattices acting on boundaries of semisimple Lie groups. Ergodic Theory and Dyn. Syst. *1* (1981), 489–494.

SPRINGER, T.
[1] Linear algebraic groups. Birkhauser, Boston 1981.
[2] Galois cohomology of linear algebraic groups, in Algebraic groups and discrete subgroups. Proc. Symp. Pure Math. IX. Amer. Math. Soc. 1966.
[3] Reductive groups, in Automorphic forms, representations, and L-functions, ed. A. Borel and W. Casselman. Proc. Symp. Pure Math., Vol. 33, 1979, AMS, Providence.

STEIN, E.M.
[1] Singular integrals and differentiability properties of functions. Princeton Univ. Press, Princeton 1970.

STEINBERG, R.
[1] Conjugacy classes in algebraic groups. Springer Lecture Notes, No. 366 (1974).

SUTHERLAND, C.
[1] Orbit equivalence: lectures on Krieger's theorem. University of Oslo Lecture Notes.

TITS, J.
[1] Algebraic and abstract simple groups. Annals of Math. *80* (1964), 313–329.

[2] Classification of algebraic semisimple groups, in Algebraic Groups and discrete subgroups. Proc. Symp. Pure Math. IX. Amer. Math. Soc. 1966.

[3] Free subgroups in linear groups. J. Alg. *20* (1972), 250–270.

VARADARAJAN, V.S.

[1] Geometry of Quantum Theory, Vol. II. van Nostrand, Princeton, N.J. 1970.

VASERSTEIN, L.N.

[1] Groups having the property (T). Funct. Anal. Appl. *2* (1968), 174.

VEECH, W.A.

[1] Minimality of horospherical flows. Israel J. Math. *21* (1975), 233–239.

[2] Unique ergodicity of horospherical flows. Amer. J. Math. *99* (1977), 827–859.

VINBERG, E.B.

[1] Discrete groups generated by reflections in Lobachevski spaces. Math. USSR Sb. *1* (1967), 429–444.

WANG, S.P.

[1] On isolated points in the dual spaces of locally compact groups. Math. Ann. *218* (1975), 19–34.

[2] On density properties of S subgroups of locally compact groups. Annals of Math. *94* (1971), 325–329.

WEIL, A.

[1] Adeles and algebraic groups. Progress in Mathematics, No. 23. Birkhäuser-Boston.

ZIMMER, R.J.

[1] Extensions of ergodic group actions. Illinois J. Math. *20* (1976), 373–409.

[2] Orbit spaces of unitary representations, ergodic theory, and simple Lie groups. Annals of Math. *106* (1977), 573–588.

[3] Amenable ergodic group actions and an application to Poisson boundaries of random walks. J. Func. Anal. *27* (1978), 350–372.

[4] Induced and amenable actions of Lie groups. Ann. Sci. Ec. Norm. Sup. *11* (1978), 407–428.

[5] Curvature of leaves in amenable foliations. Amer. J. Math., *105* (1983), 1011–1022.

[6] On the cohomology of ergodic groups actions. Israel J. Math. *35* (1980), 289–300.

[7] On the cohomology of ergodic actions of semisimple Lie groups and discrete subgroups. Amer. J. Math. *103* (1981), 937–950.

[8] Strong rigidity for ergodic actions of semisimple Lie groups. Annals of Math. *112* (1980), 511–529.

[9] Orbit equivalence and rigidity of ergodic actions of Lie groups. Ergodic Theory and Dyn. Sys. *1* (1981), 237–253.

[10] Ergodic theory, semisimple groups and foliations by manifolds of negative curvature. Publ. Math. IHES. *55* (1982), 37–62.

[11] Equivariant images of projective space under the action of $SL(n, \mathbb{Z})$. Erg. Theory and Dyn. Sys. *1* (1981), 519–522.

[12] Ergodic Theory, Group Representations, and Rigidity. Bull. AMS *6* (1982), 383–416.

[13] Ergodic actions of semisimple groups and product relations. Annals of Mathematics *118* (1983), 9–19.

[14] Algebraic topology of ergodic Lie group actions and measurable foliations, preprint.

[15] Volume preserving actions of lattices in semisimple groups on compact manifolds. Publ. Math. IHES. Vol. 59 (1984), 5–33.

[16] On the Mostow rigidity theorem and measurable foliations by hyperbolic space. Israel J. Math. *43* (1982), 281–290

[17] Kazhdan groups acting on compact manifolds. Invent. Math. *75* (1984), 425–436.

[18] Semisimple automorphism groups of G-structures. J. Diff. Geom. *19* (1984), 117–123.

[19] Actions of lattices in semisimple groups preserving a G-structure of finite type, preprint.

Subject Index